PHI BETA KAPPA
IN AMERICAN LIFE

PHI BETA KAPPA IN AMERICAN LIFE

The First Two Hundred Years

Richard Nelson Current

New York   Oxford
OXFORD UNIVERSITY PRESS
1990

Oxford University Press

Oxford   New York   Toronto
Delhi   Bombay   Calcutta   Madras   Karachi
Petaling Jaya   Singapore   Hong Kong   Tokyo
Nairobi   Dar es Salaam   Cape Town
Melbourne   Auckland

and associated companies in
Berlin   Ibadan

Copyright © 1990 by Phi Beta Kappa

Published by Oxford University Press, Inc.,
200 Madison Avenue, New York, New York 10016

Oxford is a registered trademark of Oxford University Press

Library of Congress Cataloging-in-Publication Data
Current, Richard Nelson.
Phi Beta Kappa in American life: the first two hundred years /
Richard Nelson Current.
p. cm.
ISBN 0–19–506311–2
1. Phi Beta Kappa—History.   2. United States—Intellectual life.
I. Title.
LJ85.P27C87   1990
378.1'9852'0973—dc20   89-28200   CIP

2 4 6 8 9 7 5 3 1

Printed in the United States of America
on acid-free paper

To the memory of
my children's grandmother
TERRESSA METCALF BONAR
Northwestern University 1898
the first woman student
elected to Phi Beta Kappa
in the state of Illinois

Author's Preface

This is not exactly an official history. The Phi Beta Kappa Society, which commissioned me to write it, has permitted its publication but has done so with some reluctance. The Society's history committee, consisting of five Phi Beta Kappa senators, unanimously disapproved large portions of the manuscript.

According to the committee's report, these senators "were troubled by the negative impression with which they were left at the end of their reading—an impression of an organization beset with such problems as internecine squabbles, dubious ethical practices, arbitrary granting of chapter charters, thoughtless distribution of awards, continuing financial difficulties, organizational ineffectuality, and attacks on its worth both from within and without its ranks."

In the committee's view, more attention should have been given to the Society's various programs and awards, to the scholars who have participated in them, and to those who have been honored by them. More attention should also have been given to the details of the Society's official proceedings. For example, "there should be a full presentation of how the Committee on Qualifications proceeds and what it looks for." But much less attention should have been given to such topics as the infringement of the Society's name and emblem or the relation (if any) between membership and material success—topics that, in the committee's opinion, were of little relevance and of "minimal interest."

The committee wanted the history to be brought down to date (instead of being terminated at 1976) so as to emphasize the Society's recent achievements. "What sounds in the manuscript like the end of the story is not the end of the story," the committee's report stated.

The report also pointed out a number of errors and oversights. These I have been happy to correct, but I have resisted other changes. The book is finally being published in essentially the form in which it was written. Readers will have to decide for themselves whether it is too "negative" and "depreciatory" and whether it devotes too much space to irrelevancies.

My purpose has certainly not been to muckrake. I have been a proud member of Phi Beta Kappa for more than half a century—ever since my election as a junior at Oberlin College in 1933. In the present project, I have learned and have written nothing that, taken in the context of the history as a whole, has diminished in the slightest my pride in membership or my respect for the Society.

The history of the Society, it seems to me, corresponds with Arnold J. Toynbee's theory of history in general. According to Toynbee, institutions (as well as states and civilizations) flourish or decay as a result of the adequacy or inadequacy of their responses to challenges. From time to time, Phi Beta Kappa has had to face threats, attacks, and difficulties of various kinds. I do not see how anyone could write an honest account without giving considerable attention to such "negative" phenomena. As this book shows, the Society has survived and flourished in spite of them. It has responded adequately or more than adequately to all its challenges.

In my understanding of the assignment, as I reported it to the Phi Beta Kappa Senate at the outset, the book was to include the constitutional history of the Society: the development of its organization, rules, policies, and functions. But the book would also deal (as the title indicates) with the Society's role in American life: its reputation, its influence, its response to changing standards and methods of education, and its acceptance (grudging at times) of women, blacks, and other minorities.

Often the subjects least interesting to the Senate committee were precisely those that seemed to me to have both the greatest human interest and the greatest historical significance. To me, it would be tedious (and invidious) to list awards and the scholars conferring or receiving them. This would add nothing to an understanding of the Society's development. But such an understanding can be furthered—and at least a little drama can be sparked—by a recounting of conflicts over the making and enforcement of the Society's basic policies.

As I explained to the Senate at the beginning, the book would attempt to cover systematically only the two centuries from 1776 to 1976, the more recent years being considered too close for either perspective or detachment. Inevitably, the Society keeps on growing and changing. Since my manuscript was completed, the central organization has even changed its name—from the United Chapters of Phi Beta Kappa to the Phi Beta Kappa Society. So long as the Society exists, there will be no "end of the story," but at some point there must be an end of this book.

Research for the book was done in the libraries of the University of Chicago, the University of Illinois at Chicago, and the College of William and Mary; in the manuscripts division of the Library of Congress; and in the Phi Beta Kappa Society's archives at its headquarters in Washington, D.C. (to which the backnotes of the book refer, now somewhat archaically, as the "United Chapters archives"). I am grateful to the staff at each of those repositories for the generous assistance provided me.

South Natick, Massachusetts R. N. C.
September 1989

Foreword

In December 1976, the 31st Triennial Council of the United Chapters of Phi Beta Kappa met in Williamsburg, Virginia, where two hundred years earlier five young men who were students at the College of William and Mary had come together to establish a secret society. The founders of this new organization chose as their symbol a medal bearing on the one side the Latin letters *S. P.* (for *Societas Philosophiae*) and on the other the Greek letters ΦΒΚ (for Φιλοσοφία Βίου Κυβερνήτης) and adopted as their motto: "Love of Wisdom the Guide of Life."

The occasion and location of the Council meeting in 1976 were felicitous; the representation of chapters impressive. By the time of the bicentennial Council meetings, the Society that began with a single chapter, Alpha of Virginia, had 225 chapters, and some 400,000 persons had been elected to membership. By the time of the 35th Council meeting (1988), there were 240 chapters.

In 1983, the centennial year of the formation of the United Chapters of Phi Beta Kappa (as a consequence of which the basic structure and functions of the Society as we know it today were adopted), the Executive Committee of the Senate began giving serious consideration to the desirability and feasibility of producing a new history of Phi Beta Kappa. Because of the increase in activities of the Society as well as the growth in chapters and membership, there was general agreement that the time had come for such a project. The Society will be forever grateful to the late Oscar Voorhees, author of the official *History of Phi Beta Kappa*, published in 1945. His *History* carried the story through the Council of 1940, recovering and recording evidence and details, especially with reference to individual chapters, which might otherwise have been forgotten or lost.

The Executive Committee sought the counsel of a number of historians who were also members of the Senate. Senators Charles Blitzer, Gordon Craig, LeRoy Graf, and Catherine Sims, along with the historian of the United Chapters, Irving Dilliard, encouraged the commissioning of a new comprehensive history that would place Phi Beta Kappa in the context of the social, intellectual, and educational history of the nation and carry the story forward through the bicentennial celebration of 1976.

Kenneth M. Greene, secretary of the Phi Beta Kappa Society from 1975 to 1989, sought and secured a generous subvention for the project from the Andrew W. Mellon Foundation. The Society acknowledges its profound gratitude to the Foundation, and especially to its president, John W. Sawyer, for this assistance. Then, with the concurrence of the Senate, the Executive Committee invited Richard N. Current, formerly Distinguished Professor of American History at the University of North Carolina at Greensboro, to write the new history. Professor Current's volume is not, nor was it intended to be, an "official history"; rather, it is an account written for the general reader, especially for those who might wish to know more about the origin and development of Phi Beta Kappa. And as the reader will see, the author has told not one but several stories as he traces the gradual evolution of the organization into an honor society with an emphasis upon academic performance.

There is, first and foremost, the story of the growth and expansion of Phi Beta Kappa. There was little in the early years of its history to indicate later growth. In fact, during the first seventy years of its existence, only eight charters were granted, with all but one of those located in New England. In the next thirty-five years, twice as many new charters were created. Then beginning in the 1880s and continuing to our own day, new charters were approved for institutions in the South, the Midwest and the Far West, transforming Phi Beta Kappa into a truly national organization. The post–World War II era was one of almost explosive growth and expansion, one that continues to the present time.

Growth in numbers was accompanied by a significant increase in activities. The Phi Beta Kappa Foundation was chartered in 1924 and the Phi Beta Kappa Associates were established in 1940. The first issue of the *American Scholar,* a quarterly devoted to "the promotion in America of liberal scholarship," was published in 1932; three years later the *Key Reporter,* envisioned as a regular means of communicating with the membership, was created. During the 1950s, the Society broadened its scope of activity by sponsoring the several book awards that have since become closely identified in the public mind with Phi Beta Kappa.

This growth and development did not occur without considerable stress and turmoil. In its formative period, the organization came under attack, particularly by the Anti-Masonic movement, because it was a secret society. Throughout its existence, Phi Beta Kappa would continue to face charges of elitism and discrimination from those who were opposed to its selectivity and would be exposed time and again to the criticism and ridicule of those giving expression to the persistent and pervasive anti-intellectualism that runs like a red thread through the course of American history. In addition to those external forces, the Society faced a number of internal problems as well. Over the years, it would have to struggle to survive its own disunity, diversity, and periodic decline in prestige.

Yet another story told by Professor Current deals with a series of nagging policy questions and problems that are a recurrent theme in the history of the Society. The lack of cooperation among chapters, for example, was so serious that it threatened to destroy the organization. In fact, not until 1883, when a new constitution was adopted establishing the United Chapters of Phi Beta Kappa and creating both a National Council and a Senate, was this problem resolved. The qualifications for awarding charters to new chapters was another of those durable and divisive issues. Indeed, as one reads the early history of Phi Beta Kappa, it becomes clear that the continuing quarrel over the procedure for granting new charters seemed likely to disrupt the Society. The creation by the Senate of a Committee on Qualifications did much to improve consistency and to reduce tensions with regard to the issuance of new charters. The election of new members of Phi Beta Kappa is yet another of those persistent themes in the history of the organization. From the beginning, there was considerable confusion surrounding this issue: Who? Who elects? When? How many? Using what criteria? What should a student study to be eligible for election? To anyone even remotely familiar with the workings of Phi Beta Kappa, there will be quick recognition of the fact that these very topics continue to be debated at present-day meetings of the Senate and of the Council.

And while the Society, in the view of some of its critics, has been slow to accept or adjust to change, it did adopt and implement a policy of inclusiveness with regard to membership. The election of the first woman to membership in 1875 (in the year 1918, more women than men were elected members in course) and the first black in 1877 gave some substance to the expressed wish of the founders that the Society "not be confined to any particular place, men, or description of men. . . ."

Phi Beta Kappa in American Life: The First Two Hundred Years, which tells all these stories and many more, is presented in honorable remembrance of John Heath, Thomas Smith, Richard Booker, Armistead Smith, and John Jones, to whose "happy spirit and resolution of attaining and the important ends of society" we owe the foundation of Phi Beta Kappa.

Lexington, Kentucky
October 1989

Otis A. Singletary
President
Phi Beta Kappa Society

Contents

PHI BETA KAPPA
IN AMERICAN LIFE

1

Beginnings in Revolutionary Virginia

It was a time of new beginnings. Having declared their independence, Americans of the thirteen states were fighting to establish it. After winning the war, they would face the question whether the United States of America were to be one—or thirteen. If there were to be truly one nation, it would need new institutions, cultural as well as political, to reinforce its unity. So, at least, a few Revolutionary-era college students in one of the states came to believe, and they thought they had a cultural institution of precisely the kind needed. This was Phi Beta Kappa.

The College of William and Mary was close to Revolutionary events. It stood at the edge of Williamsburg, the capital of Virginia as a colony and then as a commonwealth. Though the town numbered fewer than two thousand—except at "publick times" when the legislature or the courts were in session—it had been one of the most important centers of the agitation culminating in the Declaration of July 4, 1776. It became a center of military resistance when the British, having shifted their main campaign from North to South, invaded Virginia and were penned in only a dozen miles away at Yorktown, where they made their final surrender on October 19, 1781. By that time Williamsburg had lost its importance, the capital having been moved to Richmond the previous year.

William and Mary was, after Harvard, the oldest of the nine American colleges in operation by 1776. Church of England men had founded it in 1693 "to the end that the church of Virginia may be furnished with a seminary for ministers of the gospel, and that the youth may be piously educated in good letters and manners, and that the Christian faith may be propagated amongst the Western Indians, to the glory of the Almighty." Originally the curriculum, based on that of Oxford, was an outgrowth of the medieval trivium and quadrivium, with much of the reading in Latin or Greek. Even before the Revolution, however, the college had begun to depart from the classical curriculum, and lecture courses had begun to replace textbook recitations.

In the academic year 1779–1780 Thomas Jefferson further secularized and modernized the instruction, using his influence as governor of the state and member of the college's board of visitors. Eliminated were courses in divinity and oriental languages (such as Hebrew); added were courses in modern languages, law and public administration, international law, chemistry, anatomy, medicine, natural history (life and earth sciences), natural philosophy (physical science), and the fine arts. "The Doors of ye University are open to all, nor is even a knowledge of ye ant[ient] Languages a previous Requisite for Entrance," the William and Mary president wrote to the Yale president in 1780. "The Students have ye Liberty of attending [any of the six professors] whom they please." In its deemphasis of Greek and Latin, its introduction of practical courses, and its adoption of a free-elective system, William and Mary was at least a century ahead of Harvard and even farther ahead of Yale.

To William and Mary went most of the Virginia planters' sons who sought a higher education. Just before the Revolution, more than a hundred were enrolled in the college proper and about fifty in what was called the "philosophical school," but the numbers fell off during the war years. Some of the college students found lodgings where they could in and around Williamsburg, while the majority roomed in the college's main building. This had been an attractive edifice when first erected on the plans of Christopher Wren. After its reconstruction following a fire, Jefferson described it as "a rude, misshapen pile, which but it had a roof would be taken for a brick-kiln."[1]

As early as 1750, a group of students had formed an exclusive secret society—so exclusive that it limited its membership to six at a time, and so secret that it let the outside world know it only by the letters F. H. C. on the polished-coin medal which each member carried. These were presumably the initials of a Latin phrase, perhaps something like *Fraternitas, Hilaritas, Cognitioque,* since the members claimed devotion to friendship, conviviality, and knowledge. Outsiders came to refer to the F. H. C. clique, disparagingly and no doubt enviously, as "the Flat Hat Club."

Jefferson, an F. H. C. member during his student days in the 1760s, recalled that the society "had no useful object" at that time. Later the members, who included some of the best scholars in the college, undertook to accumulate a library, and they obtained from one of the professors a list of "the most useful and valuable books with which it would be proper to begin." But the chief activity of the "little select friendly Set" remained socializing at one of the local taverns. At such a "temple of

Mirth and Hilarity" the F. H. C. boys enjoyed many a drunken "Debauch."

A secret society aspiring to rival the F. H. C. appeared in 1773. "The initials on its medal (P. D. A.) were understood to indicate Latin words," too, according to the recollection of a William and Mary alumnus. By 1776 the new fraternity had become much like the older one. "The P. D. A. society had lost all reputation for letters, and was noted only for the dissipation & conviviality of its members."[2]

John Heath thought there ought to be yet another student society at William and Mary, one that would also be secret and select but would be much more serious-minded. The son of a prominent Virginia gentleman who had served in the House of Burgesses, Heath was only fifteen years old (as were most entering students in those days) when he got the idea. But he was excellently tutored in Greek as well as Latin, took pride in his reputation as a Hellenist, and possessed precocious qualities of leadership. He soon found four fellow students willing to join him in the project. The five met, as one of them recorded a few weeks later, on "Thursday the 5th of December in the year of our Lord God one thousand seven hundred and seventy-six, and the first of the Commonwealth, a happy spirit and resolution of attaining the important ends of Society, entering the minds of John Heath, Thomas Smith, Richard Booker, Armist[ea]d Smith, and John Jones."[3]

At that first meeting the youths agreed on a name, a motto, and a medal. The name, in Latin, was to be *Societas Philosophiae*, which could be translated as "Philosophical Society" or perhaps better as "Society of Lovers of Learning." The motto, in Greek, was to be Φιλοσοφία Βίου Κυβερνήτης, "Philosophy (or love of wisdom) the guide of life." The medal was to be a silver square engraved on one side with the initials *S. P.* and on the other with the initials ΦΒΚ set in the "philosophical design" of an "index" pointing at three stars. These stars would stand for Friendship, Morality, and Literature (literature in the antique sense of scholarship or "book learning" rather than the present-day sense of belles lettres).

A month later, on January 5, 1777, the Philosophical Society met again, with nine present, to "adopt a mode of initiation," provide for "better security," and organize by electing officers. The nine initiated themselves by taking an oath of "fidelity" in which they swore to be true to the "growing fraternity" and keep all its secrets. They then chose a clerk, a treasurer, and a president, who fittingly enough was John Heath. At the next meeting, on March 1, they adopted twenty-four bylaws.

The society was now completely organized, possessing a constitution that, with a few additions and alterations, was to serve throughout the four years of the society's early existence at William and Mary. One of the additions was a sign by which members could recognize one another—if a time should ever come when they would be so numerous and so widely scattered as to need such a sign. "For the better distinction of the Fraternity, between themselves, in any foreign country or place, it is resolved that a salutation of the clasp of the hands, together with an immediate stroke across the mouth with the back of the same hand, and a return with the hand used by the saluted, be hereby established and ordained."

The clerk duly wrote this resolution in the minutes. But suppose the minutes should somehow be exposed to the prying eyes of an outsider? To keep the salutation secret, the clerk crossed out the words describing it, so as to make them illegible. To protect the secrecy of the name and motto, he also erased them from the record, leaving only the initials. In later entries he referred to the fraternity as "the S. P. alias ΦBK Society." Members called it by either set of initials, but more and more they used the Greek letters, so that eventually the motto became the name. They pronounced the letters "Fie Beeta Cappa."[4]

A revised ceremony (adopted on February 27, 1779) made the initiation more elaborate and more impressive than it had previously been. According to the new ritual, the sponsor of a candidate for membership brought him blindfolded into a room where the brethren had gathered. The president then asked him not only whether he would respect the principles of the society and keep them confidential but also whether he would regard every member as a brother and would assist a brother in distress with "Life and Fortune." Having received affirmative answers, the president administered the oath to the candidate, now with the blindfold off, and explained the society's arcana and aims. Finally, the president welcomed him with due formality, and the secretary introduced him separately to each of the members, "all taking care to use the mode of shaking hands peculiar to the ΦBK."[5]

Eligible for membership under the original bylaws were solely "Collegians" who had "arrived at the age of sixteen years" (older than the founding father at the society's inception) and who received the "unanimous approbation of the Society." Election—as well as expulsion—was at first by voice vote and later by ballot, marked *pro* or *con*. There is no record of a member having been expelled, and there were very few cases of a nominee being blackballed. Usually members nominated only those sure to be acceptable to all the brethren, but not always. "Mr. John Page and Mr. Henry Blunt were severally recommended as worthy Members of this Society, and on a Ballot taken Mr. Page was unanimously elected and initiated in due form," the minutes of one meeting say. "Mr. Blunt on a Ballot taken for him was rejected, three Ballots being returned with the word *con*."

When the fraternity was two years old, the brothers voted to relax the membership rule. On December 14, 1778, "it appearing that the State of the Society was declining through want of Members," they appointed a committee to consider the problem, and six days later they took this significant step: "Resolved that in future, Admission to this Society be not confined to Collegians alone." The society would admit, if otherwise qualified, "irregular" students who were "not entitled to wear the cap and gown" and were "not held in estimation by the pedantic & often thickheaded cap & gown students." It would even admit men who were not enrolled in William and Mary at all.[6]

This change was eventually to make possible the perpetuation of Phi Beta Kappa and its development as a nationwide organization. But the new rule had no immediate effect on the fraternity's rate of growth. The ΦBK Society did grow faster than the F. H. C. Society, which with its strictly limited membership was socially, though not intellectually, more exclusive. The ΦBK in its first four years admitted a total of fifty, but there were not nearly that many members present at any given moment. Students came and went. John Heath himself was no longer in college by the time the society reached the end of its second year.

Unusually brief was the stay of John Marshall, who was elected to the ΦBK on May 18, 1780, when he was twenty-four years old, much older than most members, still teenagers as they were. As a captain of Virginia troops, Marshall had fought under George Washington in the New Jersey and Pennsylvania campaigns, enduring the terrible winter of 1777–1778 with the Continental Army at Valley Forge. When the enlistments of his company expired, Captain Marshall went to Williamsburg to ask the General Assembly for replacements. While there, he attended the great George Wythe's law lectures at the college, but only for two or three months. Then he returned to Washington's army.[7]

According to the original bylaws, there was to be "a regular meeting of once a month, *unless* a necessity of sooner convening should *interpose.*" There could be a special meeting whenever the president deemed it expedient to call one. In 1778 the society changed the rule three times, so as to meet regularly every two weeks, then for a while, every week, and finally every two weeks again. During the four years of the society's early existence, the regular and "call" meetings totaled more than seventy. Normally these were held in the evening—by candlelight during the darker months—and in student rooms or other rooms at the college.[8]

Attendance was seldom noted in the minutes. The largest number recorded was thirteen, on December 4, 1779, but a society document of the same date was signed by nineteen, which suggests that this many members were then residing at the college or elsewhere in Williamsburg. At each of the final two meetings, on December 5, 1780, and January 6, 1781, only five were present.

Absenteeism had been a matter of concern all along. Fines for unexcused absences were repeatedly increased but remained so seldom paid that an interest charge of 25 percent a month was imposed on those in arrears. All this seems to have had little effect. Even officers and performers sometimes failed to appear. One of Heath's successors as president, living outside the college, was absent so often that the society appointed a vice president who roomed at the college and who therefore would presumably be on hand to take charge in the president's absence.

The main business of the regular meetings was the "literary business," that is, the holding of discussions and debates. Essentially the ΦBK Society was a forensic club. Unlike its predecessors, the F. H. C. and the P. D. A., it concentrated on a serious purpose, which was self-improvement in the art of public speaking. At its sessions the members were expected to "behave with a becoming decency" and to "declare their sentiments vicissively [alternately], preventing confusion." The "least appearance of intoxication or disorder of any single member, by liquor, at a session," would subject him to a heavy fine.

The programs consistently followed a pattern laid down in the bylaws. On each occasion, two members presented written papers on an assigned topic, and two others argued the same issue from opposing sides. Three judges decided the winners of the debates and selected the best compositions, those that would "do honour to the Society," for preservation in its archives (none of these manuscripts is known to exist). It was "strongly recommended to the other Members, as an additional and improving *Exercise,* to give their Sentiments ex tempore on the same Subject, after hearing the others." Whether or not the listeners tried this exercise, they usually discussed the subjects to some extent after the set programs were over.

In the society's meetings the students enjoyed a freedom of speech that under the "scholastic Laws" they did not enjoy in the college classes. This—and not merely a taste for the mysterious—accounted for the emphasis on secrecy. As the president said in welcoming each initiate: "Now then you may for a while disengage yourself from the scholastic Laws and communicate without reserve whatever reflections you have made upon various objects; remembering that everything transacted within this room is transacted *Sub rosa,* and detested is he that discloses it." Under the protection of confidentiality, members felt free to take up any subject that interested them, no matter how controversial it might be. They were willing to look at both sides of the question, and they were able to maintain a spirit of rationality and detachment in the midst of bitterly troubled times.

In their set programs they did not take up current events, but they did discuss some issues of immediate concern to Americans in general and Virginians in particular. For example, while Jefferson as governor was advocating religious freedom and the separation of church and state, they devoted one of their sessions to "the Advantages of an established Church," another to "Whether Religion is necessary in Government," and still another to "Whether a General Assessment for the Support of Religious Establishments is or is not repugnant to the principles of republican Government."

They did not go so far as to debate the validity of religion itself. Their original oath of initiation required them to "swear on the holy Evangelists of Almighty God, or otherwise as called the Supreme Being," and the very first of their bylaws called upon them to "invoke the Deity" in all their efforts to promote "fraternal prosperity." The ritual made no reference to Christianity, though the society kept on hand a copy of the New Testament. The society was by no means atheistic or antireligious, but apparently it did incline, in the intellectual fashion of the period, toward the deistic.

Two programs were given over to the basic issue of the ongoing Revolution—the issue of rule by a king versus self-rule in a republic. On one occasion the question was "Whether Commonwealths or Monarchies are more subject to seditions & commotions"; on the second, "Whether any form of Government is more favourable to public virtue, than a Commonwealth." Other contemporary problems on the agenda included "the justice of African slavery," a standing army in peacetime as a threat to civil liberties, advantages and disadvantages of allowing "Theatrical exhibitions," the prohibition of duelling, the promotion of "Agriculture or Merchandize," the permissibility of remaining neutral in a civil war (like the Revolution itself), public versus private education, and "Whether hath a wise State any Interest nearer at heart than the education of Youth."

Those were, even by present-day standards, debatable questions, since they had a bearing on contemporary policy. Some of the others related to past events—"Whether Brutus was justifyable in killing Caesar," "Had Wm the Norman a right to the Crown of Great Britain"—and were as speculative as they were historical. Still others were even more whimsical: "Whether all our Affections & Principles are not in some measure deducible from self Love"; "Whether the rape of the Sabine women was just"; "Whether Poligamy is a dictate of Nature or not." No doubt topics of this kind provided opportunities for fun as well as fancy.

But members of the society had the greatest fun when they gathered each year "on the 5th of December in Order to celebrate that glorious Day which gave Birth to this happy Union." They celebrated three anniversaries in the Raleigh, Williamsburg's leading tavern, which contained not only the ballroom and banquet hall known as the Apollo but also smaller meeting rooms suitable for the society's annual get-togethers. For the 1779 observance, the brothers met on December 4 (the 5th being a Sunday) in Davenport's tavern on the opposite side of the Duke of Gloucester Street, the Raleigh apparently being unavailable that Saturday night.

Even on these festive occasions the members put business before pleasure. First they sat through a declamation or two on some such edifying theme as "the progress of the arts & sciences." Then "the remainder of the evening was spent in sociability & mirth," or it was spent in "Jollity & Mirth." Once, in April 1779, a member who was about to depart for Europe invited his brothers to the Raleigh for a farewell party. He "gave them a very elegant Entertainment," and they drank "many Toasts suitable to the occasion." But the ΦBK indulged in no such tippling as did the F. H. C. and the P. D. A.

The ΦBK briefly considered sponsoring something besides forensics and festivities. "Whereas the Members of this Society are willing to take under their Care, Objects worthy of Charity," the clerk noted on November 21, 1778, "Resolved that Messrs. Hardy & Cocke be appointed to look out for some Orphan, likely to receive Advantage from being put to a proper School, & make their Report of the same to the ensuing Meeting." But Hardy and Cocke made no report (or if they made one, the clerk did not record it), and the society never went into such charitable work as the education of poor boys, nor did it take on any other new activity—except for the historically most important one of establishing ΦBK branches.

In their decision to expand the society, as well as in the development of their ritual, the ΦBK leaders were acting at least partially under the influence of Masonry. A Masonic lodge had existed in Williamsburg as early as the 1750s, and in 1773 it received a charter from the grand lodge in England. In 1778, as citizens of an independent commonwealth, the Masons of Virginia set up their own grand lodge, with authority to charter other lodges within the state. Eventually Masons in other states did the same.

John Heath himself was not a Mason while a student at William and Mary, but Thomas Smith belonged to the Williamsburg lodge before joining Heath as one of the five ΦBK founders. Smith served as the first clerk of the ΦBK Society and became its president on May 3, 1777. Nine other members of the society joined the Masonic lodge during the next year. At least a dozen of the fifty men admitted to ΦBK during these first four years were associated with both groups at one time or another.

The Masons in the ΦBK did not transfer to it intact the arcana of the Masonic order. In those arcana, nevertheless, they found models they could adapt for the society's grip, sign, and other rites. In the grand lodge's practice of chartering lesser lodges, they discovered a procedure by which their own organization could expand. The F. H. C. failed to follow the same example of expansion, though some F. H. C. members also belonged to the Williamsburg lodge. Apparently the F. H. C., unlike the ΦBK, professed no ideals that its members aspired to propagate.[9]

On May 8, 1779, the ΦBK brothers made the epochal decision to branch out. Their clerk recorded:

> It being Suggested that it might tend to promote the designs of this Institution, and redound to the honor and advantage thereof at the same time that others more remote or distant will be attached thereto, Resolved that leave be given to prepare the form or Ordinance of a Charter Party to be entrusted with . . . brothers of the ΦBK . . . with delegated powers . . . to con-

stitute, establish and initiate a Fraternity correspondent to this; and that a Committee be appointed of Mr President, Mr Stuart and Mr Beckley to prepare a draught of the same and report at the next Meeting.

The term "Charter Party" (from the French *chartre partie* or "divided charter") referred to an agreement to be written in duplicate on a single sheet of paper, which would then be torn or cut in two. One part would be kept by the William and Mary society, and the other would be given to the new branch. The chairman of the committee to draft the agreement, William Short, Jr., was the last of the society's presidents during its early existence at William and Mary. He was also a Mason, having joined the Williamsburg Lodge in November, 1778.

The committee was not ready to report for nearly two months. When it finally presented the draft of a charter, the society promptly adopted it. Charters were granted to three members, one after another, authorizing them to establish branches. These, located within the state, were to be subordinate to the original society—just as the Masonic lodges within the state were chartered by the grand lodge and were subordinate to it. The original society would be known as the Alpha (the equivalent of Number One in Greek), and each branch would be designated, in the order of its authorization, by one of the successive letters of the Greek alphabet: Beta, Gamma, Delta, etc. Membership would not be limited to college students or graduates; a branch was authorized for Richmond, where no college existed.

None of the projected locals actually came to life, and the charter plan would have amounted to nothing if it had been permanently confined to Virginia. The idea of broadening its scope occurred first to Samuel Hardy, another Mason and an early advocate of expansion, who had already obtained a charter for the Beta, the very first of the proposed Virginia branches. President Short was much impressed by Hardy's new idea, which had no precedent in Masonic practice. "I remember yet my surprize when he communicated to me his plan for extending branches of our society to the different States," Short was to write more than half a century afterward. "He expatiated on the great advantages that would attend it in binding together the several States."

Expansion beyond the Virginia borders would be feasible because the society, having ceased to limit its membership to William and Mary collegians, had admitted an alumnus of both Harvard and Yale. This was Elisha Parmele, a Connecticut native who had studied at Yale until its closing on account of the war and then had graduated from Harvard. At twenty-four, while hoping for a career in the ministry, Parmele went south to regain his health (he was to die an early death from tuberculosis). He found employment as a tutor in the family of a friend of Short, who introduced him to the ΦBK.[10]

Short and Hardy saw in Parmele an agent for carrying the society to Massachusetts and Connecticut. On December 4, 1779 (at the third anni-

versary gathering), Parmele petitioned for a Massachusetts charter, and
the brothers granted him one to "establish a Fraternity of the ΦBK" at
the "University of Cambridge." The Harvard branch would be denomi-
nated the Epsilon. This implied that, like the three Virginia locals already projected, the Massachusetts one would be inferior to the Virginia
original.

At a meeting the very next evening, Parmele requested certain modifications in the plan, and Short presented them to the members for their
consideration, "suggesting the necessity of making some innovations in
the form of Charter Party to introduce it properly and give it an extensive footing, in the State of Massachusetts Bay." Phi Beta Kappa would
certainly get a better footing there if the branch was named the Massachusetts Alpha and was assured equal status with the Virginia Alpha.
But Parmele wanted more than that. At Yale he had belonged to a
student literary society that was open in both membership and proceedings, and he now proposed to make ΦBK at Harvard resemble that
society at Yale. "Resolved that so much of Mr. Parmele's petition as
relates to the Establishment of a Phi. Society to be conducted in a less
mysterious manner than the ΦBK be not agreed to, as the design appears to be incompatible with the principle of this Meeting [that is, of
this group]." Nor was it agreed at that time to change the Massachusetts
Epsilon to Alpha.

Four days later, "desireous that the ΦBK should be extended to each
of the United States," the brothers granted a second charter to Parmele
"for establishing a Meeting of the same in the College of New Haven in
Connecticut, to have the same Rank, to have the same Power, and to
enjoy the same Privileges with that which he is empower'd to fix in the
University of Cambridge." The Connecticut branch was to be called the
Zeta. Thus, while it might enjoy the same rank, power, and privileges as
the projected "Meeting" at Harvard, it would not be quite equal, in name
at any rate, to the society at William and Mary.

Before long the society amended the charters so as to authorize an
Alpha in Massachusetts and another in Connecticut. The revised forms
began with this preamble:

> Whereas it is repugnant to the liberal principles of Societies, that they
> should be confined to any particular place, Men or Description of Men, and
> as the same should be extended to the Wise and Virtuous of every degree,
> or of whatever Country.—We the Members & Brothers of the ΦBK an
> Institution founded on literary principles, being willing, and desirous to
> propagate the same. . . .

The charters went on to specify that the original society's oath, initiation,
medal, salutation, and entire "Code of Laws, herewith transmitted,"
must also be the new fraternities'. But the new Alphas would be on a par

with the Alpha at William and Mary. The prospective Harvard and Yale brothers were to be assured:

> That you by this Charter be invested with the privileges of the Meeting Alpha of Virginia, in granting Charters, for the establishment of other Meetings any where within your State of . . . which Meetings are to stand in the same Relation to you, that the junior Branches of this Society [within the commonwealth of Virginia] stand in, to the Meeting [Alpha] here.

Equipped with his two charters, Parmele went back north to inaugurate Phi Beta Kappa at both Harvard and Yale.[11]

While the Massachusetts and Connecticut Alphas were getting a good start, the Virginia Alpha was nearing its demise. At the approach of the redcoats, the College of William and Mary had to close. The ΦBK president, William Short, called a meeting on January 6, 1781, "for the Purpose of securing the Papers of the Society during the Confusion of the Times." Only he and four others attended on that final day. "They thinking it most advisable that the Papers should not be removed, determined to deliver them sealed into the Hands of the College Steward to remain with him until the desirable Event of the Society's Resurrection," the clerk noted in the very last entry in the minutes. "And this Deposit they make in the sure & certain Hope that the Fraternity will one Day rise to life everlasting & Glory immortal."

Except for Short, the remaining ΦBK members, along with the rest of the students, soon left town. Short stayed on to study French in preparation for a diplomatic career. In a couple of weeks, a messenger brought him a packet of letters and documents from the newly organized Connecticut Alpha, and he sent its president a reply in which he reviewed the recent development of the society's aims:

> I am sure, my worthy brother, your own reflection and Mr. Parmele's information have made you sufficiently acquainted with the origin of the ΦBK here. Branching it out into its present form never occurred to its founders; this could not be expected, since no human institution ever attained its highest Degree of Perfection at once; for the great improvement we are indebted to some modern members. No sooner did I hear it proposed, than the advantages of the plan crowded into my view, and made me exert every nerve to carry it into execution. I saw that a little attention to the proposition would lead into an ample field of speculation, in which perhaps the Foundation of Continental Harmony and Union may be laid. A committee was immediately appointed to bring into Maturity this inestimable Embryo. At first the most extensive idea was of spreading it through the different parts of this State; but here we despise provincial distinctions—Virginia, Pennsylvania, Connecticut are terms admitted only for particular pur-

poses—they are merged in the comprehensive Idea, America. We know no reason for extending this Institution here only; all [reasons] operated equally for extending it to the other States. Those who carried their views beyond the present period and developed the hidden events of futurity, saw the immense fabric rising upon this Foundation.[12]

Amid the uncertainties of war, some of the Virginia brothers persisted in seeing a grand future for both their fraternity and their country, with the fraternity assisting the country to achieve greatness. One who shared the vision was John Beckley, now clerk of the House of Burgesses in Richmond. Beckley wrote to the president of the Connecticut Alpha in the summer of 1782, when the last battle had long since been fought but enemy troops remained on American soil and a peace treaty was yet to be made:

How long the madness of the human race will continue, and the ravages and distress of destructive war be permitted, is only for Superior Wisdom to determine; but in the event of a return of peace I trust to see the extended influence of the ΦBK in its numerous branches; and at no distant period, produce a union through the various climes and countries of this great continent, of all the lovers of literary merit, founded on the broad basis of a virtuous emulation, and which shall be no less happy and important in preserving the future peace and grandeur of these United States, than that confederacy which has led America to the present possession of national glory and independence, through innumerable difficulties and distresses; and to this great end, give me leave to hope, all our future exertions will be pointed.[13]

During the formative years of the republic, the Virginia brothers made collectively a remarkable contribution to its success. Three of the fifty saw action as lieutenants, two as captains, and two as majors in the Continental Army. Two served in the Continental Congress, two in the United States Senate, two in the House of Representatives (one of them being John Heath), and seven in the Virginia legislature. One was a delegate to the Philadelphia constitutional convention of 1787, and nine were delegates to the Virginia convention that ratified the Constitution. William Short was Minister Jefferson's secretary and then chargé d'affaires in France. John Beckley served as clerk of Congress under both the Articles of Confederation and the Constitution. Bushrod Washington, a nephew of George Washington, sat on the United States Supreme Court as an associate justice, and John Marshall was the greatest of all chief justices, the one who did the most to make the American nation a reality.[14]

But the society at William and Mary fell far short of meeting the high expectations of the more enthusiastic brothers. When the college reopened in 1782, those returning were too few to constitute a quorum, and they held no meetings. One of them, Landon Cabell, retrieved the society's minutes and took them home for safekeeping. (These records remained with the Cabell family until 1848, in the Virginia Historical Society at Richmond until 1895, and in the William and Mary library from then on.) The college had entered upon a long period of poverty and neglect, and Phi Beta Kappa was not to be reactivated there for seventy years.[15]

Meanwhile the branches did not grow nearly numerous or strong enough to form "the Foundation of Continental Harmony and Union," not even a union of "all the lovers of literary merit," as the Virginians had so grandiosely anticipated. Nevertheless, time was to bring close to realization, in a way utterly unforeseen, their hope that the fraternity would "one Day rise to Life everlasting & Glory immortal."

2

New Beginnings in New England

For thirty-six years (1781–1817) Phi Beta Kappa remained strictly a New England institution, existing only at Yale, Harvard, and Dartmouth. Well rooted at all three colleges, the transplants from Virginia thrived despite some chilly blasts in the more northerly clime. At various times students and college officials were jealous or suspicious of the societies, yet on each campus they succeeded in gaining general acceptance as well worth cultivating. Political and religious reactionaries on and off campus grew hysterically hostile, but the society withstood their assaults and emerged stronger than ever.

While in Williamsburg, Elisha Parmele had received a charter for Harvard first and then one for Yale. On his return to the North, however, he reached New Haven before Cambridge, and so he established the Connecticut Alpha before the Massachusetts Alpha. He initiated his brother, two other Yale students, and a Yale graduate in April 1780. These men held an organizational meeting later the same year, and by commencement time in September 1781, the membership of the Yale branch totaled thirty-two.

At that time the Harvard branch was just getting started. Parmele had delayed going on to Cambridge until the summer of 1781, when he was ready to take an examination for the degree of master of arts. In July he communicated to four Harvard students, who were just completing their junior year, "a plan of correspondence with a Society at New Haven in Connecticut, and Williamsburg, Virginia, by the name of 'the ΦBK'; instituted for the purpose of making Literary [i.e., scholarly] improvement." When the four seniors-to-be expressed a willingness to join, he gave them the bylaws to read, administered the oath to each, and made a presentation of the charter. The four immediately elected three of their number as officers, then adjourned, to meet again after the summer vacation. On September 5, 1781, they held their first regular meeting.[1]

Three weeks later the Harvard members, not realizing that the Virginia Alpha had already held its last meeting, instructed their president, Samuel Kendal, to send a letter to the president at Williamsburg. When no reply came from that direction, Kendal turned toward New Haven. "Our earnest desire to cherish those seeds of friendship, already planted by the mutual communication of good offices, has engaged us, though a younger society, to ask of you the favor, and invite you to the advantages of a literary correspondence," he wrote to the Yale brothers on March 23, 1782. "As we have the interest of the two Universities, as well as those of the brothers of the ΦBK, in view, we wish you to communicate whatever curious or important in the literary way may fall within your observation."

The president of the Connecticut Alpha, Henry T. Channing, responded in cordial agreement. He also sent, for the approval of the Harvard brothers, a copy of some new regulations that the Yale brothers were adopting. This cooperativeness moved Kendal to declare: "I conceive that the institution of the ΦBK will have a happy tendency to destroy the prejudices that too frequently subsist between different Universities; to make them act upon more liberal principles, and seek the mutual advantages of the several societies with which they may, by this institution, be connected." Before the Harvard society had reacted to the Yale amendments, Channing attended one of the Harvard meetings and decided that his own society ought to reconsider the changes. "I am clearly of the opinion," he wrote to Kendal when back in New Haven, "that the uniformity of the different branches is carefully to be preserved, and that the established Code of Laws is not to be materially altered without the concurrence of the respective branches."[2]

Thus, quite early, the prospects and problems of interbranch cooperation arose. For a few years the presidents of the Harvard and Yale societies lagged in the "epistolary correspondence" they had pledged to keep up. Then, in the 1786–1787 academic year, while the thirteen states were cooperating to form a new association, the Harvard and Yale societies resumed their cooperation—to form, on a much more modest scale, a new association of their own.

This came about upon "application being made by a Gentleman from Dartmouth College for the Introduction of the ΦBK into that seat of Literature." The application, which a recent Dartmouth graduate had brought to Cambridge, caused the Harvard members to discuss whether they "had a right to give a charter out of the State." They decided to consult their fellows at New Haven and Williamsburg. From Williamsburg they again received no reply, but they heard from New Haven that a Dartmouth man had also made application there.

The president at Cambridge now wrote to the president at New Haven:

> ...we would observe: 1st. That we think it expedient that a charter be granted to certain select characters at Dartmouth College. 2ndly. That we know of no immediate method to a communication with our Brethren at William and Mary's. And 3rdly. That a Charter cannot, consistently with the principles of this institution, be granted to a college in a state where the society is not already established, without the consent of all the branches. We think, notwithstanding, that if your sentiments upon the first point coincide with ours, we might venture, in the present instance, jointly to grant a charter, without waiting for the consent of our brothers at W. & M. . . . We have not the least reason to suppose that the branch of the society in Virginia would raise any objection to the proposal; on the contrary, we have no doubt but the same benevolence which first prompted them to extend the usefulness of the society beyond the limits of their own College, still warms their breasts, & would induce them to rejoice at this acquisition. . . .
>
> But if, at any future time, application should be made by another college for a Charter, it is our opinion that it should not be granted but with the concurrence of all the branches; and we should wish that this might be one condition in the charter to be granted at Dartmouth.

The president at Cambridge went on to suggest a procedure for authorizing the new branch. Either the Yale or the Harvard society could grant the charter; the person receiving it would take it to the other society, whose president would sign it and would "communicate the laws and secrets." In addition, he recommended that each branch ought to pass a law requiring the circulation of an annual letter that would inform the other branches of members admitted, officers elected, and "any remarkable events."

The president at New Haven answered that he and his brothers gave their complete assent to all this. Further, they were "under the painful necessity" of reporting that, as they had just learned from one of their brothers tutoring in the South, the society had "ceased to exist at W. & M." Though this was sad news, it meant that there could "certainly be no impropriety" in the Harvard and Yale branches going ahead on their own. As for the suggested annual letter, the Connecticut Alpha had already passed the requisite law. We "beg you to accept our warmest wishes that you may long continue to flourish, and diffuse the influence of Literature, Morality, and Friendship throughout Massachusetts and the neighboring States."

The Harvard brothers proceeded to diffuse Literature, Morality, and Friendship to the neighboring state of New Hampshire. After adopting their own annual-letter law, they instructed one of their number to tell his friend, a Dartmouth senior, about the joint Harvard-Yale decision.

He introduced his friend to the brothers at Cambridge, and they presented him with a charter, dated June 1, 1787, authorizing him along with two other "persons of honor, probity, and good demeanor" to institute the society at Dartmouth. After receiving confirmation from the Connecticut Alpha, the Dartmouth man went back to Hanover and "proceeded to admit to the secrets of the Society, as the law directs," four of his classmates. On August 20, 1787, the five met and declared that this "was to be considered as the foundation meeting of the Alpha of New Hampshire at the University at Hanover."[3]

The New Hampshire, the Massachusetts, and the Connecticut Alphas did not continue to observe for long the law requiring the circulation of an annual report of admissions, elections, and other news. By the time the Dartmouth branch was nine years old, some of its members had become concerned about its falling out of touch with the other branches—or so the president averred. "All the Brethren of our Alpha are doubtless solicitous that the Fraternal attachment in which the different Branches of our institution were first united should be maintained and cherished," the Dartmouth society's president wrote to the Harvard society's president in 1796. "I lately attended the Phibetian anniversary at Cambridge, and looking over the records in which the names are preserved I found none from Hanover later than those who graduated in the year 1789." This was a state of affairs that "ought not so to be."

Nevertheless, the correspondence soon lagged again, and the Dartmouth brothers hardly knew what was going on in the Harvard branch, except when one of them happened to visit it. After another ten years, in 1806, a Dartmouth man was temporarily much alarmed by what he observed at a Harvard meeting. A committee there reported that, whereas "the parent branch of the society no longer existed," whereas "the three offsprings had no mutual pledge for adhering to fundamental principles," and whereas apparently "the other Alphas had considered themselves at liberty to transgress . . . *constitutional barriers*," therefore "the *federative principle* was *extinguished*, & . . . their [Massachusetts] Alpha had a right, at pleasure, fundamentally to alter their constitution and laws, to erect new Alphas, or abolish their own."

"The avowal of such principles was abhorrent to every idea I had cherished of the mode of our existence," the visitor from Dartmouth wrote to his fellows back in Hanover. "I considered it as a stab at the vitals of the Society, or an aim at so total a disunion that, at no distant period, so little resemblance would be left between us that they would wonder we ever called them brothers."

At the Cambridge meeting the Dartmouth representative explained to his Harvard brothers that his branch had not, as some of them seemed to think, adopted new "internal regulations" of an unconstitutional sort. The Harvard members then voted to recommit the disturbing report and to "open a correspondence" with respect to certain proposed alterations in their own bylaws. "We are as averse to unnecessary & wanton innovation as any of our friends and brethren," the Harvardians assured the Yalees and the Dartmouth men, "but we believe that some changes in the original arrangements, received from William & Mary, may be made in perfect consistency with our obligations, & greatly promotive of our advantage."

> Altho the Parent Society is extinct, & tho some have their doubts whether the *children* retain a strictly & indissolubly *federative* character; yet we, at Harvard, hold as a primary and inviolable principle the affiliation or confraternity of each with every branch, & that every motive of decorum & expediency enforces an adherence to some general system, no part of which is to be altered except by the joint consent of all.[4]

So passed the 1806 crisis in Phi Beta Kappa federalism—a crisis reflecting, at least in the language used, the contemporary issue of state rights under the new Constitution of the United States. Thereafter the three branches of the society managed to cooperate from time to time, but despite repeated protestations of fraternalism, they still failed to maintain close and steady contact. The problem of interbranch relations was to grow increasingly complex with the chartering of additional branches from 1817 on, and the problem was not to be solved until after the establishment of the United Chapters in 1883.

The New England Alphas quite early came to fairly close agreement on rules regarding the election of new members. This was a matter on which they could find little guidance in the bylaws they had received from the Virginia Alpha. But soon the Harvard and Yale branches and then the Dartmouth branch were following more or less uniform election procedures.

At first the elections at Yale and at Harvard, like those at William and Mary, had been rather haphazard. They took place at irregular intervals, drew from various college classes, and filled no definite quota. During the first year at Harvard, for example, elections formed a part of the business at ordinary meetings whenever someone had a candidate to propose, and they brought in new members one, two, or three at a time. These initiates that first year included a freshman, a sophomore, eight juniors, fifteen seniors, and two graduates. The total merely happened to come to thirty-three.

As early as 1782, the Yale society decided to elect from the junior class, to do so at the end of the junior year (on or near the first Monday in July), and to limit the number to approximately one third of the class. Though the Harvard brothers declined to adopt the same rule, they actually followed a similar practice, electing a total of forty-six undergraduates during the five years 1783–1787, about a third as many as received the bachelor's degree during that period. The Dartmouth branch, at its founding in 1787, set a policy of choosing up to a third of the juniors on a day shortly before commencement.

Elections, as earlier at William and Mary, had to be unanimous. At Yale each member made a list of candidates acceptable to him, and the society admitted only those whose names appeared on everyone's list. At Harvard the brothers were "determined to admit none but such as support a good character in the literary [scholarly] as well as moral line." After a visit to Cambridge, the president of the Connecticut Alpha got the impression that the Massachusetts Alpha was "composed of the reputable, judicious, and amiable part of Harvard." He advised his brothers at Yale: "We cannot be too particular in refusing admission to the ignorant and immoral, if we wish to prove ourselves attached to the interest of our institution."

Only "select characters" with a high reputation for scholarship were to be considered eligible at any of the three colleges. But it was not always easy to decide just which of the juniors met this requirement. It was not possible to go by average grades, for in none of the three colleges were students yet graded in courses. The electing members could ask an instructor for his opinion and sometimes did so. For the most part, however, they had to rely on their own judgment, and this became increasingly difficult as classes grew in size. After an election meeting in 1816, the secretary of the Massachusetts Alpha noted: "The entire evening was consumed in unavailing debate, canvassing and balloting."[5]

All three branches assumed they had the right to elect alumni of their respective colleges in addition to the juniors. Whether they could also elect graduates of other colleges, or graduates of no college at all, as honorary members remained a question for some time. There was a precedent in the 1779 decision of the Virginia Alpha to admit others besides William and Mary collegians. This decision was recorded in the minutes—which lay forgotten at the Cabell plantation in Virginia—but was not reflected in the bylaws that Parmele had carried to New Haven and Cambridge.

The Connecticut Alpha finally resolved in 1790 that "a fair and rational construction of the Constitution warrants the admission of honorary members." At the same time, however, the Alpha resolved "That no person shall be admitted who has not received a degree at some University." Phi Beta Kappa in Connecticut proceeded cautiously with honorary elections and did so in Massachusetts, too, but not in New Hampshire. The Yale brothers were upset when the Dartmouth society welcomed into its membership a Yale alumnus whom they themselves had excluded as an undergraduate. They now raised the question whether any branch ought to admit a person who had failed of election at another branch.

Through their hospitality to honorary members, the Dartmouth brothers brought upon themselves much greater embarrassment than that. In 1806 they elected John Henry, a bright and plausible young Irishman living on a farm near Hanover. Six years later, just before the War of 1812, Henry gained sudden notoriety when President James Madison, in his war message, referred to him as a source of information about British deviltry. Henry, claiming to be a British agent, had passed on to Madison (for $50,000) a set of letters purportedly documenting a plot to detach New England from the United States and restore it to the British Empire. The Dartmouth society, upon learning of Henry's scam, promptly and unanimously voted to expel this man who had "forfeited all pretensions to moral character." Henry was one of three members whom the Dartmouth branch expelled in its early years. Another had turned out to be a drunkard, and still another a thief.[6]

According to the bylaws that had come with the Yale and Harvard charters, the new branches were to provide the original society with annual membership lists, but nothing was said about the new branches sending such lists to one another. Despite their repeated resolves, the New England Alphas failed to keep these exchanges up-to-date. From the start, they maintained their own manuscript rosters, and eventually they began to put out printed catalogs listing members' names either alphabetically or by class. These catalogs appeared irregularly, once every two or three years. The first one, Dartmouth's of 1806, gave the current membership totals for the Massachusetts and New Hampshire Alphas, but had no figures more recent than 1803 for the Connecticut Alpha.

Over the years, the numbers elected declined in proportion to enrollments, except at Yale. There, during the successive decades 1788–1797, 1798–1807, and 1808–1817, the new ΦBK members averaged 12, 17, and 20 per year, while the new B.A.s averaged 32, 48, and 53. At Harvard the figures were, for ΦBK, 14, 16, and 14; for B.A.s, 38, 47, and 55—at Dartmouth, 11, 12, and 11 in comparison with 33, 35, and 36. By 1817 the three New England branches had elected a grand total of about 1,500.

By that time the vast majority of these—close to 1,450 of them—were graduates. Some had been made honorary members after graduation; most had been undergraduates at the time of their election. A certain number remained on or near the campus as graduate students, tutors or professors, or interested alumni. Men of this kind formed a core that gave stability and continuity to the organization. They often became the officers—the presidents, vice presidents, secretaries, and treasurers— the men who set basic policies, arranged for anniversary celebrations, and corresponded with other branches.

The student members, totaling no more than fifty or so at any given time (after 1788), constituted the immediate society. During their senior year they were responsible for carrying on its principal activity, which continued to be discussion and debate.[7]

To college students in that period, forensics offered both a recreational and a vocational exercise. While undergraduates, they could enjoy an oratorical contest as, in the absence of organized athletics, the only available competitive sport. After graduation they could use a well-developed skill in public speaking whether they went into the ministry, the law, or politics.

They could not get this kind of entertainment or training from their college courses, even though professional training was the colleges' avowed aim. Congregationalists had founded Harvard (1636), Yale (1701), and Dartmouth (1769) for the same reason Anglicans had founded William and Mary (1693)—that is, to produce pious and learned men, especially ministers of the gospel. The curriculum of the New England colleges continued to reflect this purpose. At Yale, for example, it included not only four years of Latin classics but also four years of the New Testament in Greek, besides parts of the Old Testament in Hebrew and courses in philosophy, theology, and ethics. Other subjects taught were logic, mathematics, geography, and English grammar and rhetoric (composition).

At the time when Phi Beta Kappa appeared at Yale, every student there was required to take this entire set of courses, no matter what career he might contemplate. A student at Harvard or Dartmouth was compelled to follow a program similar to the one at Yale. During the 1780s Yale made Hebrew voluntary, and Harvard began to excuse from the divinity course those students not preparing for the ministry. None of the three colleges allowed the students many, if any, other choices during the next few decades. Educators justified requiring the standard courses, particularly Greek and Latin, on the theory that these would develop and discipline the mind and ready the student for any kind of intellectual work.

Believing as they did that memorization provided the best mental exercise, professors seldom were inspired to anything beyond repetitious and monotonous drill. Generally they used the recitation method, assigning a text and calling upon the student to regurgitate it. They allowed little opportunity for discussion and gave still less encouragement to original thought. For that, the student would have to look outside the classroom to the kind of extracurricular activities sponsored by such a literary society as Phi Beta Kappa.[8]

The undergraduate ΦBK members met at least one evening a month and, for some periods, as often as every Wednesday night—at first in a student's room, but later, more commonly in a college hall or a local tavern. Programs followed quite faithfully the pattern inherited from William and Mary. Usually two members presented papers, two argued extemporaneously on the same subject, and one or more served as judges. The topics varied in much the same way as they had done at the

Virginia Alpha. Some, fairly specific and timely, gave the participants a chance to express themselves on issues of the day. Others, more abstract and speculative, allowed scope for imagination and occasionally humor.

Reflecting current events, one of the societies discussed "Whether Benedict Arnold can be considered as a traitor" in 1781, the year that Arnold went to England to collect his reward from King George III after deserting from the American army and leading British troops on raids in Virginia and Connecticut. In 1787, when the convention met in Philadelphia to form a new constitution, some of the Phi Beta Kappa brothers were at least willing to consider the question "Ought females to share in Civil Government?" While, during the 1780s and 1790s, the Northern states were abolishing slavery, a topic for discussion was "Ought the slaves of America to be emancipated?" When the Southern states refused to go along with abolition, and antislavery leaders proposed a boycott of Southern goods, New England brothers debated "Whether it be lawful in us to consume the produce of slaves."

In that same period Alexander Hamilton proposed funding and maintaining (rather than paying off) the debt of the new federal government, and debaters asked: "Is a national debt a national benefit?" Noah Webster urged Americans to assert their cultural independence by spelling in a simplified and distinctively American way—not only "honor" for "honour" and "plow" for "plough" but "frend" for "friend" and "laf" for "laugh"—and society members discussed the merits of Websterian orthography.

Anglo-American relations remained touchy for some time after the Revolution. In 1794, the year of the unpopular Jay Treaty with Great Britain, Phi Beta Kappas considered "Whether it would be just in the Government of the United States to sequestrate British property." But in 1798 relations with Britain suddenly improved when the United States became involved in a "quasi war" with France. The topic for discussion now was the advisability of an Anglo-American alliance. During the undeclared war Congress passed the Alien and Sedition Acts, which made it difficult for foreigners to be naturalized and for newspapers to criticize the government. Contemporary subjects for debate were "Whether in a republic any ought to be admitted to the rights of citizenship but natives" and "Ought the Press to be unrestrained?" After Napoleon made a gesture toward improving Franco-American relations by selling Louisiana to the United States in 1803, students weighed the pros and cons of the purchase from the New England point of view.

On these and other occasions the Phi Beta Kappa members looked for discussion topics in the government's domestic or foreign policy of the moment. Sometimes they turned, instead, to religion, and when they did, they maintained the open-minded attitude of the Revolutionary period long after that attitude had ceased to prevail among the American people. Sample topics of the 1790s and early 1800s: "Whether the Mosaic account of the creation be consistent"; "Was the flood universal?"; "Whether Christianity, if fabulous, ought to be supported"; "Whether a law prohibiting deistical publications in a Christian land be just."

Meanwhile the brothers also concerned themselves with issues of higher education—issues that were to persist for two centuries or more: "Whether a student in the University ought to confine his attention chiefly to the study of his profession in life"; "Whether the study of Greek and Latin classics be at the present day a judicious part of the education of youth." Obviously these students did not take for granted the instructional regimen that was imposed upon them.

Nor did they hesitate to question the careers that many of them were planning to pursue. They inquired "Whether lawyers be beneficial to society" and "Is political ambition consistent with moral integrity?" Good questions, no doubt, for the 1980s and 1990s as well as the 1780s and 1790s. Another timeless issue considered was "Does intoxication palliate a crime?"

There was room for the facetious in the discussion of lawyers' and politicians' worthiness—and even more in the discussion of such frivolous topics as women's dress. After the Revolution, "licentious" French styles replaced the comparatively prudish English ones. The skirt was "fitted closely to the figure," the neck was "cut indecently low," and "the fewest possible underclothes were worn." This provoked the brothers to wonder "Whether the wearing of low bosoms, in compliance with custom and fashion, be any impeachment of a lady's modesty."[9]

At each of the three New England colleges, the ΦBK Society at large, including graduates as well as undergraduates, concentrated its activities mainly on the anniversary day. This was not December 5, as it had been at William and Mary. At Harvard it was originally September 5, and at Yale, the first Thursday in December, but at Harvard, Yale, and Dartmouth it eventually came to be the day before or the day after commencement, the time of which varied.

At all three there developed the practice of holding both a public and a private celebration. "We celebrated our anniversary in the Chapel with open doors," the president at Harvard wrote in regard to his branch's very first observance on September 5, 1782. "The students generally attended as well as some of the first characters in the State who were on business in the town." At each of the public gatherings, a ΦBK alumnus, like as not a tutor in the college, delivered an oration, and on most occasions another graduate presented an original poem. John Quincy Adams, Harvard '87, was the Harvard orator in 1788; he addressed an audience of forty. Daniel Webster, Dartmouth '01, was the Dartmouth orator in 1807, when he was a rising New Hampshire lawyer.

At the private anniversary meetings the societies initiated new members, selected future orators and poets, and decided other important matters. Then the brothers abandoned themselves to sociability, most often at a tavern. No doubt most of them thoroughly enjoyed the get-togethers, but a few were unenthusiastic. Benjamin Silliman, a seventeen-year-old Yale student who was to become the most famous American scientist of his time, noted in his diary on December 21, 1796:

> After supper I went to attend the anniversary of ΦBK Society. Almost all the principal gentlemen of the town were present. Mr. Tutor S—— delivered an oration upon the importance of virtue. The oration was short and I think was [spelling backwards] ton a yrev yranidroartxe eno. For my part I can say that I was detnioppasid ni a hgih eerged. After the oration we were served with Sherry wine and plum cake. A number of toasts were drank, some humorous stories were told etc., and the invited gentlemen retired. After transacting some business the Society drank the remainder of the wine, sung songs, etc. and broke up about ten. I am very glad that anniversaries do not return oftener than once in a year. After such surfeits I am always sick. I got into bed at eleven, and think that my evening might have been spent more agreeably and more profitably, at least as to my purse.

Year after year the Harvard branch scheduled its anniversary dinner in the same place it held its regular meetings, that is, in Porter's tavern. In 1801 a committee for arranging the annual affair "reported that Mr. Porter would provide a good dinner at one dollar for each member, that the price of wine should be Madeira $1. pr bottle, Lisbon & Sherry 50 cents, and Port 66⅔ cents." But the expenses could run higher than that, as an alumnus of '93 discovered in 1805. He recounted:

> Met the ΦBK in the Chapel of Harvard University. At twelve the members walked in long procession to the meeting-house where Brother Harris of Dorchester prayed. Brother Benjamin Whitwell of Augusta delivered an Oration on Philosophy. Although there was such a numerous concourse of Brethren, but thirty of us dined together at Porter's. I asked the blessing. Brother Dehon returned thanks. Such was the extravagance exhibited in giving $70 for the theatrical musicians, who were a nuisance, and in paying for twenty-five dinners which were not eaten, that I now feel resolved never to dine with the Society again. Expenses to each one was $5.70.

At the annual dinner of the Harvard branch, a collection was taken up for the "relief of indigent brethren." The initiation oath, after all, pledged every member to assist with "Life and Fortune" any brother in distress. Both the Yale branch and the Harvard branch set up funds for the purpose. A report of Harvard's 1802 anniversary stated in regard to its fund: "This has been raised by an annual contribution from the Society only, & has amounted to 250 dollars."

Another item of business at Harvard was the maintenance of a society library. This originated in 1785 and, financed by gifts and by a tax on members, grew fairly fast. Before long the annual meeting had to appoint a committee to make regulations for the use of books. By 1816, however, the library was said to be "in a state of alarming diminution" because of "the abuse of the privilege allowed honorary members from a distance taking books home." The ΦBK societies at Yale and Dartmouth did not maintain libraries; they did not need to, for other student societies at those colleges were already doing so.

Still another ΦBK undertaking was the sponsorship of a scholarly magazine. As early as 1800, the brothers at Harvard proposed to those at Yale and Dartmouth that they cooperate in bringing out such a periodical. The Yale and Dartmouth men showed little interest in the project, but the Harvardians continued to talk about it. With regard to Harvard's 1802 anniversary meeting the secretary wrote: "The subject of a literary publication in the name of the ΦBK was discussed, & the discussion terminated in the choice of a 'committee to consider the expediting of a plan for a literary publication from the Society,' which committee is to report at the ensuing anniversary."

The next year the *Monthly Anthology* appeared in Boston, and for nine years it continued to appear, changing its name from time to time. Three of its successive editors were Phi Beta Kappa—as were eleven of the fourteen founders of the Anthology Club, which soon took responsibility for the publication. Meanwhile, from 1804 to 1806, the Massachusetts Alpha itself edited and published eight numbers of the *Literary Miscellany*, each issue a hundred pages long. Then, in 1806, three Yale seniors, all members of the Connecticut Alpha, announced the *Literary Cabinet*, to be "published once a fortnight, on a half sheet in the octavo form," for the purpose of "improving the youth of this College in the art of writing." This, the earliest known undergraduate periodical, lasted less than a year.[10]

At the three New England colleges, as at William and Mary, Phi Beta Kappa was by no means the earliest of student societies. It had to find a place among other societies that were at least a few years and as much as a few decades older.

At Harvard a debating club had been organized as early as 1719, and in subsequent years the faculty encouraged the formation of other such groups. They came and went. The first one to endure was the Speaking Club, which originated in 1770. Its members listened to orations on such subjects as "The Odiousness of Envy" and, at the time of the Boston Tea Party, "The Pernicious Habit of Drinking Tea." The Speaking Club was apparently the only student group of the kind in existence at Harvard when Phi Beta Kappa arrived there in 1781.[11]

After only a year the ΦBK Society had already won a place on the Harvard campus as a most prestigious and influential organization—at least in the opinion of its leaders. "I have the pleasure to inform you," the president at Cambridge boasted to the president at New Haven, "that our society commands the attention of all who have any knowledge of it . . . and I suppose it does and will have great influence upon the conduct of the students, as each one seems desirous of being a member." At the public ceremonies on the first anniversary, the presence of distinguished Massachusetts citizens "seemed to give dignity to the society" and so did the behavior of the students in the audience. "The students paid us the highest respect, standing in their places till we walked out in order."

But five years later, after the anniversary meeting of 1787, John Quincy Adams recorded in his diary: "The students attended very generally except those of the Senior Class; who kept off, from a spirit of envy, all except D." The next year the ΦBK brothers resolved "that by reason of the dissentions in the Senior Class on account of the election of members, no more than ten members be chosen previous to the anniversary." Presumably the society, by postponing elections, could give the disappointed seniors less time to complain. Unappeased, a group of seniors petitioned the university government to prohibit the discriminatory practices of Phi Beta Kappa.

In 1789 the Harvard Board of Overseers appointed a committee, with John Hancock as chairman, to consider the seniors' petition. Hancock reported back to the board: "There is an institution in the University, with the nature of which the Government is not acquainted, which tends to make a discrimination among the students." He urged "the propriety of inquiring into its nature and design." To head off the threatened investigation, the brothers promptly voted "That the President of the ΦBK society be requested to make known to the President of the University, the laws & constitutions of said society." The overseers took no further action against it. Josiah Quincy, then a recent Phi Beta Kappa graduate and later the university president, recalled: "The influence of the Society was upon the whole deemed salutary, since literary merit was assumed as the principle on which its members were selected; and, so far, its influence harmonized with the honorable motives to exertion, which have ever been held out to the students by the laws and usages of the College."

The Harvard brothers congratulated themselves on having "the kind and active patronage of the Government of the University," yet they failed to get the full and official recognition that they desired. True, the officers of the university along with the state governor and other important personages graced the anniversary celebrations with their presence year after year. But those officers attended merely as individual guests and not as university representatives. When the society invited these men to be present in an official capacity in 1791, they declined on the grounds that "a compliance with this request would prove prejudicial to the society & the peace of the college." In other words, such favoritism would intensify the hostility to Phi Beta Kappa and increase the likelihood of violence against it.

Some students continued to complain, and in 1793 one of them took their grievance to the public press. He wrote a letter to the *Columbian Centinel* of Boston, signing himself Socius. Were it not for Phi Beta Kappa, Socius intimated, life at Harvard would be very nearly idyllic:

> The members of a class generally live in harmony, the first three years—by which time they have opportunity to lay the foundation of a lasting union—when lo! just as they have become ripe for friendship, and have entered their *last year,* the daemon of discord exerts her sway, and lets loose the spirit of faction and party. Now, this she effects by the instrumentality of a certain *supposed honourable institution,* to which she has not opportunity to admit members, until their senior year. The necessary qualification of candidates, for more effectually attaining her end, she artfully conceals, lest being known, no one should feel ambitious of becoming a member—or, lest all should have equal claims, and thus union be preserved. . . .
>
> Is it not time that an abuse of this kind be reformed? That the natural rights of society should be restored in the *literary republick?* And that superior merit and abilities should form the only distinction between Harvard's sons?

Such criticism at Harvard declined somewhat during the next few years. "There is not, at this time, that opposition to the Society, which I could wish," the society's president declared in 1795. "Violent opposition rouses to action, & we more sensibly feel the advantages of our institution." A year later the anniversary orator Timothy Bigelow, an alumnus of ten years, could boast of the society's achievement:

> No expectations were formed that the connexion [with Phi Beta Kappa] would afford a dispensation from any of the duties of scholastic life, or independently assure success in the world. But it was well understood that respectability here, and eminence and good fortune through life, were exclusively attached to ability, information and merit. To improve and increase these important qualities, give complete effect to the endeavors of our instructors, more fully accomplish the purpose for which we were placed here by our friends, as well as to cultivate harmony and establish a durable friendship, the plan of the Institution was approved and carried into effect. Experience has confirmed theory; and the utility of the scheme has equalled the expectations of its advocates.

"Brother Timothy Bigelow delivered a very pertinent oration," the society secretary at Harvard informed the brothers at Dartmouth. "In this the principles of the Society were explained in such a manner that we think its enemies will have reason to admire it." The secretary also said that since the "disorders" of 1793—disorders that "seemed for a short time to check the free exercise of the friendly principle, which is the Soul of our fraternity"—"the reputation of the Society has been rapidly increasing." And he reported to the Yale brothers: "Opposition has nearly ceased." Nevertheless, he cautioned both his Dartmouth and his Yale correspondent: "If you should write you will please direct the letters to myself as librarian of Harvard University which office I now hold, & not as corresponding secy. of the ΦBK which latter title would expose them to violence."[12]

Violence to ΦBK documents remained no more than a possibility at Harvard, but it had already become an actuality at both Yale and Dartmouth.

At Yale, as at Harvard, Phi Beta Kappa experienced no trouble in the beginning. It posed no serious threat to the existing Yale societies— Linonia, dating from 1753; and Brothers in Unity, founded in 1768— even though it sponsored much the same kind of activities. "The object of our fraternity," the Linonians proclaimed, is to provide the intellectual improvement of its members by the study and practice of forensic debate, by exercises in composition and elocution, and by the delivery at stated times of written orations and poems, and also to cultivate friendly and social feelings among members of all classes in college." Members of *all classes*! Both the Linonians and the Brothers in Unity welcomed freshmen along with other students. The two societies competed for membership, the more the merrier, and practically all the undergraduates joined one or the other. A student could not belong to both, but he could belong to either one and also, if he were lucky, to Phi Beta Kappa.

The president of the Yale society predicted in 1782, two years after its establishment, that so long as the society adhered to its principles "in the choice of members," it would "never sink into disrepute" but would "furnish an additional incentive to a regular demeanor, and also give an additional spring to literary pursuits." In 1786, however, three students broke open "the Secretary's door, in his absence, entered his study and feloniously took, stole, and carried away the Society's trunk with all its contents." Before long the three were caught and compelled to return everything they had taken, to pay for the damage they had done, and to swear they would "confine within their own breasts all the knowledge of the secrets of the Society which they had criminally obtained." Then, the very next year, the records disappeared again, and this time they were gone for good.

The minutes having been lost, one of the brothers reconstructed the history of the Connecticut Alpha as best he could from memory and from correspondence that had been spared. "This Branch continued to increase and become more respectable in the view of all the students of the College," he wrote. "The younger classes were sensibly ambitious of recommending the members [i.e., themselves] to the Society by regular behavior and uncommon exertions in scholarship. At the same time the candidates in the successive senior classes who finally failed of the honor of an election, were mortified and irritated by disappointment"—hence the burglary of the society's archives.

The Yale brothers were willing to recognize and to discuss the criticisms of their fraternity but remained confident of its worth. In 1790 they debated the question "Is the Society beneficial to this University?" The decision for the affirmative was unanimous.[13]

A year after that, some students at Dartmouth similarly debated "Whether the society known by the name of Phi Beta Kappa is advantageous to this University." The listeners had no particular reason to favor the affirmative, however, for they and the debaters belonged not to Phi Beta Kappa but to the United Fraternity, which had been in existence only a year when the Greek-letter society appeared in 1787. At that time the "Fraters" already faced bitter rivals in the "Socials," the members of the three-years-older Social Friends. "But," a historian of the New Hampshire Alpha was to relate,

> as the Phi Beta Kappa made no pretensions to rivalry in regard to a library, and drew its members on the basis of personal merit indiscriminately from both the other societies near the end of the course, and by its aristocratic position and connections abroad held out hopes of both honor and advantage, it was with little delay received by the others into fellowship; and yet the same reasons that made it tolerable by the older societies gave occasion for individual jealousies and rancor, that repeatedly came near destroying the entire system [of undergraduate societies].

During the first year at Dartmouth, student hostility to Phi Beta Kappa was so intense that its leaders wondered whether it could survive. "Till our late election it remained doubtful whether we should be able to resist the violent opposition made to our establishment, or whether we must have suffered the effects of envy and ignorance in ceasing to exist," the vice president wrote in 1788 after the first anniversary meeting. "We were fearful of a combination among the students and an agreement not to join the Society, but not one refused who was invited." Its success seemed to equal the brothers' "most sanguine expectations."

Before the second anniversary, however, the New Hampshire Alpha became, like the Connecticut Alpha, the victim of a burglary. Three seniors entered the room of one of their number while his roommate, a ΦBK member, was absent. According to their confession, they "in a clandestine and scandalous manner broke open the chest . . . took from it papers belonging to the ΦBK Society . . . and likewise published them, not only among the students of Dartmouth College, but at Cambridge." The ringleader had "taken unwearied pains to undermine and overthrow said Society," his motives being admittedly "envy and malice."

As late as 1803, a few of the Dartmouth students were still trying to overthrow the society. "There was a general outbreak of a spirit of destruction in a clique of reckless dissipated boys, occasioned by jealousy of elections made by Phi Beta Kappa," as the society's historian was to put it. Daniel Webster's brother Ezekiel, a Phi Bete by virtue of the most recent of those elections, wrote excitedly to Daniel about the "little affair," which had so "discomposed" him he could hardly hold his pen. According to him, "the conspirators aimed at the abolition of every Society in College."[14]

At Dartmouth, however, Phi Beta Kappa was, of all the societies, the most frequent object of student jealousy, and at Harvard and Yale it was the only such object. The students who were excluded from it would hardly have been provoked to anger if membership had not become a mark of considerable prestige. Hence this very hostility, persisting as it did, betokened the high reputation that the fraternity had achieved by the early 1800s. But the fraternity was already facing other attacks of a much less flattering and much more threatening kind.

The threat arose from a peculiar conjunction of politics and religion. A presidential election was coming up in 1800. While waging their undeclared war against France, the Federalists under President John Adams charged that the Republicans under Vice President Thomas Jefferson favored the enemy and subscribed to the radical, godless doctrines of the French Revolution. Conservative churchmen blamed Jefferson and the French for the decline of religiosity in the United States. Evangelists were about to start a recurring wave of revivalism that would put the Devil and the Deists on the run. And Phi Beta Kappa became an object of the politico-religious zeal.

Of all the zealots, none aroused hotter indignation among Federalists than did the president of Yale College, Timothy Dwight. In his baccalaureate address of September 9, 1797, Dwight warned the students and others in his audience about "The Nature and Dangers of Infidel Philosophy." Religion (of the proper Christian sort), he said, brought the believer "progress, life, wisdom, virtue, happiness, and glory." But philosophy consigned its devotees to "annihilation and despair." And by philosophy he meant the teachings of Voltaire and Voltaire's followers, among whom he included Jefferson.

Dwight thought the peril imminent. He could cite as an incontrovertible authority the just-published book by the University of Edinburgh's Professor John Robison, *Proofs of a Conspiracy against All the Religions and Governments of Europe, Carried on in the Secret Meetings of Free Masons, Illuminati, and Reading Societies.* The Illuminati were presumed to be underground agents of a mysterious organization headquartered in Bavaria. In a 1798 Fourth of July speech, "The Duty of Americans at the Present Crisis," Dwight said that true Masons in the United States were not "implicated" in the worldwide conspiracy, but he insisted: "Illuminatism exists in this country." He demanded: "Shall our sons become the disciples of Voltaire, and the dragoons of [the French revolutionary extremist] Marat; and our daughters the concubines of the Illuminati?"

Good Federalists among Dwight's listeners were deeply impressed. One of them acknowledged that the speech had opened his eyes to

Jefferson, whom he had previously considered nothing worse than a designing politician. Now he realized that Jefferson was "the *real Jacobin*, the very child of *modern illumination*, the foe of man, and the enemy of his country."

Good Federalists among the Phi Beta Kappa members, listening to Dwight's harangues or reading them in pamphlet form, could hardly avoid twinges of concern and even of guilt. Plainly a secret society could be a devilish thing, and *they* belonged to a secret society, which had originated in Virginia at the College of William and Mary—the state and the college of the Jacobinical Jefferson himself. And the society's very name stood for "philosophy the guide of life," which was precisely the satanic error that Dwight warned against. Perhaps God-fearing, right-thinking members of the thing ought to terminate it before it developed its potential for mischief.

Such considerations led to the publication of the following notice in the *Connecticut Courant* of Hartford on July 29, 1799:

> A number of the members of the ΦBK Society, convinced by an attention to the history of modern times, that secret societies have been improved as engines of intrigue and iniquity; and considering the *time, place,* and *circumstances* which gave birth to this society, and the probable designs of its institution which are further deducible from the import of its initial letters, and apprehensive that this society, tho at present harmless, is liable to be abused to the infidel and seditious purposes of other secret associations, do request a general meeting of said society, at the State House in New-Haven, on the day succeeding the approaching Commencement at 7 o'clock A.M., to take into consideration the expediency of abolishing the Alpha of Connecticut, and of corresponding with the other branches of the society with design to effect the dissolution of the whole institution.
>
> *The several Printers in the state are requested to insert this Advertisement in their papers.*

ΦBK undergraduates at Yale did not wait to see what ΦBK alumni throughout the state would do about this advertisement. The undergraduates chose as their very next question for debate: "Is it expedient to abolish this Society?" The negative won by a unanimous vote.

As for the brothers at Harvard, they promptly held a special meeting to consider what to do in response to the proposal that the Connecticut Alpha had made "on account of the suspicions to which all secret societies are liable in the present dangerous & critical situation of the country." A committee appointed at that Harvard meeting prepared a "remonstrance," which ten days later the society approved and ordered sent to the Connecticut and New Hampshire Alphas. The remonstrance read:

> A strict examination of the publication, which has led to this address, has we trust enabled us to discover the motive, that induces its authors to wish for

our dissolution. We think it is a fear that public opinion may attack or general prejudice may depress our society. We however strongly doubt if even in this era of jealousies and mistrust we shall suffer from general censure, supported as we are by the virtue, talents, & eminence of many of our members, & shielded by that propriety of conduct, which has obtained us the flattering meed of public approbation. But let us suppose the contrary, & we ask shall public opinion be respected farther than is just? Shall it not when degenerated to unmanly suspicions & patronizing ungenerous surmises meet with our contempt? Yes—Virtue will not be frightened by popular clamors from the performance of its duties. Philosophic ardor shall not hence be intimidated to restrain its inquiries, nor friendship to break its silken cord. Animated by a consciousness of right the noble mind rises superior to opposition. Should it be our fate then to be traduced, let us as individuals boldly profess our attachment to our society;—let us declare to teeth of clamor, that it is not only harmless, but virtuous in its objects, & useful in its effects;—that the circumstances of its origin here [whatever the circumstances of its origin at William and Mary!] indicated, not a design to sow infidelity with sedition, but a benevolent wish to enlarge the heart & improve the mind; & that our initials are only expressive of a submission to true wisdom from a love to true virtue. Should we meet the rude shock of persecution let us stand firm & undaunted, steady in our resolutions, & more energetic in our exertions.

No doubt the same line of reasoning occurred to ΦBK members in Connecticut. At any rate, the result was anticlimactic when they finally held their anniversary meeting in the New Haven State House on September 12, 1799. "Nothing material was transacted," according to the minutes. "There was some conversation on the expediency of abolishing the Society, but the Society determined to do nothing respecting it."

This was perhaps a victory of sorts for academic freedom, and yet it did not mean complete freedom for Jeffersonians to express their views. To give the oration at their next anniversary on September 8, 1800, the predominantly Federalist brothers at Yale invited Abraham Bishop, a past president of their society but also an ardent Jeffersonian. They disinvited Bishop after they learned he was going to talk on "political subjects to advance the interests of a party by abusing the confidence of the Society and involving the members in that political turmoil which disgraces our country." Friends of Bishop then hired a hall for him in New Haven, and on the evening of the ΦBK exercises, he held forth to a large audience of his own. Speaking on "The Extent and Power of Political Delusion," he charged the Federalists with misusing the pulpit for partisan ends.

Later, on March 11, 1801, a week after Jefferson's inauguration as president of the United States, Bishop reminded his fellow Republicans at a rally in Wallingford: "Some pious good men have been made really to believe that all the meeting-houses and bibles were to be burnt, and that Jefferson was to be the high Priest of Infidelity." In a pamphlet containing the Wallingford speech, Bishop recounted his treatment at the hands of his erstwhile brothers. "Why am I not expelled from the P. B. K. society?" he challenged them. "You will not do me that honor. But from this moment I cease to be a member . . . and join myself to the great community of unprivileged men."[15]

While favoring the Federalist party, most of the New England brothers continued to expect their meetings to be free from partisan politics—by which they meant Republican politics. This was as true at Harvard as at Yale, as Alexander H. Everett learned when, addressing the Harvard branch in 1813, he raised some questions about Edmund Burke's conservative view of the French Revolution. "These intimations, though moderately expressed, and with no application to our politics," Everett recalled long afterward, "gave dissatisfaction to the Federal portion of the audience, which comprised, probably, nine-tenths of the whole." The society nevertheless voted, as usual, to publish the speech, but to no avail. "The chairman, a strong Federal partisan, was so much scandalized by the heterodox character of the politics, that he abstained from performing his duty, and the address has consequently never been printed."[16]

Politics threatened to wreck both Dartmouth College and the New Hampshire Alpha when the Republicans got control of the state government in 1816 and tried, in effect, to convert the private college into a state university. For a time the "university" authorities contested with the college authorities for the custody of the campus and the loyalty of the students. Undergraduates in the ΦBK Society continued to meet, but on May 1, 1817, the date for choosing candidates from the junior class, the minutes said "the peculiar state of the College rendered it impossible to elect at once."

Some of the men on each of the rival boards of trustees and all members of the two contending faculties were ΦBK brothers. When the case of *Dartmouth College* v. *Woodward* went to the United States Supreme Court, the chief attorney for the college was Brother Daniel Webster, and the defendant personifying the "university" was Brother William H. Woodward, recently president of the New Hampshire Alpha. Three brothers—Joseph Story in addition to Bushrod Washington and John Marshall—were among the judges who finally decided in favor of the college.

After the Supreme Court's decision, the Dartmouth Phi Beta Kappas on the winning side celebrated at an anniversary dinner, where Webster himself deigned to appear and to speak as the orator of the evening. "A large number of our most distinguished brothers met each other with mutual congratulations for the deliverance of our present institution from the arbitrary interposition of legislative power," the minutes read; "and while the numbers present seemed to give strength and respectability to the Society, a luster was thrown around it by the presence of him, who had so dexterously and successfully wielded the Aegis of the Law."

The ΦBK members on the losing side took no part in these festivities.[17]

Phi Beta Kappa had survived all the attacks upon it. What the Massachusetts Alpha had asserted in 1799 was even truer by 1817: the society could stand up against "jealousies and mistrust" because of the "virtue, talents, & eminence" of many of its members.

By this time a remarkable number of them (in addition to Webster, Marshall, Washington, and Story) held positions of local and even national prominence, though ever since the demise of the Virginia Alpha there had existed only the three New England branches. While the largest number lived in New England, others were scattered throughout the rest of the states, the District of Columbia, Canada, and several other countries. These men were mostly lawyers, clergymen, or schoolteachers, but also physicians or businessmen, and they were very heavily concentrated in the comparatively few available posts as college presidents or professors. One of the best known was Eli Whitney, inventor of the cotton gin and pioneer in the technology of mass production. Two were leading members of the incoming James Monroe administration: John Quincy Adams, the secretary of state, and John C. Calhoun, the secretary of war.[18]

3

Limiting the Fraternity's Growth

The Virginia founders of Phi Beta Kappa had once dreamed of extending it throughout the United States and even overseas. Their New England successors revived the idea from time to time. Though "our institution is now confined to a few Seminaries in this country," an orator declaimed at Harvard in 1796, "what obstacle exists to its further extension, or even a communication of it to foreign universities?" One obstacle did exist—the high standard that existing branches maintained. They refused to charter a branch at any college they considered inferior to their own. And they chartered so few that by 1847, when the number of colleges and universities in the country had mushroomed to more than 150, the number with ΦBK branches had grown to only 8, all except one of these in New England.

Even if ΦBK leaders had been more eager than they were to expand their fraternity, they would have been hard put, in evaluating sites for branches, to keep up with the proliferation of institutions of presumed higher learning. Several of the original states chartered public universities soon after achieving independence, and most of the newer states did so soon after gaining admission to the Union. Meanwhile the states chartered several times as many private colleges. These attracted the greater share of both students and resources.

Most of the private colleges arose from the waves of evangelicalism that swept the country, especially the newly settled areas of the South and West, during the last decade of the eighteenth century and the first decades of the nineteenth century. The various Protestant denominations competed with one another in an effort to direct religious life on the advancing frontier. Some Protestant leaders discovered a new and

37

more dangerous rival when Roman Catholic immigration from Germany and Ireland rapidly began to increase in the 1840s and 1850s. They feared that the Jesuits, through the "unobtrusive, unobserved power of the College," were scheming to give the Catholic church control of the Mississippi Valley. "There, Brethren, there our great battle with the Jesuit, on Western soil, is to be fought," one missionary declared. "We must build college against college."[1]

In this contest the Presbyterians, Methodists, and Baptists would seem (by 1860) to have come out ahead. At any rate, they had founded, respectively, the largest numbers of colleges. The Catholics were in fourth place, though they could claim less than 15 percent of the total of church-sponsored institutions. In proportion to church membership, the Episcopalians, Presbyterians, and Congregationalists, in that order, led the race.

Colleges had sprung up much more frequently in the less populous than in the more populous regions. Between 1800 and 1850 the number in New England increased from eight to only fifteen; in the Middle Atlantic states, from eight to thirty-two; and in the South Atlantic states, from eleven to thirty-seven. But in the Old Southwest the number grew from five to sixty-one; and in the Old Northwest (the Midwest), from zero to sixty-seven.[2]

This disproportionate increase was due not only to the zeal for evangelizing the newer settlements but also to the booster spirit of the frontier. Promoters wanted their infant town to grow into an Athens of the West, or at least they wanted it to grow, so as to bring about a rise in land values. They assumed that a college would attract people, and people of the finest type. Hence they and the leading citizens in general gave their support to the establishment and maintenance of a denominational college when one was offered, whether they belonged to the sponsoring denomination or not. The college proceeded to draw students from the community as a whole.

To attract students—when the available ones were so few and the competing institutions so numerous—the college typically kept admission requirements low and broadened its aims beyond the original one of preparing members, ministers, and missionaries for a particular church. Sometimes it offered practical courses, intended to train young men for occupations besides the ministry or the other professions. But it could seldom afford to provide much variety, for the student body was usually quite small (averaging much fewer than one hundred) and the faculty still smaller in proportion. The professors were mostly clergymen, giving only part of their time to teaching. Financing such an institution was difficult, even with the motivation of serving both God and Mammon. Many colleges, having received their charters, never materialized. Many others, having opened their doors, failed to keep them open longer than a decade or two.[3]

Among the surviving institutions in the South and West, there was considerable variation in quality. Some were colleges or universities only in name, consisting merely of a preparatory department while intending to add a more advanced program or remaining permanently nothing more than an academy or preparatory school. Others provided a fairly high level of what passed for higher education at the time, but none could match, in prestige at any rate, the well-established colleges of the Northeast. Nor could any of the state universities, for they as yet received little tax support and, in their faculty and offerings, differed little from the ordinary denominational colleges.

Having chartered these colleges, the states seldom made any effort to regulate them. In New York, however, the State Board of Regents as early as 1811 set minimum standards for financial backing, then asserted a right of visitation and later demanded annual reports. The president and trustees of Union College refused to comply, maintaining that the Supreme Court's decision in *Dartmouth College* v. *Woodward* protected Union from interference by the state government. The Dartmouth decision, incidentally, encouraged the founding of private colleges by making their charters inviolable. The decision also encouraged the establishment of new state universities, since it indicated that existing private institutions could not easily be converted into public ones.

At least in respect to courses taught, the denominational colleges tended toward uniformity as a result of the influence of Yale. This was the largest institution of higher learning in the country (having in 1840 an enrollment of 410 as compared with 341 for Dartmouth, 270 for Union, and 236 for Harvard). Many Yale graduates went out to the West and the South to help found colleges and to teach in them.[4]

Yale exerted an extremely conservative influence. Members of the faculty there reaffirmed their time-honored educational philosophy in a famous 1828 report. Why favor recitations over lectures? Lectures, the report explained, do not develop the same sense of responsibility in the student. "To secure his earnest and steady efforts is the great object of the daily examinations or recitations."

> But why, it is asked, should all the students in a college be required to tread in the same steps? Why should not each one be allowed to select those branches of study which are most to his taste, which are best adapted to his peculiar talents and which are most nearly connected with his intended profession? To this we answer, that our prescribed course contains those subjects only which ought to be understood, as we think, by every one who aims at a thorough education.

The required subjects, especially the Greek and Latin classics, were the best adapted to the college's main purpose, which was "to discipline the mind."

Yale dominated the American Education Society, which from 1815 on solicited contributions from benevolent Christians throughout the country and used the money for scholarship aid to "hopefully pious" and deserving students in the various colleges. The A. E. S. insisted that the colleges, to be eligible for its philanthropy, must conform to the Yale standard of classical instruction. In 1838 the A. E. S. cut off funds to students at the Oberlin Collegiate Institute after finding the institute deficient in Latin and Greek. The society resumed the funding after Oberlin had mended its ways.

Thus the American Education Society functioned as a kind of normative or accrediting agency so far as the content of instruction was concerned. The society—and the attitude it represented—discouraged changes such as those that Professor George Ticknor proposed at Harvard. Ticknor had been at Göttingen in 1815, when the great German university boasted 840 students, 40 professors, and 70–80 lecture courses, among which the students were free to choose. His field being modern languages, he was particularly eager to see an increased number of them taught as electives at Harvard. The university did add a few electives. Some lesser colleges boldly experimented with alternative curricula, such as a "scientific" program paralleling the "literary" one but not requiring Latin or Greek and not leading to a Bachelor of Arts degree.

In his book *Thoughts on the Present Collegiate System in the United States* (1842) President Francis Wayland of Brown University, while advocating curricular expansion and variation, noted that this idea had made little headway. The courses of study, he wrote, "in all the Northern Colleges are so nearly similar that students, in good standing in one institution, find little difficulty in being admitted to any other." But while there was considerable uniformity in curriculum (and Wayland no doubt had in mind only the well-established colleges of the Northeast), there was far less uniformity in financial stability, faculty salaries, admission requirements, and the adequacy of libraries and other facilities.[5]

Which of these multifarious institutions ought to qualify for hosting a unit of such an elite organization as Phi Beta Kappa prided itself on being? The existing ΦBK branches themselves refrained from taking the initiative in expansion. Applications for charters had to come from individual ΦBK members interested in starting branches at other places. In considering the applications, the society developed at least rudimentary qualification standards.

The Yale and Harvard branches early set a precedent against accepting applications from men not connected with any college or university. In 1783 two Yale and two Harvard graduates living in Albany, New York, asked their brothers at Yale to charter a "meeting" that would be "known by the name of the Alpha of the State of New York." The Yale brothers took no action. Seven years later a similar foursome, two with Yale and two with Harvard degrees, met in

Augusta, Georgia, and petitioned the Harvard brothers for a Georgia Alpha. The brothers referred the petition to a committee, which declined to report on it. Unattached groups, such as those in Albany and Augusta, could have formed Phi Beta Kappa associations without securing a charter, but almost a century was to pass before such associations began to appear.

When the application came from students at Dartmouth in 1786, the Yale and Harvard branches approved it with no serious question except, for a while, a question as to the propriety of going ahead without the authorization of the men at William and Mary. But the Yale, Harvard, and Dartmouth branches could not agree to grant any of the next several requests coming from college students or faculty.

In 1789 Abel Flint applied on behalf of a group at Rhode Island College, a comparatively small institution which the Baptists had founded in 1764 and which had operated in Providence since 1770, except for a few years during the war when American and French troops used its buildings as barracks. Flint, a young tutor at Rhode Island College, held A.B. and A.M. degrees from Yale. His Yale brothers readily voted to approve the chartering of a Rhode Island Alpha on condition that the Harvard and Dartmouth brothers should concur.

When the Yale men met for their 1790 anniversary, Flint was among them, expecting to receive his charter or at least assurance that it would be forthcoming. With him were two recent graduates of Rhode Island College, one of them Moses Brown, whose older brother was to provide money and a new name for the college. Flint proposed the election of his two companions as honorary members of the Yale society. The society obliged him by electing them but disappointed him by reversing itself and rejecting his application for a charter. A letter had just arrived with news of the Harvard branch's nonconcurrence and with persuasive reasons for it.

After receiving Flint's petition "for extending the fraternity to Providence College," the Harvard branch had referred the matter to a committee consisting of four youthful tutors and a new graduate. At the next anniversary meeting the committeemen reported "that they thought it inexpedient to transmit a charter to that College as it would lessen the dignity of the society & multiply similar applications from literary institutions of still less importance." The Harvard brothers approved this report "pretty unanimously," and their corresponding secretary sent letters to Hanover and New Haven to explain their action.

The letter to New Haven read in part:

> The grounds of our determination were a number of facts . . . which compelled us to conclude that students at Providence college were by no means possessed of literary qualifications equal to students of the same standing at this college, Yale or Dartmouth. The facts were these. The government of Providence college had admitted into their Sophomore class, one of our

students in the middle of his freshman year, who was admitted into our
freshman class with difficulty and when he went away was by no means
equal to many of his classmates. The same government has admitted into
their Sophomore class, young gentlemen, whose Preceptors told us that at
the time of their admission they were not qualified to enter our freshman
class, according to our laws and regulations. The same government has
been known to admit a person who was rusticated at this college to a bachelor's degree before the class in this college to which he originally belonged
was graduated here.

Obviously Rhode Island College did "not require so great literary qualifications for admission" as Harvard, Yale, and Dartmouth did. Considering the "spirit of our [ΦBK] charter and institution," the Harvard brothers wondered "whether it was compatible with that to transmit a charter
to any college the members of which could not be presumed to possess
literary qualifications equal to those of the members of those colleges in
which our fraternity was already established." The Harvard men "could
not avoid concluding in the negative."

> Another reason reported by the committee, and which had its weight with
> us, arose from the small number of students at Providence college. We tho't
> that, should a charter be transmitted in compliance with the application,
> there would be danger that many might be admitted to our fraternity,
> merely to make its numbers respectable, without sufficient regard being
> had to their literary merit.

Members of Phi Beta Kappa had to bear in mind both their objective to
"encourage knowledge and benevolence" and their "dignity as a band of
literary brothers."

> Much of the utility of such a fraternity as ours depends, as you must well
> know, upon its respectability in the eyes of spectators. Should those of the
> students at Harvard, Yale or Dartmouth whose literary merit does not
> entitle them to admission amongst us, see a branch of the Φ B K established
> at Providence, and at the same time justly feel their equality and perhaps
> superiority in literary merit to the generality of students at Providence of
> the same standing, the Society would in our opinion lose very much of its
> influence in promoting emulation, and have a direct tendency to breed ill
> will and animosity.

The Massachusetts Alpha had voted, "from motives of delicacy," to
make no response whatever to the application from Providence; "but, if it
should be repeated, to reply in the negative, as politely and tenderly as
possible, provided the determination of the Alphas of Connecticut and
New Hampshire should coincide." These determinations did coincide.
The decision was not kept secret from the applicants, however, since their
agent Abel Flint was in New Haven with his Yale brothers when they
considered the report from Cambridge and voted to concur with it.[6]

Still, the application from Providence remained unanswered, and "a partial request of the same kind" soon came to Yale from the College of New Jersey at Princeton. An expression of interest also came from Williams College in Williamstown, Massachusetts. There was no formal application from either Princeton or Williamstown.

In 1797 the Connecticut Alpha nevertheless "Voted that the corresponding Secy be directed to correspond with the other Alphas upon the policy of extending this society to Princeton, Rhode Island & William's colleges." Princeton, founded by the Presbyterian church in 1746, had been a college of some distinction for more than half a century, much longer than Dartmouth. Williams, beginning as an academy in 1791, had been operating at the college level for only four years. But—what was more to the point—eight of the Yale brothers had been serving as tutors there. The Yale secretary assured the Dartmouth and Harvard branches in regard to Williams: It is "flourishing and bids fair to be a very useful Institution."

The Dartmouth secretary, William H. Woodward (he of later *Dartmouth College* v. *Woodward* fame), replied that his brothers favored giving charters to Princeton and Williams as well as Providence "provided regular application be made." "It is our opinion that it would not be consistent for us to grant a Charter to Williams College and continue to refuse Rhode Island," Woodward added. "The objections which existed some years since in regard to Rhode Island, it is thought are not so weighty now. That Institution is supposed to be growing into respectability, and considering the age of Williams College we think that we ought to grant the request from Rhode Island if the Society be any further extended."

When the Dartmouth letter arrived at Yale, there was still no regular application from either Princeton or Williams, and there were "some doubts being expressed respecting the willingness of Princeton College to accept of a charter." In any event, the Yale brothers at their 1798 anniversary found themselves in utter disagreement with the Dartmouth brothers. "The Alpha of Connecticut appeared almost unanimous," as its secretary recorded, "that it would be best to grant a charter to Williams college, but not to Providence or Princeton colleges." The reasons were "some facts injurious to the character of these two Colleges."

As yet, no reply had arrived from Cambridge. Apparently the secretary there had never received the inquiry from New Haven about establishing the three new branches. But the Harvard brothers did get mail from the Williams tutors, who had at last deigned to make formal application. (No one at Princeton bothered to do so.) When the Yale brothers finally heard from Cambridge in 1799, they learned only that the Harvard brothers had received the Williams application "but had not come to any determination on the subject." If the members of the Massachusetts Alpha had wished, they could have obliged the Williams group without even consulting the Connecticut or the New Hampshire

Alpha, since their charter from William and Mary empowered them unilaterally to grant other charters anywhere within the Commonwealth of Massachusetts.

The upshot was that no new charter was issued for the time being, nor was one forthcoming when some men at Union College in Schenectady, New York, applied in 1803. No college, with men desiring to organize a branch, could yet qualify to the satisfaction of all three existing branches.[7]

All three subscribed to certain qualifying standards, which were neither sharply defined nor exactly measurable but which nevertheless served as general guides to judgment. The curriculum itself did not enter in, since this was assumed to be pretty much the same from place to place. Taken into consideration were the difficulty of admission, promotion, and graduation and the size of the student body. The students must be numerous enough and their progress difficult enough that they could be expected to compare, in the distribution of scholarly ability, with the students at Dartmouth, Harvard, and Yale.

These standards were hard to meet, at least in the opinion of the men who imposed them. No other college qualified until thirty-six years after the granting of the charter to Dartmouth.

A fifteen-year-old student had taken the lead in founding Phi Beta Kappa at William and Mary. Seniors, recent graduates, and youthful tutors were responsible for extending the society to Yale, Harvard, and Dartmouth. Such people were also responsible for the unsuccessful atempts to extend it to Rhode Island (Brown) and Williams. Then men of greater age and higher rank— college presidents and professors—took the initiative. They succeeded in getting branches for Union, Bowdoin, and Brown.

Union College had obtained its charter of incorporation from the state of New York in 1795 (a charter that prevented denominational control by providing that a majority of the trustees must not belong to one church). Thus Union, slightly younger than Williams, had been in existence only eight years when some of its friends sought a ΦBK branch for it in 1803. But the college grew fast, especially after Eliphalett Nott took over its presidency the following year. Though Nott had never been a college student, he had received an honorary M.A. degree from Rhode Island College after passing with distinction all the tests given to seniors there. He became a powerful Presbyterian preacher and then a very successful college administrator. By 1820, Union was to have exactly as many students as Harvard—234—though it was to have only four professors and two tutors in comparison with Harvard's twenty-one professors and seven tutors.

In 1813, ten years after the first try, President Nott encouraged some of the students to make another attempt to get a ΦBK charter for Union

College. Fourteen seniors signed a letter to be sent to each of the existing branches. Nott added a postscript certifying that the signers were "young gentlemen of unblemished moral characters and of respectable literary acquirements," were "worthy to become members of the Society in question," and had "obtained permission to forward the foregoing application."

The New Hampshire and Massachusetts Alphas gave prompt approval to this petition from Schenectady, but the Connecticut Alpha delayed action for three years. Finally, in 1816 a committee at Yale reported that they had "examined the records of former proceedings on similar applications from Brown University and Williams College" and were satisfied that "the reasons which prevented grants to those institutions" would not "operate to prevent a grant to Union College." The committee also noted that earlier "a similar request was preferred by Union College to the Alpha of Massachusetts and by that Alpha rejected on the ground that the literary attainments made at Union College were unequal to those usually made at colleges where Alphas were established." Still, since the other two branches had already approved the more recent petition, the committee recommended that the Yale branch do the same, and it did so.

But the charter, which the Yale committee drafted, did not go to the fourteen seniors at Union who had petitioned for it. Instead, it went to three friends of the college, the foremost of whom was James Kent, Yale '81, a member of Yale's first ΦBK class and now chancellor (the highest judicial officer) of the state of New York. Kent and his fellow "commissioners" met in 1817 at his home in Albany and organized the "New York Alpha of Union College." They elected several members, starting with the college president, Eliphalet Nott, whose newly created brothers then elected him fraternity president.

Bowdoin College, the next to receive a ΦBK charter, was somewhat younger than Union, having begun in Brunswick, Maine, in 1802. In 1820, the year Maine became a state, the four existing Alphas received from Bowdoin a proposal for a fifth, the Alpha of Maine. The application was signed by fourteen friends and faculty members of the college, all of them already "Brethren of the ΦBK," including the college president, William Allen, who recently had been president of the ill-starred "Dartmouth University." Eleven of the fourteen signers were members of the Harvard branch.

The Harvard brothers were understandably enthusiastic, and at their annual meeting in 1822 they voted overwhelmingly in favor of acceding to the Bowdoin request. But the Yale members disapproved it. They believed that Bowdoin "was not superior to many Colleges in the United States to which the ΦBK could not, consistently with the design and interests of the Society, be extended." The Harvardians urged the Yalees to reconsider, and the next year the Yalees reversed themselves by a unanimous vote. The Dartmouth men having failed to act, the Harvardians also appealed to them, and they "voted to come over with the other Alphas of New England." For a time, the Alpha of New York still held out, postponing consideration because its members were unacquainted with the "merits and standing" of Bowdoin and unfamiliar with the qualifications usually required. Finally, after a four-year wait, the Bowdoin men obtained their charter, and in 1825 nine of them—eight of whom were Harvard graduates—organized the Maine Alpha.[8]

A Rhode Island Alpha soon followed. At the providence college in 1797, after the failure to obtain a charter for a ΦBK branch, some of the alumni and professors had formed a local society on a similar pattern, the Federal Adelphi Society. This was the first of what in time were to be a great many imitations of Phi Beta Kappa, each confined to a single campus. At this one college—known as Brown University from 1804 on—the Federal Adelphi took the place of Phi Beta Kappa for more than thirty years.

At its 1826 anniversary meeting, the Federal Adelphi voted that a committee "take into consideration the expediency of this Corporation applying to become a Branch of the Phi Beta Kappa Society at Cambridge." Months later the Adelphi secretary wrote diffidently to the Massachusetts Alpha to say an application would recently have been made "had it not been that, thirty years ago, a similar application . . . was made and rejected." He now inquired "whether an application in the present state of things might be successful." He got no encouragement from Cambridge.

By this time Brown University had a new president, who soon arranged to have Phi Beta Kappa dispose of Federal Adelphi, not by incorporating it but by replacing it. The thirty-one-year-old Francis Wayland, a member of the New York Alpha, inherited at Brown a fractious faculty that had been responsible for the resignation of the previous president. Wayland provoked the faculty further when he undertook to carry out various reforms, among them a rule that every professor must reside in college (like an Oxford don) and give his entire time to teaching and proctoring. The leaders of the resistance to Wayland were also leaders of the Federal Adelphi. He determined to weaken the resistance by destroying the society and to do that by bringing in a branch of Phi Beta Kappa. Shrewdly he laid his plans.

First, in 1828 Wayland prompted a Boston friend to find out how to proceed from Edward Everett, recently a professor of Greek at Harvard and now the president of the Massachusetts Alpha while also a member of Congress. "With regard to the establishment of new branches of the Φ B K.," Everett responded, "I believe the course to be pursued is not marked out, by any written rule of procedure nor any distinctly ascertained usage." In the case of Bowdoin, he went on, "the brethren of the Phi Beta Kappa resident at Brunswick" had sent a written request to each of the branches, and all of them had given their approval. "There is a good deal, which is loose in this mode of proceeding, & I think it would be well to have a regular course, adopted by the consent of all the branches." Meanwhile, the thing to do would be to follow the Bowdoin example.

Next, Wayland obtained a petition bearing the signatures of ΦBK friends of Brown, including Harvard graduates living in the Boston-Cambridge area. Then he wrote a letter of application in which he stressed those features of his institution he thought most likely to qualify it in the judgment of the existing branches. He pointed out that Brown was, "after Harvard University and Yale College, the oldest college in New England." He also declared "that its requirements for admission are equal to those of the oldest colleges in New England; and its course of study is as elevated; and that its officers devote themselves exclusively to the business of instruction." Their success, he intimated, could be seen in the large number of Brown's distinguished alumni. (He did not mention Brown's enrollment, which in 1830 was only 101 in comparison with Yale's 346, Union's 268, Harvard's 248, and Dartmouth's 172.)

Wayland sent copies of this letter to Harvard, Yale, and Dartmouth but not to Union or Bowdoin. Everett, who earlier had advised getting the consent of all the branches, now wrote him: "I agree with you in opinion, that it is not necessary to obtain the assent of more than the three older branches, for the erection of a new one."

Finally, following up his epistolary appeal, Wayland enlisted powerful friends to present his case in person—among them the eminent Justice Joseph Story, who was about to become a Harvard law professor while keeping his seat on the Supreme Court; and the distinguished Judge James Gould, who ran a famous law school in Litchfield, Connecticut. Story assured Wayland: "I will certainly attend our annual meeting at Cambridge . . . for the purpose of urging your Claims; & I cannot doubt, that they will be successful." Gould informed Wayland: " . . . your college is as justly entitled to a Branch of the Ph. B. K. society, as other colleges, which already have branches. And tho' I shall not be able to attend the next anniversary meeting of the Alpha of Connecticut; you shall not want the *benefit,* (if you esteem it so), of this opinion, fully & decidedly expressed, to those, who may attend."

Under this kind of pressure, the Harvard branch readily gave in, and soon the Yale branch and then the Dartmouth branch followed suit. In 1830 Wayland and two other "commissioners" met in his office and elected to the Rhode Island Alpha the entire faculty of Brown University. Like Nott at Union, Wayland at Brown became the branch's first president.

Wayland had succeeded in introducing Phi Beta Kappa before his enemies in the Federal Adelphi realized what he was up to. Some of the F. A. members on the faculty declined election to the new organization

and, along with other members, many of them prominent Rhode Is-
landers, undertook to keep the old one alive. The Adelphi is "to Brown
University precisely what Phi Beta Kappa is to Cambridge," declared the
Providence Journal, pleading for Brown's own society as against the "foreign Institution." In 1831 the ΦBK branch met on commencement day
to hear Wayland orate in observance of the first anniversary, and the
F. A. assembled the following day to listen to a different orator. But after
that, Phi Beta Kappa had everything to itself at Brown, and Wayland no
longer had to contend with a hostile organization on his campus, for the
F. A. soon ceased to exist.

The brothers at Union College did not seem to mind the chartering of
the Providence branch without consultation of all the Alphas, but it
rankled the brothers at Bowdoin for a long time. In 1831 the secretary
of the Maine Alpha informed Everett at Harvard that "a communication
was received, purporting that it proceeded from a Rhode Island Alpha,"
and questioned whether such an Alpha could possibly exist, considering
the unanimity rule. As late as 1838, the brothers at Brown learned "that
some members of this Alpha had, on presenting themselves at the Alpha
of Maine at Bowdoin College, not been recognized as members of the
Fraternity at Large." In 1839 the Bowdoin brothers voted "that the
subject of the irregular establishment of an Alpha at Brown University
be recommitted to a committee." Finally, ten years after that irregular
establishment, the committee recommended that the Maine Alpha take
no further action in regard to it.[9]

As Everett had remarked about the method of granting charters,
there was much that was "loose" in the "mode of proceeding." The
procedure long since agreed upon, however, was not quite so loose as
Everett imagined it. Everett also said: "In earlier times, a single branch
admitted other branches." This was not so—except, of course, in the
instance of the admission of the Yale and Harvard branches at a time
when there existed only a single "branch," the original one at William
and Mary. But Everett was right in saying "it would be well to have a
regular course" of procedure. It would also have been well if he had
observed the precedent already set and had insisted on unanimous
agreement instead of encouraging the departure from it. Over the coming years the problem of procedure in expanding the fraternity was to
persist, growing more and more difficult before eventuating in a work-
able solution.

After the chartering at Brown in
1830, fifteen years elapsed before the establishment of any more
branches. Then two new ones appeared, both of them in New England.
As yet, not one was to be found farther west than Schenectady, New
York, which lay only about a dozen miles west of the Hudson River.

None existed even as far south as New York City. This was not due to an utter lack of applications from the city or from places in the more distant South and West.

As early as 1818, an application had come to Harvard from the College of South Carolina (predecessor of the University of South Carolina), the first state institution to express an interest in Phi Beta Kappa. The document bore the signatures of five students who had been "selected by a respectable portion" of their fellow students to make the request. Accompanying and endorsing it were letters from four men: Abraham Nott, a South Carolina lawyer and a ΦBK graduate of Yale; Jonathan Maxcy, the college president (not a ΦBK member); John C. Calhoun, the secretary of war; and John Quincy Adams, the secretary of state. Maxcy had sent the application and the Nott letter, along with his own, to his South Carolina friend, Calhoun. Calhoun had then added his testimonial and passed all the papers on to Adams, and Adams had forwarded them with his personal note to the secretary of the Massachusetts Alpha.

It was clever of President Maxcy to take this roundabout approach through men so highly placed and influential. And certainly he had persuasive arguments on his side. He himself pointed out that at South Carolina College "a regular course of instruction similar to that pursued in the Northern colleges" was "supported by a President, four Professors, and two Teachers; and a strict course of discipline maintained through the classes." The five students testified: "From the unexampled liberality of the legislature in providing for the excellence and elegance of our building, their appropriations for its future support, the enjoyment of an extensive apparatus and library, and the distinguished abilities of the Faculty, we most reasonably anticipate that our Institution will ere long assume a distinguished station among the seats of Science." Calhoun declared that the college, now in its twentieth year, already was "decidedly the most flourishing in the Southern and Western section of our Country."

The South Carolina students even included an appeal to patriotism in their application. Sectionalism was growing apace as the third Southerner in a row occupied the national presidency—Jefferson, Madison, and now Monroe—and more and more Northerners complained about what they called the Virginia Dynasty. There was an increasingly precarious balance between free and slave states, a balance about to be threatened by a bill to admit Missouri as a slave state. "No bond, whether literary or political, by which the members of the American family may be united is unimportant," the Carolinians now urged the ΦBK societies of the North. "Your association, by its extent and importance, will vastly contribute towards fashioning our national character, by dissipating the sectional jealousies which now exist."

Unknowingly, the South Carolina applicants were repeating the dream which had inspired the Virginia members to expand their society decades earlier—the dream of "the extended influence of the ΦBK in its numerous branches," an influence that would tend to bind the states together and would be "important in preserving the future peace and grandeur" of the country. But this idea, along with the other arguments on behalf of South Carolina College, failed to move the New England inheritors of the society of those Virginia idealists.

Having received the South Carolina documents "through the medium of the Hon. John Quincy Adams," the secretary of the Massachusetts Alpha sent the other Alphas copies of the application itself but not of the accompanying recommendations. He also sent a reminder that "the acquiescence of every existing branch of the Society" would be necessary. The Dartmouth brothers disposed of the application thus: "Resolved, that this Society is not in possession of such information as would justify them at the present time in granting the prayer of the petitioners." That was the end of it.[10]

Almost twenty years later, in 1837, applicants at Hampden-Sydney College in Virginia noted that there was still no branch of Phi Beta Kappa anywhere in the South. There ought to be one, and their college was qualified for it, according to the petitioners, among them William Maxwell, Yale '02, who was shortly to become the college's president. The institution, they said, was "of more than fifty years' standing"; indeed, they could have traced its beginnings, as a creation of the Presbyterian church, all the way back to 1776. It had a fine reputation because its "standard of scholarship" had always been high; its faculty consisted of "very distinguished men"; and its alumni included "not a few whose influence, both in Church and State," was "felt through the length and breadth of our land."

Again, the existing branches were unimpressed, and none of them made a positive response. At Harvard, Professor Henry W. Longfellow gave an unfavorable report, and the brothers voted to pigeonhole the application. They thought "the Alpha of the State should be established in the chief literary institution of the State," and that was not Hampden-Sydney but the recently founded University of Virginia. Hampden-Sydney was "a very respectable College," a committee at Yale reported. "Owing, however, to want of funds and other untoward circumstances, its prosperity has been impeded and its number considerably reduced, its average number being only about forty." The Yale society therefore decided "to postpone any decisive action on the petition until the prospects of the Institution offer sure grounds to hope for its permanent usefulness and prosperity." (Hampden-Sydney was to wait more than a century for a ΦBK charter.)

Men at Georgetown College, which Baptists had founded in Kentucky in 1787, received even less consideration when they applied to the Harvard branch in 1841. The Harvard members routinely voted that it was "not expedient to grant the request at present."[11]

Considerable discussion, however, followed even after the Dartmouth and Brown societies had turned down the 1836 application on behalf of the University of the City of New York, an institution then only four years old. Since a New York Alpha already existed at Union College, the question arose whether the Union branch could grant the University of the City of New York a charter on its own authority and whether the new branch would be another Alpha or merely a Beta and, as such, inferior to the Alpha. The Union brothers sent a letter of inquiry to the Yale brothers. According to its own charter, the Connecticut Alpha had the power of chartering branches in the state of Connecticut by itself, and these would presumably be dependent Betas, Gammas, and so on.

The Yale brothers now (in 1837) decided that applicants were to be granted charters "on the same uniform and impartial principles, whether the said applicants be the first from any State or not, and the united consent of all the Alphas to be obtained." This meant that all future Alphas, Betas, Gammas, etc., were to be created equal. It also meant that the Union brothers could not by themselves issue a charter to the University of the City of New York or to any other institution in the state. Neither could the Yale brothers act independently with respect to institutions in Connecticut.

Just six years after that, in 1843, the Yale brothers took under consideration a petition "received from Washington College praying to become a Beta of the Phi Beta Kappa that they might see and taste its glorious fruits under the shadow of their own vine and fig tree." Washington College, already in operation for two decades under Episcopal auspices and shortly to be renamed Trinity College, was located in Hartford, Connecticut. It had (in 1840) only seventy-seven students. The Yale members voted at their next annual meeting to give a charter to the college without referring the petition to any other Alpha—despite their previous decision that in such a case as this, "the united consent of all the Alphas [was] to be obtained." The Yale men justified their action by noting in this 1844 charter for the Connecticut Beta that in the 1779 charter for the Connecticut Alpha, "authority was delegated and given to the said Alpha to grant Charters for the establishment of Branches of the Society in the State of Connecticut."

At the same 1844 meeting, after granting the application from Trinity College, the Yale members appointed a committee to consider one from Wesleyan University, which the Methodists had launched in Middletown, Connecticut, as recently as 1832. The committee reported in 1845 that Wesleyan already held "such a rank among the Literary Institutions of the Country" and had "such prospects of permanent usefulness and respectability" that a branch of the fraternity might "suitably be planted there." Actually, Wesleyan's prospects were not yet very good; its enrollment fell from 147 in 1840 to 104 in 1850. No matter. The Yale brothers voted to establish at Wesleyan a Gamma of Connecticut "with all the rights and privileges enjoyed by other Branches." The people at Wesleyan had sent copies of their application to the rest of the Alphas too, but action or inaction on the part of these was now irrelevant.[12]

Clearly, Phi Beta Kappa was yet to develop consistency or system in qualifying colleges or universities for branches of the society

4

Transition to
an Honor Society

From the very beginning, while primarily a social fraternity and a forensic club, Phi Beta Kappa had also been something of an honor society. In the course of time, it lost all but a playful pretence to secrecy and the mysteries of fraternalism, and it put increasing emphasis on academic performance, as faculties took control from students and based elections more and more on grades. Meanwhile, it inspired the establishment of new Greek-letter fraternities, which took over its abandoned function of keeping the arcana of collegiate brotherhood. And for the most part, it ceased to carry on its undergraduate discussions and debates, leaving these to the existing local and unexclusive campus literary societies. Eventually, by the 1840s, there was little or nothing left for Phi Beta Kappa to do—except to honor scholarship.

Strictly speaking, Phi Beta Kappa was never a secret society. Its members always prided themselves on belonging to it; they certainly did not pretend that it was nonexistent, nor did they attempt to conceal their membership in it. Like Freemasonry, Phi Beta Kappa began as a *society with secrets,* but unlike Freemasonry, it eventually gave them up in response to public clamor against what people called secret societies.

Hostility had died down but had not died out since the excitement of the 1790s, when President Dwight of Yale warned Americans about the (imaginary) international conspiracy of the Illuminati. Suspicions about Phi Beta Kappa persisted. Was it foreign in origin, radical in tendency, and perhaps somehow related to the worldwide antireligious plot? Even members of the fraternity wondered about these questions.

In 1819, only two years after the founding of the Union College branch, its corresponding secretary wrote to Thomas Jefferson: "The Brothers of the New York Alpha of the Phi Beta Kappa Society have directed me to request of your excellency the communication of any information you may be in possession of, in relation to the Introduction of said Society into this country, as it has been told to the Society, that you first brought the charter from Oxford or elsewhere to William & Mary College in 1776 or thereabouts." Jefferson could only say: "I am an entire stranger to the P. B. K. society, its history and its objects."[1]

No doubt the suspicions regarding Phi Beta Kappa arose in part from an anti-intellectual strain in the evangelicalism of the period. As James Gould, the Litchfield law-school proprietor, told his Yale brothers in his 1825 oration: "It has been supposed by many that deep research, especially in physical and metaphysical science, or abstract learning in general, has a tendency to promote scepticism and free thinking, and consequently licentiousness in morals." Or as another of the Yale brothers, class of 1821, recalled long afterward about the society: "There was no iniquity in its secrets; they were merely puerile—except that those mystical initials, *Φ. B. K.* 'writ large,' were susceptible, though not necessarily, of an infidel or a heathen sense."[2]

Hatred of secret societies flared to new extremes after the disappearance of a man named William Morgan, who was last seen in 1826 in the vicinity of Niagara Falls. Morgan had been about to publish a book purporting to expose the secrets of Masonry, and his friends and sympathizers quickly concluded that vengeful Masons had done away with him. More Americans than ever became convinced that the "brethren of the mystic tie," favoring one another in law and politics, formed a dangerously undemocratic clique. Indeed, they seemed to run the country.

The president of the United States himself, Andrew Jackson, was one of them, a loyal member of the lodge. Some politicians opposed to Jackson took up the Antimasonic agitation and made it the basis for the Antimasonic party, the first third party in American history. One of its Pennsylvania leaders, Thaddeus Stevens, had been embittered when, as a Dartmouth student a dozen years earlier, he failed to win election to Phi Beta Kappa. Of his more fortunate classmates, he wrote at that time: "Those fawning parasites . . . must flatter the nobility, or remain in obscurity; . . . they must degrade themselves by sycophancy, or others will not exalt them." As for himself, he would not defer to the "patricians"; he was proud to remain merely one of the "poor plebeians." Now, as an Antimasonic politician, he had a chance to get back at the patricians of all kinds.[3]

One of the Dartmouth ΦBK members, Samuel L. Knapp, did the society no good when he replied to the Antimasons in *The Genius of Masonry, or a Defence of the Order* (1828). Knapp praised Phi Beta Kappa by associating it with the Illuminati! "To these men [the Illuminati] we are indebted for the spirit of philosophical investigation of the present age," he argued. "A branch of the illuminati is now found in this country under the name of the Phi Beta Kappa." Brought here by Jefferson, it "exists only as connected with seminaries of learning in the United States."

With such notions in the air, it was no wonder that the foes of secret societies should include Phi Beta Kappa as one of their targets. "The political Anties are extending their plan of attack. They have now pointed their artillery at the Odd Fellows and Phi Beta Kappa societies," the *Rochester Republican* observed in 1829. "It is expected that . . . an 'excitement' [will] be raised against the Phi Beta Kappatarians—as, whether this society be diabolical or not, it certainly has a very 'hard name.' "[4]

The most damaging attack came with the publication, in Boston in 1831, of Avery Allyn's *A Ritual of Freemasonry . . . to Which Is Added a Key to the Phi Beta Kappa.*" Allyn said he was including Phi Beta Kappa in his exposé because, though not yet guilty of the crimes of Freemasonry, it too was "a *secret* society" and was "as susceptible of being perverted to unholy and dangerous purposes." More than that, he charged: "It is a species of Freemasonry, and bears a strong affinity to it; and for aught I know, may be a younger branch of the same tenebrous family." It had been "imported into this Country from France," supposedly by Jefferson. Its letters *ΦBK* meant that philosophy was the guide of life, whereas in truth the Bible ought to be the guide. "But here a vain, imported, and infidel philosophy is exalted into the place of divine revelation, and that Holy Book, which contains the words of eternal life, is superseded and set aside."

Allyn managed to reveal the deepest secrets of Phi Beta Kappa. He explained not only the meaning of *ΦBK* on one side of the medal but also what *S. P.* on the other side stood for—*Societas Philosophiae.* (These words, *Societas Philosophiae,* were in time to be forgotten again, and for many years the initials would remain mysterious to the most scholarly members of the society itself.) Allyn even provided sketches illustrating the society's grip and sign: two figures clasping each other by the wrist with the right hand, and another figure passing the back of his right hand across his mouth. After these revelations, Phi Beta Kappa was no longer much of a secret society or even a society with secrets.[5]

Edward Everett, president of the Massachusetts Alpha, was seriously troubled when he read the part of Allyn's book bearing on Phi Beta Kappa. He promptly wrote to William Short, the last president of the Virginia Alpha and at that time living in Philadelphia, to ask about the society's pedigree. "Do you happen to know whether it came to William and Mary from abroad, and if so from what Country? Had it in its origin any specific object further than literary improvement, in general? And what? Why was it made a secret society?" Short immediately replied: "I can assure you that it was purely of domestic manufacture." It was founded "in opposition" to another student society that had "lost all reputation for letters," and presumably it was made secret "merely to follow the example of that which preceded it."[6]

Everett was not satisfied. He thought the society, at the very least, ought to "drop the affectation of secrecy." But perhaps "it would be best to abolish it altogether . . . unless such a liberal change can be made in the terms of admission as to make it a comprehensive Fraternity of the children and friends of the College." By means of a newspaper advertisement, he hastily called a special meeting at the Academy of Arts and Sciences in Boston "to consider the propriety of revising the charter and fundamental laws of the Society."

This was an illustrious gathering. Among the fifty to sixty members present was Everett himself, the congressman with a Ph.D. from Göttingen, who had a long career in politics, diplomacy, and public speaking ahead of him (he was to share the platform at Gettysburg in 1863 with Abraham Lincoln). Joseph Story, the Supreme Court justice and Harvard law professor, now at work on the first of several monographs that were to make him an unrivaled authority on American jurisprudence, was also present, as was John Quincy Adams, the sixty-four-year-old ex-president of the United States, about to begin a new and notable career as a congressman from Massachusetts. At the moment, some Antimasons were mentioning Adams as their candidate for the presidency at the next election.

Once Everett had opened the discussion in the Academy of Arts and Sciences, Adams "enquired what other branches of the Society were known to exist," as he recorded in his diary, "with a view to propose some communication with them before adopting any definitive measure." But the rest of the brothers seemed inclined to go ahead on their own, so Adams "moved a resolution that no oath should be administered to any member of the Society to be hereafter elected, and no promise of secrecy required." The group voted to lay this motion on the table and to adopt instead Story's motion for a committee of nine to consider the matter. Everett appointed Story as chairman and Adams as one of the committeemen.

These men twice took four hours from their busy schedules to meet at the Boston Atheneum and discuss Phi Beta Kappa affairs. Story proposed what "was in substance a new constitution for the Society, making several essential changes in its laws, of which the abolition of all secrets and promises was one." He produced "two volumes of the records of the Society," which showed "that in 1806 the oath had been abolished and the solemn promise of secrecy substituted in its place." Adams objected to any basic change except the complete removal of secrecy. The committee, however, adopted Story's report.

When the Massachusetts Alpha met again to consider the report, one of those present argued "that the secrets of the Society ought not to be removed, because it would be yielding to a mere momentary popular clamor; that there was nothing worth retaining in the secrets themselves, but that to abolish them in deference to an artificial excitement got up among the people was unworthy of the Society." Adams replied that this excitement was the very reason he wanted the secrets abandoned. He "considered the excitement prevailing in the public mind on this subject just and well founded."[7]

Before the end of 1831, the Harvard brothers passed Adams's motion to require no oath of secrecy—but still to keep confidential the proceedings in regard to the election of new members. (Having abolished the oath, the brothers thereafter discontinued the initiation ceremony.) They also revised their constitution to read in respect to the aims of the Society: "its motto is intended to indicate that Philosophy, including therein Religion as well as Ethics, is worthy of cultivation as the guide of life." Avery Allyn, in a second edition of his book, took credit for the changes at Harvard.

Having brought about reform at home, Everett made a trip to New Haven to advocate the cause among his brothers there. "He used a tender tone, stood half-drooping as he spoke," a Yale undergraduate afterward recalled, "and touchingly set forth that the students at Harvard had such conscientious scruples as to keep them from taking the vow of secrecy, and the society life was thus endangered. There was stout opposition, but the motion prevailed, and the missionary returned to gladden the tender consciences of the Harvard boys." (Of course, it was not the Harvard boys but their elders who had had the conscientious scruples.) Later the Yale branch reimposed secrecy with respect to the election of new members, since those rejected "ought never to know the manner in which their claims have been discussed."

In 1832 the brothers at Brown similarly removed the obligation of secrecy except for discussions and votes "respecting the qualifications or character of any candidate proposed for admission." The Dartmouth brothers continued to exact the formality of a pledge for nearly twenty years, until a professor moved that, "as the Society had no longer any secrets," it was unnecessary for new members to promise to keep them. The men at Bowdoin clung to the pretence of secrecy for almost thirty years, and those at Union for much longer than that. But it was only a pretence.[8]

Scholarship was always one of the qualifications for election to Phi Beta Kappa, but not necessarily the most important one. Until the 1820s the seniors generally elected the juniors who would succeed them, and the seniors ostensibly based their choices on considerations of "morality" and "friendship" as well as "literature." Literary merit could refer to forensic skill rather than book

learning, while moral character could mean congeniality, thus reinforc-
ing the element of friendship or fraternalism, which found practical
expression in the rule allowing a single member to blackball any candidate. From the 1820s on, scholarship became more and more the
desideratum.

So many candidates were blackballed at Harvard that some members
wanted to abolish the unanimity rule in 1822. The rule was retained
when the bylaws were revised in 1825, but it came under severe attack in
1831, when the Antimasonic agitation led to proposals for getting rid of
it along with the oath of secrecy.

At that time John Quincy Adams defended the unanimity rule as
resolutely as he denounced the secrecy oath. When someone advocated a
two-thirds vote for admission, Adams said: "I think it will change entirely the character of the Society, and make it less elect: for one im-
proper exclusion which it will prevent, it will secure the admittance of
ten pale-colored candidates, of little or no value to literature or to the
reputation of the Society." And when the motion seemed about to pass,
he sarcastically "moved as an amendment to it that everything having
reference to *friendship* as being one of the objects of the institution
should be expunged, leaving it a mere literary society." The principle of
friendship had already been weakened, he had just then learned, for
"the promise of assisting a brother with life and fortune had been dis-
carded" some time ago.

Simultaneously there was another crisis concerning elections at Harvard. The "members elect of the class now commencing Seniors at the
university had declined accepting the membership, burdened as it now is
with a tax of fifteen dollars each on admission, to defray the expenses of
the anniversary." In addition to this tax, each of the admitted students
had to pay for his medal and for the pink ribbon he was to wear with it.
Adams "suggested the expediency of retrenching many of the expenses,
such as the band of music, and even the dinner." In the end the elder
statesmen decided not to levy so rapaciously upon the young initiates.
They passed a motion "that the expenses of this year's anniversary celebration should be paid from the fund of the Society, with a small assessment upon the class now to be graduated"—not upon the class about to
be initiated.[9]

At the same time the Harvard elders decided to require a three-
fourths instead of a two-thirds vote for the election of new undergraduate members. Then, at the anniversary meeting the same year, they
reversed themselves and reimposed the unanimity requirement. Meanwhile, the brothers at Dartmouth and Yale kept the requirement (though
making other changes in their constitutions) and so did the brothers at
Union and Brown.

At Bowdoin the election procedure was quite different from the very
outset. According to the branch's original bylaws of 1825, there would

have to be a unanimous vote, but it would be a vote of graduates, not undergraduates. "No undergraduate shall become a member of this Society." Students would be elected at the end of their senior rather than their junior year, and they would not be initiated until the day after their graduation. This meant, in effect, that the ΦBK men on the faculty would do the electing. Here was a model that other branches were later to imitate.

With respect to qualifications for membership, the Bowdoin brothers made explicit the principles that other branches already, no doubt, were implicitly following: "in the election of members, the Society shall direct their [i.e., its] attention to the *talents, acquirements, character,* and *disposition* of the candidate." The problem—a problem becoming increasingly difficult with the growth of enrollments—was to find some fairly reliable basis for judging these traits and especially the intellectual talents and acquirements.

Help toward a solution came with the colleges' development of a grading system. Beginning early in the eighteenth century, Harvard had ranked students imprecisely on the basis of professorial opinions. In 1783 Yale began to rank students not individually but in four groups: best, second best, good, and worst. These categorizations did not necessarily depend on classwork alone, but in 1813 Yale started to keep a record of numerical grades based on oral examinations given twice a year. In 1825 Harvard announced that it would sent the parents of each student a quarterly report of his "every mark of approbation or distinction," his "every punishment or censure," and "all his absences from exercises, lectures, and publick worship." By 1838 not only Yale and Harvard but also Dartmouth, Bowdoin, Brown, and some other colleges were keeping fairly precise records of academic work along with "general deportment," "application to study," and absences. The academic ranking depended on daily marks for recitations; written examinations did not come into use until after 1850.[10]

As early as 1819, the ΦBK seniors at Harvard had looked to the university administration for assistance in rating candidates. "Not having sufficient information concerning the relative rank of others in the Senior Class," the society "voted we have a special meeting . . . and members were requested previously to procure a list of scholars in that class from the [university] government and from individuals of the class." A few years later the society sent two members to President Samuel Kirkland "to request a copy of the scaling of rank," but Kirkland was unwilling to give out confidential information of that kind, and so for the time being the society had to rely on "general literary characteristics." In 1826 the secretary went to "consult with some members of the Faculty previous to the election from the Junior class, respecting the persons whose rank entitled them, according to custom, to be candidates."

By 1840 President Josiah Quincy could say of Phi Beta Kappa at Harvard:

> In process of time, its catalogue included almost every member of the Immediate Government [the faculty and administration], and fairness in the selection of members has been in a great degree secured by the practice it has adopted, of ascertaining those in every class who stand the highest, in point of conduct and scholarship, according to the estimates of the Faculty of the College, and of generally regarding those estimates.

Thus at Harvard the undergraduate members continued to do the electing, but they generally accepted the judgment of the faculty as to which of the candidates stood highest in "conduct and scholarship."

At Yale and at Brown the students likewise retained the power of making elections but came to follow faculty advice in doing so. At Brown the idea occurred to Francis Wayland, as it had occurred to Edward Everett at Harvard, that perhaps the society ought to lessen or even eliminate its exclusiveness. In 1847 President Wayland moved that a committee "take into consideration the subject of enlarging the basis of the Society, by modifying or altogether removing the present restrictions upon admissions." A committee was appointed with Wayland as chairman, but he never got around to making a report.

At Union and Dartmouth, as earlier at Bowdoin, the entire responsibility was eventually left to the ΦBK men on the faculty. In 1838 the Union undergraduate members found themselves unable to agree on which candidates to choose, and they resolved "That the Faculty be requested to make such elections." Thereafter the faculty did so, always examining first the "merit roll" of the college. The elections were rescheduled for late in the senior year, and initiations were set for the day before graduation, so that undergraduate membership ceased to exist.[11]

In 1845 the Dartmouth members appointed a committee to "solicit recommendations from the faculty" regarding juniors to be elected. Later that same year the committee proposed "that the constitution be so altered that the election be made from the Senior class instead of the Junior class" and that it be made "on the day preceding commencement." The constitution was so altered. Henceforth there would be no undergraduate members, and so the graduate members—the professors and tutors for the most part—would have to take charge of student elections.[12]

Thus, by the 1840s, the faculty indirectly influenced or directly determined the choice of students for Phi Beta Kappa membership. In theory, the choice was still based on "character" as well as "acquirements" and on "conduct" as well as "scholarship"; in practice, it depended essentially on the record of grades in courses, since the professors were better acquainted with this than with evidence of the student's standing among his peers. To the extent that the undergraduates' role in elections had declined or disappeared, the society had lost its claim to being a true fraternity. As John Quincy Adams had suggested, the society might as well expunge all references to *friendship* if a time should ever come when each of its members no longer possessed the right to exclude uncongenial candidates. Phi Beta Kappa might continue to call itself a fraternity, but undergraduates would have to look elsewhere on campus for a real sense of brotherhood.

For comradeship, students could look to new Greek-letter fraternities and other campus clubs that sprang up in great profusion from 1825 on. These featured the characteristics that Phi Beta Kappa was losing. They kept secrets. They emphasized sociability and allowed each member to exclude anyone he considered undesirable. They were student-centered and student-controlled.

The emerging Greek-letter societies resembled the primordial one in still other respects. Each of them took a name with initials that stood for some motto in Greek. Each adopted a more or less close imitation of the Phi Beta Kappa "key"—which had evolved from the original small, thin, square medal when this was attached to the kind of tiny key once used to wind a watch. Some of the new fraternities remained confined to a single campus, but most of them sooner or later followed the example of Phi Beta Kappa and established branches, which they called "chapters."

In one important respect, however, these proliferating fraternities were quite different from Phi Beta Kappa. Generally, they were glad to get high-ranking students—students of the kind that would also be elected to Phi Beta Kappa—and occasionally a chapter even prided itself on the scholarship of its members. But scholarship was not the prime requisite for belonging to one of these newer societies. Their raison d' être was social, not intellectual.

ΦBK students themselves started the new fraternity movement, their object being to keep what they liked and to gain what they lacked as brothers in the existing "fraternity." Nine seniors at Union College met in 1825, in a room that two of them shared, to organize under the name of Kappa Alpha. Seven of the nine were or became members of Phi Beta Kappa. But the Union ΦBK branch, from its very beginning eight years earlier, had always been in the hands of its older members, particularly those on the faculty, and its undergraduate members found themselves with little or nothing to do in its activities.

Kappa Alpha, though at first extremely small, soon aroused the envy of the Union students it excluded, and in 1827 they founded two competitors, Sigma Phi and Delta Phi. In 1831 Sigma Phi, designating itself as the Alpha chapter of New York, chartered the Beta chapter of New York at Hamilton College. In 1833 Kappa Alpha and in 1834 Sigma Phi placed chapters at Williams College. Fraternities with Greek-letter names had appeared on other campuses as early as a dozen years before the first of the three at Union, but these three, the "Union Triad," set the pattern for the system of social fraternities in the United States.

These early fraternities resisted the Antimasonic hue and cry, and they remained secret societies, as did nearly all their successors. Student-led as they all were, they asserted their independence from college faculties and administrations. They were not going to submit to dictation from cautious and conformist elders in the way the Harvard branch of Phi Beta Kappa had given in to such men as Everett, Adams, and Story. A few students did yield to the clamor, however, and formed anti-secret fraternities. At Williams there was O. A. (*Ouden Adelon,* Greek for "nothing secret"); at Union, Equitable Union; and at Amherst, Delta Sigma. These three combined in 1847 to form Delta Upsilon.[13]

Before the new secret societies appeared at Yale, the ΦBK Society there was already declining in sociability. As early as 1820, the faculty complained about "noises which had in times past occasionally passed the bounds of decorum," and so the society decided on "a curtailment of the usual preparations" for "the entertainment customary on the admission of new members." Soon the brothers resumed the usual toasts, in one of which they drank to the success of the local temperance society. Then in 1826 the entertainment committee "prohibited the introduction of ardent spirits." For the time being, wine was still permitted, and there was "every variety" of it at the annual meeting that year, but at the next big affair "wind was excluded" also. Other refreshments were "handed round," and according to the secretary, these made for "a cheerful entertainment."

Apparently, seniors who had failed to make Phi Beta Kappa were responsible for the founding of the first and foremost of the social fraternities at Yale in 1832, one open to seniors only. Its founders did not follow the ΦBK example in choosing a name; they called their new society "Skull and Bones." Regarding it, a Yale graduate wrote long afterward: "Some injustice in the conferring of Phi Beta Kappa seems to have led to its establishment, and apparently it was for some time regarded throughout college as a sort of burlesque convivial club"—a burlesque of Phi Beta Kappa, that is. After Skull and Bones, there appeared a rival society for seniors, "Scroll and Key."

There also arose societies for juniors, then for sophomores, and in 1840 for freshmen. These took Greek-letter names. Certain of the Greek-letter fraternities came to Yale from elsewhere, as Alpha Delta Phi did in 1837 from Hamilton. Others branched out from Yale, as Sigma Epsilon did, with chapters at Amherst, Rensselaer Polytechnic Institute, and Dartmouth. The fraternities of the first three years at Yale had "a general resemblance to one another and to those at other colleges."

But Skull and Bones and Scroll and Key were "peculiarly Yale institutions." They "are the only Yale societies whose transactions are really secret. Their members never even mention their names, nor refer to

them in any way, in the presence of anyone not of their own number."
These societies, in their own way, were more highly selective than that
older secret society, Phi Beta Kappa, which accepted about a third of the
class. The two new ones admitted, between them, barely more than a
fourth of the seniors and were "composed exclusively of 'big men' . . .
preeminent above their fellows in college repute."[14]

At Harvard, as at Yale, Phi Beta Kappa eventually became less noted
for sociability than were other student societies. For many years the
Harvard brothers had tippled a lot, especially at anniversary meetings.
The Reverend John Pierce, '93, grew more disgusted as he grew older,
though he continued to attend regularly. He complained that the 1838
meeting was "far too Bacchanalian" for his taste. Some meetings were
even worse: "my judgment and feelings equally revolted at the quantities
of wine drunk, among others by clergymen, and of these by one who not
long since delivered an eloquent lecture in many places on total absti-
nence from all which can intoxicate."

Dr. Pierce expected the worst of the 1841 affair when "the room
became dark and nauseous by the tobacco smoke, and consequent ex-
pectoration which it occasioned." But he was relieved when time came
for the toasts and the president "did not preface the sentiments to be
uttered with calling on his brethren to fill their glasses." He noted with
some satisfaction: "Comparatively little wine was drunk, and consequently there was the less boisterous mirth. It is my earnest wish yet to
witness a ΦBK dinner, at which there shall be no unnatural excitement
from alcoholic liquors, and at the same time 'a feast of reason and a flow
of soul.'" Before long, he was to get his earnest wish.

Another group at Harvard was more forthrightly devoted to food
and drink, though actually Phi Beta Kappa often spent more per person
on wine at dinner than the other group did. This was the Porcellian Club
(the "pig club"). From its founding in 1791, many of its members were
also ΦBK members and attended the ΦBK as well as the P. C. dinners.
But the Porcellians ceased to do so after 1846, when the Massachusetts
Alpha, in response to the persuasive New England temperance movement, banned from its meetings all kinds of alcoholic beverages, including wine. (The ban remained in effect until 1946).

Social fraternities multiplied despite the opposition of college officials—opposition of a kind that Phi Beta Kappa no longer faced. By
1850 there were sixty-three chapters on seventeen campuses, with at
least one chapter on every campus having a ΦBK branch. A member of
one social fraternity could not belong to another, but he could belong to
Phi Beta Kappa if he met its scholarly requirements. Indeed, sixty-five of
the first one hundred men initiated into Kappa Alpha were or became
members of Phi Beta Kappa as well. Though some of the new societies,
like Skull and Bones, may have originated as competitors of Phi Beta
Kappa, most if not all of them soon came to look upon it as occupying an
entirely different sphere.[15]

As Phi Beta Kappa became less and less a fraternal society for undergraduates, it also became less and less a forensic club for them. Some of the newer branches sponsored no undergraduate activities whatever from the outset. Even Harvard, Dartmouth, and Yale sooner or later discontinued them.

Harvard was the first to do so. As early as 1813, after having noted for some time and with increasing frequency the large number of absences and "unprepareds," the secretary recorded: "though this is the regular evening for assembly, yet there was not a member present." During the next several years, attendance did not improve very much. By 1818 the "literary exercises" were "altogether neglected," the secretary wrote. "There were but few who condescended to attend the most necessary meetings, and some who did not attend at all." Too many other things were going on. The society finally made the decision to dispense with literary exercises entirely, a decision that could be "traced in great measure to the late surprising multiplication of lectures and so forth." The undergraduates held their last debate in 1820.

From time to time, the undergraduate society at Harvard tried to revive the discussions and debates but without success. On June 20, 1831, the members passed a resolution for scheduling fortnightly meetings once again. Then, before it took effect, the minutes read: "Abolished the preceding resolution on Sept. 13, 1831—as the Secrecy of the Soc. was now destroyed. It was thought that no other motives than vanity of a *Secret* would keep up exercises in a promiscuous assemblage of individuals little interested in each others' *peculiar* success."[16]

At Bowdoin, from the founding of the branch in 1825, there were no undergraduate meetings, since there were no undergraduate members, the elected seniors not being initiated until after their graduation. The founders of the branch preferred not to go into competition with the two strong literary societies, the Peucinian and the Athenean, already flourishing on the campus. That the new society was expected to play a very different role was early indicated when the *Portland Argus* editorialized: "The establishment of Phi Beta Kappa at Bowdoin will . . . have the beneficial effect of allaying the rivalship at present existing between the two literary societies in that place, which is productive of great inconvenience, since it induces them to seek recruits among the idle as well as the industrious and takes away one great motive to exertion by making literary distinction cease to be the reward of literary exertion."

At Brown the university president and professors made no provision for undergraduate activities when they organized the branch there in 1830. For several years, meanwhile, the undergraduates at Union met fairly often for discussions and debates, but discontinued them entirely in 1838.

At Dartmouth such activities were dying out by that time. The under-graduates there used to meet every other Thursday afternoon for decla-mation and dispute in Society Hall, which the college provided for the use of Phi Beta Kappa and the rest of the student societies. But as at Harvard, interest and attendance fell off, especially after the college began to offer courses in rhetoric (public speaking) in the 1820s. Again and again the members managed to revive the old programs, but not the old enthusiasm, and finally they decided to give them up for good. They scheduled the last ones in 1845, the year before postponing elections to the end of the senior year and thus ruling out all undergraduate ac-tivities by eliminating undergraduate membership.[17]

At Yale, too, the Phi Beta Kappa members long used, on different days, the same college hall as the other debating societies, the most important of which were the Linonian and the Brothers. "The exercises of the ΦBK were generally of a higher order than those of the other societies," a Linonian of the class of 1821 recalled. "They ought at least to be so, since the members [of ΦBK] from the higher classes only, are selected for their talents and attainments."

The same Yale graduate saw Linonian and Brothers as performing, in the 1840s, functions that Phi Beta Kappa had originally prided itself upon. "I regard the Debating Societies as a very valuable means of improvement," he wrote. "The recitation and lecture rooms may make scholars; they cannot make debaters." Phi Beta Kappa at Yale might get the best of the debaters, but it certainly had no monopoly on self-im-provement in the art of public speaking.

This Yale alumnus also saw Linonian and Brothers as, potentially at least, serving the high cause to which the William and Mary students had once dedicated Phi Beta Kappa—the cause of fostering the greatness of the country by keeping it united and at peace with itself. He related that at the time of the Missouri controversy in 1819, the Linonian society had split over an event "of the same kind as that which, as some predict, will one day sunder the American Union—the election of a President." When a Northern candidate defeated a Southern one for president of Linonian, the Southern members all seceded and formed a new society, the Callio-pean. Yet, Northerners and Southerners, "these young men are to meet . . . on the great stage of public life . . . and it will be well then that they can recognize each other as old acquaintances and friends." "One might imagine a case in which an intimacy formed at college saved the Union."

As late as the academic year '1842–1843 at Yale, the seniors in Phi Beta Kappa were still debating questions of the traditional kind: "Ought a National University to be established?" (decided in the affirmative) and "Are the intellectual capacities of Females equal to those of Males?" (decided in the negative). Already the members were concerned about the health of the society, however, and they had decided on a self-exam-ination, setting up a committee to "take into consideration the condition of this venerable Institution." The committeemen reported "that they had examined the patient and found . . . that he was sick." They recommended and the society decided that it should hold fewer meetings and invite ΦBK members from the faculty and from the community to attend and participate. So, separate undergraduate exercises finally came to and end at Yale. That was in 1846—seventy years after the founding of Phi Beta Kappa, sixty-six years after the founding of the Yale branch.

The newest branches, those at Trinity College and Wesleyan University, had recently started out by patterning themselves on the new model that Yale was providing. At Trinity, from the beginning in 1845, members residing in college were to "constitute a Local Society" and were to be responsible not only for elections and initiations but also for "such literary exercises . . . as the Society shall think fit." The meetings were to be addressed by members "of the ΦBK Society in the vicinity, either undergraduate or otherwise." These "literary" meetings did not continue very long at Trinity. At Wesleyan from 1845 on, such meetings were to be held on the second Wednesday of each month "in term time," and professors and local residents as well as undergraduates were to be encouraged to take part. The meetings persisted longer at Wesleyan than at Trinity, but at neither place were there ever separate undergraduate activities of the original Yale-Harvard-Dartmouth type.[18]

The "society at large," consisting of professors and other alumni, played a larger and larger role in Phi Beta Kappa affairs as the "immediate society" of undergraduates lost most or all of its activities and even its existence. The remaining purpose was chiefly, if not exclusively, to honor scholarship, which the society at large did at each of its annual meetings. It honored individual scholars by welcoming them into its company, and it honored scholarship in general by inviting learned or knowledgeable men to speak.

Phi Beta Kappa became famous for the addresses that it sponsored. The earliest orators were mostly recent graduates, men like the young John Quincy Adams and the young Daniel Webster whose reputations were yet to be made, but many of the later ones were already quite well known. Indeed, they were often chosen in the expectation that their names would attract large audiences and would generate wide publicity. Among the more notable orators were De Witt Clinton, the New York governor and Erie Canal builder; Joseph Story, the Supreme Court justice and budding jurist; and, most memorable of all, Ralph Waldo Emerson, the greatest American philosopher of his time.

For years, leading magazines such as the *North American Review* published many of the Phi Beta Kappa addresses in their entirety. From the 1820s on, though printing fewer complete texts, the magazines often ran long reviews, but after 1840 only brief notices appeared as a rule. In 1825 a member of the New York Alpha hoped to provide an alternative vehicle in his weekly, to which he gave the wordy name *The New York Literary Gazette and Phi Beta Kappa Repository*. After six months, however, he announced: "*Phi Beta Kappa*. We shall drop this part of our title with the present number. When we adopted it we had the promise of the cooperation of several members of the ΦBK Society, and of the orations and poems appertaining to the institution. But those who were most interested in this have not taken the pains to furnish us with the articles we expected." (That was the last attempt at what could be called a Phi Beta Kappa periodical until 1910.) In fact, the society did not need the *Phi Beta Kappa Repository*. The various branches were already publishing most of their annual orations—or most of those worth publishing—as separate pamphlets.[19]

The orators chose topics as disparate as "The True Wealth or Weal of Nations" (Yale, 1837), "The Influence of Scientific Discovery and Invention on Social and Political Progress" (Brown, 1843), and "The Connection Between Science and Religion" (Harvard, 1845). No matter what the subject, the speaker almost always had something to say about education and its importance for the future of the United States. He could use the occasion itself to advance the cause, or so it seemed to Governor Clinton. "When you were pleased to intimate that the deep interests of science, in exhibitions of this nature, might be promoted by my cooperation, I considered it my imperative duty to yield a cheerful compliance," he began, when speaking on "The Achievements and Responsibilities of the Age" at Union in 1823. "I endeavour to enforce those considerations which ought to operate upon us generally as men, and particularly as Americans, to attend to the cultivation of knowledge."

Orators elaborated on a familiar theme: the opportunities and responsibilities of the educated American. "Never may you, my friends, be under any other feeling than that a great, a growing, and immeasurably expanding country is calling upon you for your best services," Edward Everett said in 1824, addressing a Harvard audience that included the French hero of the American Revolution, the Marquis de Lafayette. Everett, whose topic was "The Circumstances Favorable to the Progress of Literature in America," found in the classics an inspiration and a challenge for native writers. From the "tombs of departed ages" he could seemingly hear voices. "They exhort us, they adjure us, to be faithful to our trust," he declared. "Greece cries to us . . . Rome pleads with us. . . ."

Orators might differ, though, in their assessment of the classics. Speaking on "Science and Letters in Our Day," at Harvard in 1826, Joseph Story remarked upon the perduring value of classical learning, the relevance of ancient Greece and Rome for modern America. But Charles Sumner expressed some doubts about that when he spoke to the Union branch on "The Law of Human Progress" in 1848. Sumner—a lawyer, literary dilettante, and aspiring Free Soil politician—concluded that the classics possessed a "peculiar charm" as models of composition and form. But, he said, they lacked "purity, righteousness, and that highest charm" which was to be found only in the law of God. "For eighteen hundred years the spirit of these classics has been in constant contention with the Sermon on the Mount." And "heathenism" was "not yet exorcised"; it continued to exert "a powerful sway, imbuing youth."

There was a recurring note in the familiar theme of the educated American's opportunities and responsibilities. This was the national need for originality, creativity, emancipation from the bondage of European and particularly English culture. As early as 1818, Edward Tyrrel Channing, then editor of the *North American Review* and soon to become professor of rhetoric and oratory at Harvard, addressed the Harvard society on "Literary Independence." Channing exhorted: "Let the American Scholar turn homeward a little"; let him "cultivate domestic literature." Subsequently, Channing gave instruction and inspiration to the man who was to develop this "American Scholar" idea to its fullest— the man who, indeed, was to deliver the greatest of all Phi Beta Kappa addresses.

That man was Ralph Waldo Emerson, the Harvard orator for 1837. Then thirty-four, a refugee from the Unitarian ministry, Emerson had already published one book and was on the way to fame as a free-lance philosopher, essayist, and lecturer. He was the sage of Concord, Massachusetts. Different listeners got different messages from his Phi Beta Kappa lecture, "The American Scholar," but he certainly had something to say about Phi Beta Kappa itself and about its anniversary meetings, stressing their role in the past and his hope for an even more significant role in the future. He began:

I greet you on the recommencement of our literary year. Our anniversary is one of hope, and, perhaps, not enough of labor. We do not meet for games of strength or skill, for the recitation of histories, tragedies, and odes, like the ancient Greeks; for parliaments of love and poesy, like the Troubadours; nor for the advancement of science, like our contemporaries in the British and European capitals. Thus far, our holiday has been simply a friendly sign of the survival of the love of letters amongst a people too busy to give letters any more. As such, it is precious as the sign of an indestructible instinct. Perhaps the time is already come when it ought to be, and will be, something else; when the sluggard intellect of this continent will look from under its iron lids and fill the postponed expectation of the world with something better than the exertions of mechanical skill. Our day of dependence, our long apprenticeship to the learning of other lands, draws to a close.

The hope of the future, Emerson went on to proclaim, was the American Scholar, and the scholar was "Man Thinking." Such a man required the experience of nature, of books, and of action. Though he must resort to books, to " laborious reading," he must not depend on such things alone. "Colleges, in like manner, have their indispensable office,—to teach elements. But they can only highly serve us when they aim not to drill, but to create; when they gather from far every ray of various genius to their hospitable halls, and by the concentrated fires, set the hearts of their youth on flame." Emerson concluded:

> We will walk on our own feet; we will work with our own hands; we will speak our own minds. The study of letters shall be no longer a name for pity, for doubt, and for sensual indulgence. The dread of man and the love of man shall be a wall of defence and a wreath of joy around all. A nation of men will for the first time exist, because each believes himself inspired by the Divine Soul which also inspires all men.

Not all the hearers on that August 31 were impressed. "Rev. Ralph Waldo Emerson gave an oration, of 1¼ hour, on The American Scholar," noted the Reverend John Pierce, that faithful attender who so much disliked alcoholic drinks and tobacco smoke. "It was to me in the misty, dreamy, unintelligible style of Swedenborg, Coleridge, and Carlyle." Emerson's friend Herman Melville thought that much of what Emerson wrote was "oracular gibberish." But Oliver Wendell Holmes afterward averred: "The grand Oration was our intellectual Declaration of Independence."[20]

These yearly occasions, with their eloquence and their honors, made Phi Beta Kappa unique. As President Quincy wrote in his 1840 history of Harvard, "The Phi Beta Kappa Society, from its long continuance, the number and respectability of its members, and the public interest excited by its annual celebrations, deserves a particular notice."

The society had evolved most notably along just one of the lines that its founders had intended and its early leaders had stressed. John Heath at William and Mary conceived of an organization that, while providing a sense of brotherhood and a chance to practice public speaking, would bring together the more serious-minded of his fellow students. The pioneer officers at Yale and Harvard prided themselves on admitting only those with "a good character in the literary as well as moral line" and excluding "the ignorant and immoral." The honor conferred by election was shown by the envy of those who failed to gain election. When one of them (in 1793) sarcastically referred to Phi Beta Kappa as "a certain *supposed honourable institution*," he surely meant an institution reputed to be honorary.

This reputation was confirmed with the passage of the years. Phi Beta Kappa served as "a stimulous [*sic*] to youths in College," according to an 1828 book on Masonry, "and by confining it to a minority of every class an election to this order was a matter of distinction in college, and of course no small exertions were made to deserve this honor." An 1821 graduate observed in his 1847 book on student life at Yale: "The desire to be elected to this society was hardly less than that of appointments [honors] at commencement; and for the same reason, namely, that it was regarded as a proof of scholarship." Clearly, Phi Beta Kappa had become preeminently an honor society—indeed, *the* honor society of American academe.[21]

5

Expansion, Disunification, and Decline

During the first seventy years (1776–1846), each of the local groups had been known as a "society," a "fraternity," or a "branch" of Phi Beta Kappa (which, at large, was also known as a "society" or a "fraternity"). From 1847 on, the local group came to be more and more commonly referred to as a "chapter," a term borrowed from the burgeoning Greek-letter social fraternities. During those first seventy years a total of only eight branches had appeared, in addition to the original William and Mary society, which had disappeared. During the next thirty-five years (1847–1882), a period half as long, twice as many new chapters were chartered. This growth did not mean greater strength. Instead, it resulted in increasing diversity and disunity and was accompanied by declining prestige. By the time Phi Beta Kappa reached its centennial, the society seemed to be approaching disruption if not also, as some believed, extinction.

The society now expanded from New England not only to New York and New Jersey but also to the trans-Appalachian West and even to the Deep South. Yet it was far from being widespread or strong enough to overcome the strains of sectionalism, as its founders had hoped it would do. The Southern chapters fell victims to the Civil War. Afterwards, as late as 1882, there were still no active chapters in the South and only two active ones in the West, both of these in Ohio. All the rest were located in the Northeast—ten in New England, eight in New York, and one in New Jersey.

The first of the trans-Appalachian chapters, the Alpha of Ohio, was organized at Western Reserve College in 1847. This Alpha gave its blessing to the establishment of a Beta at Kenyon in 1858 and a Gamma at Marietta in 1860. Meanwhile, Western Reserve fell upon hard times as

71

professors abandoned it and the president resigned. Its ΦBK chapter lapsed into inactivity during the war years (1861–1863) and again during 1879–1885 while the college was being moved from Hudson to Cleveland and was being transformed into Western Reserve University.

The first chapter in the Deep South, the Alpha of Alabama, resulted from the exertions of Frederick A. P. Barnard, a Massachusetts-born professor at the University of Alabama. Students and visitors at the university's 1851 commencement were mystified when observing that here and there a graduating senior wore a green ribbon on his gown's lapel. The seniors thus identified were those who had just won election to Phi Beta Kappa. But seniors were to repeat this commencement scene fewer than a dozen times. With the coming of the war, the Alabama chapter ceased to exist. It was not to be resurrected until 1912.[1]

The Virginia Alpha was resurrected in 1851, seventy years after its demise. This society owed its refounding to Northerners as the Alabama Alpha did its founding that same year. Two of the William and Mary professors were alumni of Union College and members of the New York Alpha. In 1849 one of the two wrote to William Short, the last president of the Virginia Alpha, who was still alive in Philadelphia. Short eagerly "commissioned" the men from Union to reorganize the society at William and Mary. The men had access to the society's 1776–1781 records, which the Cabell family had transferred to the Virginia Historical Society in Richmond. In these records the two found ample authority for what they were doing. "The legality of this proceeding," they informed the other Alphas, "rests upon Article 24th of the original Constitution, which declares that any number of members shall be competent to act whenever it is necessary for the preservation of the Society."

Some members of the revived Virginia Alpha, like the leaders of the original society, looked upon Phi Beta Kappa as both a scholarly and a patriotic organization. "Its annual gatherings constitute the great literary jubilee of our country," the 1855 orator declared. "Sir, I indulge the hope . . . that its re-institution here . . . appealing as it does to our love of letters and to our love of country, is an omen of cheering import." The 1860 poet had only a curse for the secessionist: " . . . should some rash and parricidal wretch Put forth his hand . . . To sever the union of these equal States . . . [Let us] damn him to the hell of scorn."

The Civil War damaged the College of William and Mary more than the Revolutionary War had done, and it hurt the society of Phi Beta Kappa there almost as badly as had the earlier war. Again, in 1861 the society went out of existence, though only for thirty-two years this time. In 1862–1863, while United States troops occupied Williamsburg, the Confederates repeatedly attacked them, and some of the fighting took place on or near the college grounds. "Upon the evacuation of the rebel cavalry and the return of our troops," a congressional committee afterwards reported in regard to one skirmish, "a body of stragglers from the

United States forces, drunken, disorderly, and insubordinate, fired and destroyed the college building, with the library, apparatus, furniture, and other property therein belonging to the institution."[2]

When a committee of three Yale professors drafted a charter for the Western Reserve branch, in 1847, they wanted to be sure that the new Alpha of Ohio would be "guarded against any and every departure from the fundamental principles of the ΦBK Society," and so they included a "brief statement of those principles." These had to do with elections to membership, the establishment of additional branches, the conduct of meetings, and other matters of mutual interest.

The Yale chapter, being the oldest one still active at the time, presumed thus to set standards for the society as a whole, or at least for its newer branches. The Yale men did so without consulting the other Alphas—not even the Massachusetts Alpha, which was practically as ancient as the Connecticut. Still, these men were setting a worthwhile objective, for Phi Beta Kappa certainly needed to counteract the accelerating trend toward heterogeneity among its various chapters. Unfortunately, the Yale members failed to adhere consistently to their own rules, and some of the other members were even less inclined to follow them. The society became more heterogeneous, not less.

With regard to elections, the charter from Yale to Western Reserve stated this basic aim: "the Society of ΦBK, by electing the best and most promising scholars, of a good moral character, in the principal Colleges, and uniting them in one great Fraternity of Scientific and Literary men, has been found, not only profitable to the members of said Society, but also conducive to the advancement of sound learning in our Country." Another statement of purpose came from a University of Alabama trustee when the university president objected to the establishment of a ΦBK chapter because he and the professors had "become painfully aware of the evils attending the existence and multiplication of secret societies formed among students beyond the inspection and control of the Faculty and Trustees." The trustee assured the president that Phi Beta Kappa was quite different from other Greek-letter fraternities. He wrote:

Its secret consists in having nothing to divulge . . . the only society in our college over which the Faculty exercise a controlling influence; the members are selected by the society entirely on the ground of college scholarship, on their standing in their classes respectively, for excellence in college studies, and this must principally be determined by the Faculty who attend meetings and hold offices, and meet the other members in consultation on equal and confidential terms since its great object is to promote among the students a generous spirit of emulation in college duties by their distinguishing those who have been most successful, and to promote order.

Still another expression of the society's criterion for membership came from College of the City of New York (CCNY) students who, appealing to the New York Alpha for a charter, referred to the society as having been "founded solely for the benefit of those high in scholarship, without regard to any other qualifications."

Thus there was a slight difference of opinion as to whether there should be no other qualifications or whether "a good moral character" should also be required. Probably most members would have agreed that scholarship ought to be the most important if not the sole consideration. But the various chapters could disagree on specific standards and procedures for holding elections.

According to the "fundamental principles" in the Western Reserve charter, no undergraduates should be elected before attaining "the rank of Junior Sophisters," that is, before reaching advanced junior standing. No more than "one-third part of any College Class and only those the most distinguished for genius, character, and good scholarship" should be admitted. Voting on new members was to be by "silent ballots," and "a single negative vote, if not withdrawn," would "prevent election."

Not all chapters observed this unanimity rule. In 1847, while Yale was instructing Western Reserve to do so, Dartmouth decided that a two-thirds vote could override a single negative ballot. In 1851 the Dartmouth chapter further relaxed the rule by providing that a single negative could not prevent election even in the first instance. Instead, there would have to be two "no" votes out of fifteen, three out of twenty-five, and four out of thirty-five. The two-thirds vote could override these negatives also.[3]

Nor did all chapters elect the same proportion of students. Most filled the one-third quota, but Marietta reduced it to one fourth. Amherst set no quota; instead, it elected those students with an average of 85 percent or better at the end of their junior year and those with an average of at least 80 percent at the end of the first term of their senior year. Marietta based its rankings on "the college averages, of one or more years next preceding," and most other chapters took the average of all previous grades. But Dartmouth again showed some originality. There—where, as at some other places, elections were held at the end of the senior year—the chapter began in 1876 to give added weight to grades in junior and senior courses. It multiplied these averages by three and multiplied those of the freshman and sophomore years by two, then divided the sum by ten.

In general, the chapters limited their election to those students who took the "full classical course," including both Latin and Greek. But the Cornell chapter, at its inception in 1882, resolved to broaden its scope somewhat by admitting students of high rank "from courses in which

Latin only" was required. The Kenyon chapter chose its members not only from the college proper but also from the affiliated theological seminary. The chapter also elected theology professors.

For professors, as for other honorary members, standards generally were lower than for undergraduates. Time and again a new chapter started out, as in earlier years, by incorporating a large proportion of the college faculty. The Amherst constitution provided that all permanent members of the faculty should also be members of the local chapter. For other honorary elections, Amherst required only a two-thirds majority, not a unanimous vote.

In one respect, all elections in all chapters were uniform for almost a century. The elections were confined to men. Women were not even admitted to college until Oberlin opened the way in 1834. Later the Midwestern state universities adopted coeducation, and women's colleges began to appear—Vassar in 1865, Smith and Wellesley in 1875.

The University of Vermont was the first institution with a ΦBK branch to admit women to its classes. In 1875 the members of the branch resolved "That in the opinion of this chapter all the graduates of this university should be eligible to membership in ΦBK without distinction of sex," and "That the Corresponding Secretary be directed to communicate the action of this chapter and a statement of the situation which occasioned it to the other chapters of the order." The "situation" was the presence of two women among the four top-ranking seniors, all four of whom the chapter proceeded to elect. None of the other chapters objected.

One of the others promptly followed the Vermont example. At Wesleyan, which had begun to admit women in 1872, the chapter decided in 1876 that the election of women would be "constitutional" and that henceforth they would be "eligible to membership on the same conditions as men." Cornell had admitted women from its founding in 1868, and when its ΦBK chapter was organized in 1882, the chapter's constitution provided that sex should not be taken into account in elections.[4]

The Yale-drafted Western Reserve charter listed the following as one of the "fundamental principles" of Phi Beta Kappa: "Each Branch of the Society, while bound by these principles, has full power to frame its own Constitution and By-Laws for the regulation of its meetings, its exercises, its forms of election and initiation, its choice of officers, its financial concerns and all its internal affairs, according to its own pleasure."

Each of the newer chapters founded between 1847 and 1882 certainly catered to its own pleasure in regard to the activities it sponsored. Some, from the very beginning, carried on practically no activities except for the election of members and officers—as a few of the older branches already were doing. Others started out with considerable ceremony and a busy schedule, which later tapered off to little or nothing. They repeated a cycle that the Yale, Harvard, and Dartmouth chapters had previously gone through and that the Yale society soon extended to the point of self-annihilation.

There was no initiation ceremony at Cornell and not much of one at New York University, where elected students were given a reception but no information about Phi Beta Kappa's history, nothing that would cause them to take much pride in their membership. At first, the Amherst graduate members were expected to meet to initiate the newly elected seniors. Later the faculty merely posted the names of the winning one third of the class and then listed as members those who signed the roll and paid the fee. At Madison (Colgate) the treasurer introduced the members-elect, and the president asked their "consent" to the constitution and the laws, explained the medal "according to his discretion," and made "such other remarks as he deemed advisable, closing thus: 'Gentlemen: We now have the pleasure to recognize you as brethren.'"

The procedure at Kenyon was quite different. There the initiation was elaborate in the old-fashioned way, and the initiate had to promise to keep "inviolate all the arcana of the Association," including the "secret" grip and sign—which were no longer secret. In 1862 the chapter held "an animated debate" on the proposition: "Everything connected with the Phi Beta Kappa Society which may be called secret ought to be abolished." When the members gave the victory to the affirmative by the margin of a single vote, the anti-secrecy men proposed to inquire of other chapters about the right to abolish the secrecy pledge. Apparently the Kenyon brothers were unaware that the Harvard chapter had abolished it thirty-one years before. In any case, the Kenyon chapter retained it and remained, on paper, a secret society.

For a time, the Kenyon chapter and several other new ones held fairly regular literary" meetings, which the older chapters had never done or had long since ceased to do. At Kenyon a literary committee prepared lists of topics for discussions and debates, which usually occurred once a month. At Western Reserve the members met for literary exercises three times a week. At Rutgers the bylaws called for reports from five divisions of the chapter—philology, ethics and metaphysics, mathematics and natural sciences, political and social science and history, theology—and the members undertook to discuss the reports at monthly meetings.[5]

Only once did Rutgers arrange an anniversary oration, one to celebrate Phi Beta Kappa's centennial in 1876. Some chapters sponsored no orations at all; others staged them annually or occasionally. People had begun to laugh at ΦBK oratory as "a type of turgid, pompous, and useless declamation," the *North American Review* observed in 1855. Still, the *Review* contended, the orations served a useful purpose by breaking down intellectual barriers. At Harvard, where the annual oration and poem still drew a great deal of attention, the revivalist Henry Ward

Beecher had appeared in 1875 to remind liberal Congregationalists that orthodox Congregationalism had much to offer.

The College of the City of New York was one of the institutions where the ΦBK chapter (1867) undertook few activities to begin with. Members gradually lost interest in even these few. For a dozen years the chapter met irregularly and did nothing except elect and initiate members, choose officers, and discuss the constitution and bylaws. Attendance fell off, and most of the alumni elected to membership failed to appear for their initiation. Between 1875 and 1880 only three meetings were held.

New chapters that started out with brave plans for regular literary exercises sooner or later abandoned them. At Western Reserve the student members found little appeal in the programs that professor members arranged for the meetings. The exercises were discontinued after three years. At Marietta the monthly meetings, at faculty homes, went on for only two years.

Even at Kenyon, where the chapter made the greatest effort to survive as a debate and discussion club, the effort ended in failure. Here, too, the club included professors along with students, not students alone as in the earlier days of Phi Beta Kappa. The professors bore most of the burden of assigning or presenting papers and arranging or leading debates. Tiring of these duties, the faculty more and more evaded them or simply stayed away. During the 1870s the officers repeatedly called off meetings because of a lack of performers or programs. Finally the chapter scheduled a debate on the question "Should Phi Beta Kappa live?" The affirmative won the decision, and the chapter remained alive, but it never quite managed to breathe new life into the literary exercises.[6]

There was a real question whether Phi Beta Kappa would survive at Kenyon or any other college, for the society was already dead at Yale, which (after the second collapse at William and Mary), had the most venerable chapter, the first among equals. The Yale chapter had also been in some ways the most influential. Its members had done the most to enlarge the society by founding new branches and had taken the greatest pains to unify the society by laying down the law to them. But Phi Beta Kappa at Yale lost out in competition with the social fraternities, as did the literary societies. "The sub-Freshman is pledged to his society [fraternity] months before he approaches the college walls," a Yale alumnus of the class of 1869 wrote soon after his graduation, "and the graduate keeps up his senior-year connection long after he has left those walls behind him." The peculiarly Yale societies—Skull and Bones, Scroll and Key—were the most prestigious of all. When these held their elections, the anticipatory excitement among the "likely men" was intense. "All college, too, is on the alert, to find out what the result may be."

Neither Linonia nor Brothers in Unity provoked any such excitement. Most students no longer bothered to join either of the two, and they were arbitarily assigned in alternating alphabetical order to one or

the other. "Nobody now cares which society he is assigned to." As for Phi
Beta Kappa, it aroused little if any more interest

> With its mystery departed its activity also; and for fifty years past it has been
> simply a "society institution" possessed of but little more life than it claims
> to-day, though membership in it was thought an honor worth striving for
> until quite a recent period. . . . Personal prejudice sometimes kept a few
> high-stand men out, and personal favoritism sometimes brought a few low-
> stand men in, but, in general, scholarship alone decided the matter,—the
> society confining its elections pretty closely to the list recommended it by
> the faculty. . . .
>   The constitution of the society required every member to wear the badge,
> but only about a third of them did so fifty years ago, when their society was
> a living, wide-awake affair; and now, as for a dozen years or more past, the
> sight of a Phi Beta Kappa key would raise a cry of derision. The last Yale
> Senior who once or twice ventured to expose such a thing to the gaze of the
> populace belonged to the class of '67.
>   This was also the last class for whose initiation a special meeting was held.

The writer obviously had no Phi Beta Kappa key of his own, either to
display or to conceal, and he no doubt wished to believe the grapes were
sour, as many a disappointed student had done before him. Nevertheless, events soon bore him out in his diagnosis of the society's condition at
Yale. For years the chapter had existed only to elect members and to
invite an orator and a poet to perform during commencement week. In
1870 the chapter sponsored its last oration and in 1871 its last poem.
Then it stopped all activity, even the electing of new members.[7]

   When there existed only the three Al-
phas of Connecticut, Massachusetts, and New Hampshire, they sometimes had difficulty in coordinating their policies with respect to the
chartering of additional branches. Those three and the newer Alphas
experienced increasing difficulty as the society expanded, especially
after 1847. They worked out no system that all of them could or would
consistently follow in the granting of new charters. Consequently, by
1880 the society approached a crisis that threatened its disruption.
   There was little or no disagreement about how to begin the chartering
procedure. The Yale charter of 1779 from William and Mary had authorized Elisha Parmele to "establish a fraternity of the ΦBK, to consist of
no less than three persons . . . and as soon as such number of three shall
be chosen . . . to hold a meeting" and organize. In later years the Yale
members took this to mean not that an emissary could anoint three
persons as Parmele had done but that three persons already members of
Phi Beta Kappa must be connected with the institution that was to shel-
ter the new branch. This nucleus of three or more members would be
responsible for the application.
   If they were seeking to form a new Alpha, they would usually make

their first overture to the existing Alpha to which one or more of them belonged. If they were planning a chapter in a state that already had an Alpha, they would be expected to apply to that particular one. The Betas, Gammas, etc., were given no part whatever in the chartering process. Yet they were presumably equal to the Alphas, as the Yale chapter had implied in 1837 when it stated that all the branches—Alphas, Betas, and the rest—were to have charters "on the same uniform and impartial principles."

Occasionally the real initiative in seeking a charter came from students, as it did at Amherst, Williams, and CCNY. To get any attention, however, the students needed the cooperation of faculty members, who would make the formal application. Most often the idea originated with professors, presidents, or trustees. Quite a few of them continued to look upon Phi Beta Kappa as a source of prestige for their institutions, even though the society was losing its attractiveness for students at colleges where it already existed. After obtaining a charter for Alabama while a member of its faculty (1851), the great educator F. A. P. Barnard secured one for Columbia while its president (1869).

The Alphas receiving the application made little effort to investigate the college concerned, and they made less and less effort as time went on. No longer did they discuss qualifications and standards of the strict if rather general sort that the Connecticut, Massachusetts, and New Hampshire Alphas had formerly imposed. Indeed, it appears that in this later and less demanding period none of the applications was denied, though a few of them were delayed.

Yale received the application from Western Reserve in 1841, and Bowdoin did not approve it until 1847. Union meanwhile appointed a committee "to inquire into the state of this College" and especially to ascertain whether it had a "prospect of permanency." The delay in granting the charter could hardly have been due to thoroughness in the investigation, since Western Reserve was approved even though the college had a rather poor prospect of permanency and was soon struggling for survival.

Supposedly, all the existing Alphas had to approve the creation of a new Alpha. When an application came from the University of Vermont (1848), a Yale committee promptly gave a favorable report—"on account of the high character of the petitioners, the prosperity and rank of the University, the fact that there is at present no branch of the Society in Vermont, and the manifest advantages of extending the influence of our Fraternity." Soon the neighboring New Hampshire Alpha drafted a charter for the new Vermont Alpha and notified its organizers: "With the concurrence and consent of our brothers at Cambridge, at New Haven, at Schenectady, at Brunswick, and at Providence, we do hereby ratify and confirm this charter." There was no mention of the brothers at Hudson, Ohio, though the Alpha of Ohio had already been established at Western Reserve.

It was Vermont's turn to be overlooked when Rutgers applied twenty years later (1868). Six of the Alphas, including the one at Western Reserve, quickly approved, and then the one at Yale did so "with some reluctance." The Union chapter sent its president from Schenectady to New Brunswick to supervise the organization of the New Jersey Alpha. He presented a certificate: "The undersigned . . . *Hereby Certifies* that he has examined the authorities by which the seven Alpha chapters of the Phi Beta Kappa Society have authorized the establishment of a Chapter of the Society in Rutgers College in the State of New Jersey and finds that they are in due form and in accordance with the constitution and usages of said Society." But things were not quite in due form, since there existed not seven but eight Alpha chapters, and the constitution and usages of the society would have required the endorsement of all eight, including the Vermont Alpha.[8]

Though the Alphas were somewhat careless and inconsistent in setting up new Alphas, they had no serious disagreements as a consequence. They became far more careless and inconsistent in establishing Betas, Gammas, Deltas, and so forth. Over some of these cases, they did come into conflict, and it was this kind of conflict that eventually threatened to disrupt the society.

The issue was this: Could an Alpha on its own authority set up new chapters within its state, or could it do so only with the consent of other Alphas and, if so, how many? From time to time, when the question came up, the successive officers of the same Alpha gave different answers to it. They must have neglected to consult their own archives.

Thus, in the role of lawgiver to other chapters, the Yale society had told the Union brothers in 1837 that they could not establish branches in New York without unanimous consent. Then, several years later, the Yale society gave charters to both Trinity and Wesleyan in Connecticut without even notifying the other Alphas. In the code of laws they presented to Western Reserve in 1847, the Yale officers forgot what their own society had so recently done with Trinity and Wesleyan. They now declared that the Alpha of each state had "power, *with the consent of the other Alphas,* to establish other Branches of the Society within the bounds of said State."

The men at Amherst were not aware of the proper procedure when in 1847 they undertook to form the Beta of Massachusetts. They applied not to Harvard but to Yale. The Yale society approved the application and advised the Amherst men to get approval from the rest of the Alphas and a charter from Harvard. Deferring to Harvard, the others gave their consent on condition that Harvard should approve. Harvard did so, but neglected to provide a charter, and six years went by before the Amherst men could organize their chapter.

The applicants at Williams (which had applied unsuccessfully in 1799) obtained their charter much more expeditiously than those at Amherst. Indeed, they set a record for speed. "A meeting of students from the

Senior and Junior classes, comprising those of highest standing in scholarship, was held July 12, 1864," according to the minutes of the meeting, "for the purpose of establishing a provisional organization of the Phi Beta Kappa Society in this college, subject to the recognition and approval of the parent Alpha of the State of Massachusetts." Without referring the matter to any of the other Alphas, the Harvard chapter granted a charter right away for the Gamma of Massachusetts, and in less than three weeks after the July 12 meeting, the Williams men could make their provisional organization a permanent one.

The much younger Vermont Alpha was far more diffident than the Massachusetts Alpha. When in 1865 an application arrived for a Vermont Beta at Middlebury College, the Vermont Alpha voted to postpone consideration. According to its minutes, "the dignity of the Society and the preservation of its high character would forbid the establishment of new chapters except by the concurrent action of several branches"— several branches, but not necessarily Alphas and not necessarily *all* of them! Confusion was confounded when the Connecticut Alpha responded to the Vermont Alpha's inquiry by saying it had "voted to suspend the rule requiring the concurrence of three other chapters." There had never been any such rule. Anyhow, the Vermont Alpha chartered the Vermont Beta after waiting three years and getting the consent of the chapters at Yale, Dartmouth, and Brown.

Meanwhile the Ohio Alpha was generous in granting charters. When it received a petition from Kenyon (1858), it appointed a "commission to organize said Beta when consent of the other Alphas shall be obtained." But the Ohio Alpha allowed the Ohio Beta to go ahead and organize while awaiting the other Alphas' consent. When a Marietta professor appeared in person in the spring of 1860, the Ohio Alpha gave him to understand that there would be "no obstacle" to his getting a charter if his faculty contained the requisite minimum of three Phi Beta Kappa members. He did not have to wait for the other Alphas to act but obtained his charter that summer and proceeded to organize the Gamma of Ohio at Marietta. Other Ohio colleges undoubtedly would have acquired chapters if they had bothered to apply.[9]

A larger number of New York colleges were interested, and the New York Alpha obliged all those that applied to it. This was the most prolific of all the Alphas, producing seven of the twelve Betas, Gammas, etc., that originated between 1847 and 1882.

From the outset of its chartering career, the Union chapter proceeded independently. The question of its authority to do so arose when New York University, which had failed to get a charter in 1836, applied to Union in 1857. A committee reported to the society at its next anniversary meeting: "The Alpha of Massachusetts and the Alpha and Beta of Connecticut—with whom correspondence had been had—thought the consent necessary only of the chapter or chapters in the State in which it was proposed to organize a new chapter, but that if their consent was deemed necessary to the organization of a Beta in New York University it would be cheerfully given." Here was another passing suggestion that chapters other than the Alpha might play a role in the chartering of additional chapters within a state. This time the suggestion emanated from those fountainheads of Phi Beta Kappa, the Connecticut and Massachusetts Alphas, but they soon forgot all about it, as did the New York Alpha. At the moment, this Alpha could feel free to go ahead on its own since there would be no Beta in the state for it to consult until it had acted.

Receiving an application from CCNY a decade later, the New York Alpha referred it to a committee, which reported: "in its [the committee's] judgment it is the province of the Alpha . . . to form other chapters in the State without consulting the junior chapter or chapters." Apparently the committee gave no thought to whether the Alpha could form new chapters in its own state without consulting the Alphas in other states.

The New York Alpha made quick work of the next three applications, though dilatory with a fourth. It enabled Columbia (1868–1869) and Hamilton (1869–1870) each to organize a chapter within a year, and Hobart (1871) within a month. Madison (Colgate) had to wait three years (1875–1878), but that was because of an oversight on Union's part, not because of time-consuming deliberation.

No doubt the Union chapter would have acted promptly and favorably on the petition from Cornell in 1869 if the Cornell men had not made the mistake of sending it to Harvard. The secretary at Harvard recorded: "A vote was passed expressing a willingness on the part of this chapter of the ΦBK that a chapter should be established at Cornell University." Jealous of their prerogative, the Union brothers at their next annual meeting "appointed a committee to investigate the alleged establishment of a chapter of this society at Cornell University by the Phi Beta Kappa of Harvard."

Not until ten years later, in 1880, did the Cornellians get around to making an overture to Union. Its secretary informed one of them that Union was willing to comply, but the Cornell man neglected to transmit the message to his colleagues. Growing impatient, they told Union they were going to approach Harvard once more. The Harvard chapter had full power to respond, according to a recent report of one of its committees. This concluded that, "with the recent extinction of the chapter at Yale, the Harvard Alpha became the parent Alpha" and, as such, possessed the authority to grant charters on application from *any state*! The Union men were incensed. Their secretary wrote to the Cornellians:

> Harvard College had no authority to found a chapter anywhere. The Constitution provides that all the Alpha chapters shall concur in the founding of a new branch. This, however, of late years has been tacitly construed that

the Alpha of each State shall have jurisdiction in this matter over its own State.

You will see the confusion that could and would be produced if any chapter could found a branch anywhere and I hope you will not find it necessary to threaten us with Harvard again. We stand ready, whenever we are notified, to establish your chapter.

The New York Alpha at Union was soon to establish the New York Theta at Cornell, and so the particular case was about to be disposed of, but the troublesome, divisive issue remained. Certainly, as the Union secretary pointed out, confusion "could and would be produced if any chapter could found a branch anywhere." But confusion would also result if every Alpha could unilaterally found a branch anywhere in its own state, as the New York Alpha claimed it could do.

Conceivably the Alphas would compete in producing their own satellites, and Phi Beta Kappa would tend to break up into a number of separate clusters, state by state. Even if it remained nominally a single society, it would come to have branches at a great variety of institutions, some of them much less select than others. Most if not all of the colleges added from 1847 to 1882 were probably as good as the previous average, but there was no assurance that all those to be added in the future would be equally good. Clearly, the society had an urgent need to do something to control its expansion and maintain standards for the qualification of new chapters.

The Harvard chapter stood ready to offer a solution to the problem. Its members had taken the following action (1880):

> *Voted,* That whenever any college shall ask for a charter of Φ. B. K. the application shall be made to the Alpha of the State in which said college shall be situated. In case no such Alpha exists, then application may be made to any Alpha of Φ. B. K., and such charter shall be granted on a vote of not less than two-thirds of the Alphas of the United States.
>
> In case the Alpha first applied to shall refuse to take action, then application may be made to any other Alpha; and the charter shall be issued by the oldest Alpha voting in the affirmative.
>
> *Voted,* That, in the event of the assent of all the existing Alphas to this proposal, it be adopted as a new article of the constitution of Φ. B. K.

The existing Alphas were never to adopt this proposal, but in trying to persuade them to do so, the Harvard chapter was to bring about a much better way of dealing with this problem as well as other problems facing them. For the Harvard effort was to lead to the uniting of the chapters.[10]

The society also needed to do something to improve both its self-image and its reputation. Even the best informed of its own members knew little about its past development or its present condition. Outsiders knew still less, and what they thought they knew was often unflattering or false.

In his book *Four Years at Yale* (1871) Lyman H. Bagg, '69, commented unfavorably on the society in general as well as on the chapter at his alma mater. The society as a whole, Bagg pointed out, had no real unity and no real life. Not since 1852 had the Yale chapter published a catalog of its members.

> Each separate chapter of the fraternity has doubtless published similar lists, but no general catalogue of the members of all the chapters has ever been issued. With each branch showing signs of life but one day out of the three hundred and sixty-five, there can of course be no tangible tie between them, and any general work in the name of the whole fraternity is clearly out of the question.
>
> The history of the chapters elsewhere has been essentially the same as at Yale, save that the younger ones have never known any active life, but have been from the outset simply society institutions. . . . In short, "Phi Beta Kappa" is, always and everywhere, a mere official compliment paid by the faculty to high scholarship. Its key, or the right to wear it, is simply a medal, or reward of merit, certifying that the owner ranks with the first third of the class. This fiction, myth, abstraction, pious fraud, or what not is naturally the object of much merriment at Yale. . . . Such are the sarcastic and derisive utterances now heard in regard to the venerable fraternity which, almost a century ago, started out upon its mission of inculcating the doctrine that "Philosophy is the guide of Life."[11]

William R. Baird likewise described Phi Beta Kappa as moribund when in 1879 he published the first edition of his encyclopedic *American College Fraternities*. This society, Baird adjudged, had "outlived its activity and almost its usefulness."

> The proceedings of the society were always stiff and formal, and lacked vitality, although elections were eagerly sought by college students, as it was in a measure a confirmation of their rank. . . .
>
> Meetings of the members are still held about commencement time, and every winter banquets are partaken of in the larger cities. At these latter gatherings papers upon education and kindred topics are presented, and usually these are the only signs of life the old fraternity exhibits.

Baird repeated the hoary tales that "Thomas Jefferson was the founder" and that "it was of Masonic origin." He also quoted "a work published in London, 1876," which asserted: "*Phi Beta Kappa*, a branch of the Bavarian Illuminati, is supposed to have been established in America about the close of the last century, endowed with the above grotesque title."[12]

Both Bagg and Baird were devotees of the social fraternity system and evaluated Phi Beta Kappa in that light. This fraternity was not carrying on the kind of activities that other fraternities were; therefore it must be defective. But devotees of Phi Beta Kappa seemed to share the widespread indifference to its history and significance. In 1875 the Harvard chapter directed a committee "to consider the propriety of a celebration of the General Fraternity of the Phi Beta Kappa in the year 1876 . . . the centennial year of the foundation of the original chapter." That was as far as the Harvardians went in their preparations for 1876. They never got around to arranging a joint observance by the various chapters or even a separate observance by their own chapter. The only one to celebrate the occasion was the Rutgers chapter, which did so by hosting an orator for the first and last time.

As for the Harvard chapter's own centennial, its past president Edward Everett Hale wanted to make some observance in 1879, since the Harvard charter was dated 1779 (though the chapter was not organized until 1781). The fifty-seven-year-old Hale was a prominent Unitarian minister and a prolific author, best known for his story "The Man Without a Country," which he had written for the *Atlantic Monthly* in the midst of the Civil War to stir the war spirit of the North. Now, to observe the 1879 anniversary, Hale wrote an article for the *Atlantic Monthly* entitled "A Fossil from the Tertiary," the first account of Phi Beta Kappa history longer than a paragraph or two. Unfortunately, he did not have access to the minutes of the original William and Mary society. "The old records cannot now be found," he explained, "but probably exist in some Virginia archives." He did not realize that they were, indeed, well preserved in the archives of the Virginia Historical Society in Richmond.

Still, Hale did have access to the records of the Harvard chapter, including the correspondence of William Short with its president, and he had access to relevant passages from the John Quincy Adams diary. Using these materials, he was able to correct the more notable errors still current regarding Phi Beta Kappa's history. There was "not a shadow of a line of evidence to show that Jefferson had anything to do with it." Also fanciful was the story "that Phi Beta was invented by the French officers in Rochambeau's army after the pattern of the German Illuminati." It was preposterous to suppose that radicalism or irreligion had had anything to do with the society's origin and spread. "Far from being unchristian in its cradle, the Phi Beta Kappa owed all that extension which has given it any renown to a young student for the Christian ministry." The "St. Paul" who had brought the word to Yale and Harvard was, of course, Elisha Parmele.

Well informed though he was about much of the society's early history, Hale knew very little about its contemporary existence outside of—or even inside of—Cambridge. "At the present moment there are nineteen chapters, connected with as many leading colleges in the Union," he wrote. In fact, there were at the moment not nineteen but twenty chapters still active. Hale probably overlooked, as others often did, the chapter at Marietta College, away down on the Ohio River, the one most remote from Harvard Yard. Still, he had in mind a much more nearly correct count than the majority of members probably did. The president of the Madison (Colgate) chapter, in his 1878 initiation talk, said "our sister Societies in Massachusetts, Connecticut, New Hampshire, Maine, and Rhode Island have rendered honorable the appellation of ΦBK." He seemed unaware that there were also sister societies in Vermont, New Jersey, and Ohio.

Hale revealed his most surprising misconception when he referred to the meetings that had begun at Harvard in 1781. "From that time to this time the society has been in regular work," he averred. "It originally held meetings as often as once a week among the undergraduates. Such meetings still continue in all the colleges where branches have been established." Actually, regular and frequent undergraduate meetings of that kind had not been held at Harvard for more than half a century and were not at that time being held at any of the colleges.

In appraising the achievements of Phi Beta Kappa, Hale was extremely modest. Its name, he allowed, was "pretty well known, even to school-boys," who (in New England at least) were called upon to recite its more famous poems or passages from its more eloquent orations. In the "infant literature of the nation" these productions marked "some noteworthy steps." And Phi Beta Kappa was "the first of the Greek-letter societies," though quite different from the rest. "The society is one of the queerest things in America. It is indeed one of the very few visible relics of the mythical age of our national history; and it is not very visible at that."[13]

6

Organizing the
United Chapters

The first century of Phi Beta Kappa
had ended on an unpromising note. Already extinct were two of the
ancestral chapters, the one at William and Mary and the one at Yale.
Most of the others seemed to have little vitality left. Certainly they
showed a lack of esprit de corps as constituents of a common society.
Some of its leaders now determined to breathe new life into the venerable institution. What it most urgently needed, they believed, was a national organization, and this they managed to create during the decade
of the 1880s.

Phi Beta Kappa leaders naturally
thought of organizing on a larger scale, for other groups of all kinds
were doing the same thing. Consolidation with standardization was an
accelerating trend of the time.

During the 1880s the typewriter and the telephone came into general
use, adding to the feasibility of large-scale organization. Railroads, already combining to fix rates and pool earnings, combined to set standard time zones for the United States and Canada. John D. Rockefeller
converted the Standard Oil Company into the first of the "trusts." To
deal with business combinations, Terence V. Powderly consolidated the
"assemblies" of the Noble Order of the Knights of Labor, and Samuel
Gompers brought scattered unions together in the American Federation
of Labor.

During that same decade, professionals sought to set standards and
provide a voice for their respective fields by forming national associations. The list is fairly long. It includes the American Historical Association, the American Society of Mechanical Engineers, the American Ornithologists' Union, the American Society of Naturalists, the American

Economic Association, the Geological Society of America, the American
Mathematical Society, and the Modern Language Association. Such were
some of the organizations taking form simultaneously with the new Phi
Beta Kappa organization.[1]

The idea had previously occurred to some members of the society. As
early as 1858, when the Amherst chapter was only five years old, its
members voted to propose "a general convention of the branches of the
Phi Beta Kappa Society, for the purpose of forming a more compact
organization," and to request that each branch appoint a delegate. The
Dartmouth chapter responded favorably, and the Union chapter even
named its delegate, but nothing came of the proposal.

In 1877 the chapters of the three Manhattan institutions—Columbia
University, New York University, and the College of the City of New
York—cooperated to set up an organization of limited scope. Columbia
sought "a plan for making the Society more useful to its members" and
concluded that the best plan would be to form an association of ΦBK
alumni of whatever college living in the city. Thirty to forty graduates of
a dozen different colleges attended a series of meetings, drew up a
constitution, organized the ΦBK Alumni in New York, and celebrated at
Delmonico's. Soon the association boasted a hundred members. Surviving and prospering, it set an example for other alumni associations that,
along with it, were eventually to become significant elements in the
overall Phi Beta Kappa organization.

The need for such an overall organization became increasingly evident to thoughtful members. At Hobart the 1880 anniversary orator, the
college's former Latin professor, Francis Philip Nash, put the case effectively. "If the Phi Beta Kappa Society will wake out of her slumber and
will grapple in right earnest with the educational problems of the day,"
Nash declared, "she will soon accustom the public . . . to look upon her
as a living force and listen reverently and obediently to her teaching." He
suggested such "an improvement in the organization of the Society as
will facilitate common discussion of living questions." He said a national
council was needed, and "general conventions should be sufficiently
frequent to attract and fix public attention upon the deliberations and
labors of the Society." Nash was to have a key role in bringing about the
creation of such a national council.

Not Hobart but Harvard provided the initiative for the first convention. Taking the lead was the author of that rather whimsical 1879 account of Phi Beta Kappa history "A Fossil from the Tertiary," Edward
Everett Hale. At the 1880 meeting—where the Harvard men drafted a
constitutional amendment to define and limit the chartering power—
they also appointed a committee "to arrange a convention of all chapters
of the Fraternity," where they could propose the amendment along with
other constitutional reforms. In addition to Chairman Hale, himself a
Unitarian minister, the committee consisted of two other reverends (one

of them seventy-five years old), a retired judge and congressman, and the Harvard librarian, Justin Winsor (not quite forty).

Their chapter being the oldest one still in active existence, the Harvard men considered themselves the logical ones to sponsor such a convention. An ideal occasion was coming up—the hundredth anniversary of the chapter to be celebrated on June 30, 1881. Accordingly, Hale and his fellow committeemen signed letters to the rest of the chapters, inviting them to send representatives to the anniversary meeting, as many as five per chapter with each delegation to cast but a single vote. You will "give such power to your delegates, that the assembly of delegates may act as a convention of Phi Beta Kappa, to determine on any changes which may be necessary in the Constitution of the Fraternity."[2]

It was a bright, fresh, beautiful June morning when, at nine o'clock, Hale called Phi Beta Kappa's very first convention to order in Gore Hall. He immediately chaired a committee to ascertain and report the names of the delegates. They represented the chapters at a dozen institutions: Bowdoin, Dartmouth, Vermont, Harvard, Amherst, Brown, Wesleyan, Union, New York University (NYU), CCNY, Hamilton, and Hobart. Six of the chapters were Alphas, the others were not, but all were being treated alike with one vote apiece. Of the total of fifty-six delegates appointed, fifteen boasted the title of professor. Another fourteen called themselves "reverend" or indicated a theology degree; some of these, too, were faculty members. Six others were politicians of sufficient stature to merit the prefix "Hon."

Hale and his fellow Harvard delegates expected the convention to complete its work that very morning. They were not planning on a new organization for the society but only on the accomplishment of a few reforms. They had their agenda ready. For one thing, they wanted the convention to adopt as the model for all future charters the charter that Yale had drafted for Western Reserve in 1847. For another thing, they wished to secure the adoption of the constitutional amendment their chapter had recently endorsed in the hope of defining and limiting the authority of each Alpha to issue charters.

But Professor Nash of Hobart still had in mind a more drastic change, one too extensive to be immediately achieved. Before the convention could turn to other business, he rose to offer the following resolution:

Whereas the Phi Beta Kappa Society has not hitherto exerted upon the intellectual life of America an influence commensurate to its true and legitimate importance, having been precluded therefrom by the lack of any regular method of ascertaining and expressing the views of the Society as a whole;

And whereas it is highly desirable that a voice and utterance should be given to the collective learning, wisdom, and experience of the Society, in order that the Society may obtain the influence and moral power which legitimately belong to it;

And whereas this object cannot otherwise be attained than by entrusting the expression of the opinions of this Society to some sufficient representative body delegated by the several chapters,—therefore,

*Resolved,*That this Convention do hereby earnestly recommend to all chapters of the Phi Beta Kappa Society to choose delegates to the number of three for each chapter, to meet together at _____, on the _____ day of _____, 1881, and from day to day thereafter, who when thus assembled shall constitute the National Council of the Phi Beta Kappa Society and shall have power to express the opinion and sentiment of the Society upon all such questions as may from time to time be presented to said Council for its consideration.

Thus Nash changed the question from reform to reorganization, a question he would postpone until later the same year. The National Council, in the form in which he proposed it, would have had rather limited power—"power to express the opinion and sentiment of this Society"— and yet it would have given the society at least the nucleus of a national organization.

Professor John A. De Remer of Union College moved to amend Nash's resolution by providing for a committee of one member from every chapter to meet at the University of the City of New York on October 18, 1881. The Nash resolution with the De Remer amendment passed. Presumably the committee would act as a new convention with authority to consider and to adopt or reject not only the Nash resolution but also other proposals. Hale now brought up the two Harvard favorites—the adoption of a model charter and the limitation of each Alpha's chartering ability—and agreed to refer them to the October convention in New York.[3]

Before adjourning the Cambridge convention, the delegates passed a resolution of thanks to the Harvard chapter for its kind reception. The chapter then invited them to accept its hospitalities for the rest of the day. As guests at Harvard, the men from lesser institutions had a chance to see how impressively Phi Beta Kappa conducted its annual affairs in the place where they were longest established and best known. These, coming on the day after commencement, at the end of "prolonged college festivities," no longer attracted large crowds as a rule, though the society had ceased to have "a real existence except on the day of its oration and poem and dinner." This year, however, the centennial attracted an unprecedented number of celebrants in addition to the delegates from other chapters.

The delegates went to Boylston Hall to attend the Harvard chapter's annual meeting. Then in the afternoon they marched in the procession to hear the obligatory oration and poem. "The pink and blue ribbon, which has replaced the square gold watch-key of other days, fluttered in every button-hole, and with pealing music leading the way, the long, long procession—a Φ. B. K. procession such as even Harvard never saw before—wound under the imposing buildings toward the beautiful college hall, the Sanders Theatre."

The orator was one to rival the greatest of the past and make this day in 1881 as memorable as that day in 1824 when Edward Everett "closed with an apostrophe to Lafayette, sitting upon the platform" beside him, or that day in 1837 when Ralph Waldo Emerson read his famous lecture on "The American Scholar." The orator was Wendell Phillips. Now going on seventy, Phillips had been a flaming leader of the antislavery movement, and after the winning of that cause he devoted himself to new crusades, most notably those for workers' rights and women's suffrage. Though less strong of voice than he used to be, he was still his eloquent self as he took up his chosen theme, "The Scholar in a Republic."

Phillips was out to shock his more conservative hearers, and he began by repeating the story of a radical, antireligious origin of Phi Beta Kappa:

A hundred years ago our society was planted—a slip from the older root in Virginia. The parent seed, tradition says, was French,—part of that conspiracy for free speech whose leaders prated democracy in the *salons*, while they carefully held on to the flesh-pots of society by crouching low to kings and their mistresses, and whose final object of assault was Christianity itself. Voltaire gave the watchword,—
"Crush the wretch."
"Écrasez l'infame."
No matter how much or how little truth there may be in the tradition: no matter what was the origin or what was the object of our society, if it had any special one, both are long since forgotten. We stand now simply a representative of free, brave, American scholarship. I emphasize *American* scholarship.

Thus, after a brief detour through fantasy, Phillips quickly returned to the ground made familiar by Emerson and many other Phi Beta Kappa orators.

But Phillips the reformer continued to say provocative things as he went along. "History is, for the most part, an idle amusement, the daydream of pedants and triflers." (The history of Phi Beta Kappa in particular was apparently not for serious thinkers.) "I urge on college-bred men that, as a class, they fail in republican duty when they allow others to lead in the agitation of the great social questions which stir and educate the age." "Never again be ours the fastidious scholarship that shrinks from rude contact with the masses." "Intrench labor in sufficient bulwarks against that wealth which, without the tenfold strength of modern incorporation, wrecked the Grecian and Roman States; and, with a sterner effort still, summon women into civil life as reënforement to our laboring ranks in the effort to make our civilization a success."

All those in the audience knew they had attended "as noble a display of high oratory" as they could ever hope to hear, or so it seemed to one of them, George William Curtis, editor of the "Easy Chair" in *Harper's Magazine*. "Ah!" Curtis heard another man exclaim, "it was a delightful discourse but preposterous from beginning to end." Curtis conceded that the speech had logical gaps. "Yet its central idea, that it is the duty of educated men actively to lead the progress of their time, is incontestable."[4]

No doubt many of the chapter delegates agreed with Curtis as they left Sanders Theatre to join the procession again and march to Massachusetts Hall, where they were to be dinner guests. Afterward they could reflect upon an inspiring day, though no more than a partially constructive one. There still remained the task of converting Phi Beta Kappa into an organization that might do much more than it was doing to encourage the educated "actively to lead the progress of their time."

On July 2, 1881, two days after the centennial celebration in Cambridge, President James A. Garfield was shot as he walked through the Union Station in Washington. He lingered through the summer at a New Jersey seaside resort and finally died on September 19, less than a month before the date set for the ΦBK committee to meet in New York.[5]

When the committee met in the NYU council chamber on October 18 at 10 a.m., sixteen members were present, each of them representing a chapter. According to the minutes of the secretary, Justin Winsor, "The following (five) chapters were not represented: Amherst, Western Reserve, Univ. of Alabama, Univ. of Mississippi, Madison University (Hamilton, N.Y.)." In fact, there had never been a chapter at the University of Mississippi. The chapter at Western Reserve was temporarily inactive, and the one at Alabama was as dead as the one at William and Mary. Besides Amherst and Madison, two other functioning chapters were unrepresented—Kenyon and Marietta—but these two were not even mentioned in the secretary's report. Obviously, the committeemen were less than fully informed about the condition or extent of the society they were preparing to reorganize.

They promptly voted to accept, with thanks, two social invitations— one from the president of CCNY for a reception at the college that evening, another from the president of Columbia and members of the Columbia chapter for a reception at that college the following afternoon. Then Professor Nash of Hobart offered a draft of a constitution, a revised form of the one he had proposed at the Harvard convention.

During the two-day session the committee managed to agree on a "Constitution of the National Council of the Phi Beta Kappa Society."

According to this plan, the National Council would consist of three delegates from each chapter plus the members of the Senate. The Senate would consist of twenty people whom the National Council would choose from the society at large. Senators would have six-year terms, which would be staggered, so that ten senators would be elected every three years. The National Council would choose a president from among the senators, and it would choose from the society in general a vice president, a secretary, a treasurer, and such other officers as it might find necessary.

The National Council would meet every third year. The Council or, between Council sessions, the Senate would have "exclusive power to grant charters" and "to prescribe the form of constitution for new chapters." The Council would also have power to elect by secret ballot "Members at Large of the Phi Beta Kappa Society," who would have all the rights and duties of members elected by a particular chapter.

During the triennial sessions of the Council, the Senate would "have no separate existence," the senators merely taking places in the Council along with the delegates. "When the National Council is not in session the Senate shall constitute an independent body charged with the duty of representing the Phi Beta Kappa Society and speaking in its name, and exercising in addition the functions of a permanent Executive Committee of the National Council."

This constitution was to take effect when ratified by twelve chapters. Having agreed on its terms, the committeemen scheduled the first National Council to meet in the Adirondack resort of Saratoga Springs on the first Wednesday in September 1882. They instructed their secretary to provide each chapter with a copy of the draft, invite each to send three delegates to Saratoga Springs, and request the chapters "to empower their delegates to ratify said Constitution with such modifications" as the Council might decide to make.

Thus, in the view of the drafting committee,the National Council at its first session would serve as both a ratifying convention and a governing body. Indeed, it might operate as a constitutional convention at any of its sessions, since it would have power to amend the constitution of the society at large as well as to prescribe the constitution for new chapters.

Before adjourning, the committeemen voted "That the consideration of a uniform form of charter, and all the other matters referred to this committee, be referred to the National Council." Thus they disposed of their unfinished business. In particular, they recommended that steps be taken to achieve the following: the drafting of "a uniform Constitution and form of initiation, as desirable for adoption by all existing chapters, and to be used in the institution of new chapters"; "the compilation of a general catalogue of ΦBK."; "the foundation of a ΦBK travelling fellowship, open for the competition of all American scholars"; and the specification of "a uniform condition of membership."[6]

Under the new constitution there would no longer be the inequality between the Alphas and the other chapters, as there had been when only the Alphas had been allowed to issue charters. There would be, however, another kind of inequality, one between the chapters antedating and those postdating the establishment of the National Council. By a kind of unwritten grandfather clause, the earlier societies would have greater autonomy than the later ones. The committeemen hinted at the difference when they spoke of preparing a uniform constitution and initiation ceremony. These were to be recommended as "desirable" for existing chapters, but were "to be used" by new ones.

The question of uniformity was to persist through much of Phi Beta Kappa history—uniformity not only with regard to chapter constitutions and ceremonies but also with regard to standards for qualifying new chapters and electing members. When the committeemen suggested the foundation of a traveling fellowship, they were touching upon another question that was to persist, one that was to become more and more urgent with the passage of time. Beside electing and initiating members and, at a few places, sponsoring an annual oration or an annual poem, what could the society do to further its avowed aims? Could it do anything besides enlarging and perpetuating itself?

The meeting at Saratoga Springs, September 6–7, 1882, proved to be a second sitting of the constitutional convention rather than the first session of the National Council. Present in the council room of the town hall were twenty-eight delegates from fifteen chapters, but only seven of the delegates had come with power to ratify the constitution—those from Amherst, CCNY, Kenyon, Trinity, NYU, and Wesleyan. Two others, those from Rutgers and Vermont, had power to do so only on condition that certain amendments be added. Since the committee in New York had set a minimum of twelve chapters for ratification, the constitution could not yet go into effect, and the delegates could not proceed to act as members of the National Council.

So they remained in Saratoga Springs to discuss and amend the constitution. After making a few changes, which did not affect the basic structure, they unanimously approved the revised draft. Then they voted that fourteen rather than twelve would have to ratify. The seven already having power to ratify did so. Finally the delegates called for the National Council to meet at the same time and place the following year if the necessary seven additional chapters should have ratified by that time.

By June 1883 the following chapters had added their approval: Harvard, Dartmouth, Union, Bowdoin, Wesleyan, Williams, Columbia, Hamilton, and Hobart. These made a total of sixteen charter members of the new national organization. The four other active chapters were slow to ratify. Brown delayed for a dozen years, until 1895. The national organization did not even recognize the existence of the Marietta chapter, which had not been invited to the Harvard centennial or to any of the organizational meetings, until 1896.[7]

Only thirteen of the sixteen chapters already having ratified were represented when twenty-three delegates gathered in Saratoga Springs on September 5, 1883. The delegates began by electing members of the Senate, awarding six-year terms to the ten candidates with the largest number of votes, three-year terms to the next ten. Then the senators and the other delegates constituted themselves the first National Council of the Phi Beta Kappa Society. For president of the society, they elected the president of Harvard University, Charles W. Eliot.

The delegates kept on with the constitution-making process by taking a step toward the designing of a standard constitution for new chapters. They instructed the newly elected senator John A. De Remer to look into the constitutions of the existing chapters and report to the next Council, three years later. De Remer was an expert in the field, having been a leader of the Alpha that had brought forth more branches in New York than any other Alpha had done in any other state.

While the Council was getting down to business, the Senate stalled. The secretary called the senators to meet in the Overseers' room at Harvard on December 27, 1883. Only seven bothered to leave their Christmas holidays and attend. These seven decided they did not constitute a quorum. They talked informally about the condition and prospects of the society but took no action, though the secretary reported that he had received applications for charters from several colleges.

By September 1886, however, the Senate was ready to respond when the second National Council requested it to report to the next Council meeting "a Form of Charter and Constitution to be used by the National Council for all Charters hereafter granted." There was, the resolution pointed out, "nothing in the character of the ΦBK Society making secrecy obligatory upon any Chapter." When the Senate next met in Boston on January 8, 1887, De Remer arrived with the draft of a model charter. The Senate quickly adopted it and immediately used it in authorizing a new chapter—the very first to be chartered under the auspices of the reorganized society. The new chapter, the Iota of New York at the University of Rochester, lay within what had been Union's and De Remer's bailiwick.

Not until 1887, with the adoption of a standard charter form, did the term "United Chapters" appear in an official document of Phi Beta Kappa. None of the conventions, councils, or committees had formally agreed that this term should be used to designate the national organization, as it was henceforth to do. The words had simply crept into the organization's title when De Remer wrote and the Senate approved the charter form.

The form began: "Whereas the NATIONAL COUNCIL OF THE UNITED
CHAPTERS OF THE PHI BETA KAPPA SOCIETY has . . . authorized the Senate
to grant a charter. . . ." It proceeded to say the Senate hereby established
"a separate and subordinate branch" having "all the powers, privileges,
and benefits" enjoyed by "other and existing Chapters." Thus the document affirmed the principle of equality, though in fact a new chapter
would not be quite equal in every respect to all the older ones. Finally,
the form enjoined "a due regard for, and strict compliance with, the
Constitution of the aforesaid National Council and all the laws, customs
and usages of the Phi Beta Kappa which have obtained in our several
Chapters in the years past." Here was a call to uniformity, though the
customs and usages of the several chapters in years past had hardly been
uniform.

In 1889 the third National Council approved a new model charter,
much like the one of 1887, and also a model constitution. This referred
to the society at large by almost exactly the title that was to become
official. The document read:

> I. This society is one of the co-ordinate branches of the body known as *The
> United Chapters of the Phi Beta Kappa,* and shall be called the _____ Chapter
> of the Phi Beta Kappa society in the State of _____.
>
> II. The object of the Phi Beta Kappa society is the promotion of schol-
> arship and friendship among students and graduates of American Colleges.
>
> III. The members of the Chapter shall be elected *primarily* from the best
> scholars of the graduating classes of the college, *secondly* from those gradu-
> ates of said college where post-graduate work entitles them to such honor,
> and *lastly* from any persons distinguished in letters, science or education;
> provided, however, that the selection from each graduating class shall not
> exceed one-fourth of the number graduated. But the Chapter may make
> further limitations or restrictions.
>
> IV. In addition to scholarship, good moral character shall be a qualifica-
> tion for membership, and any member who is found to have lost this qualifi-
> cation may be expelled from the society by a four-fifths vote of the members
> present at a regular annual meeting of the society.
>
> V. The Chapter shall send a delegation to represent it at each National
> Council of the United Chapters, shall contribute its equal part to the finan-
> cial support of the United Chapters, and shall conform to the constitution
> of the United Chapters and all the lawful requirements of the National
> Council.
>
> VI. This Chapter shall, by the enactment of suitable by-laws, provide for
> its election of officers, the initiation of members, the conduct of its meet-
> ings, and such other matters as it shall deem wise to so regulate.[8]

With this document the constitution-makers of Phi Beta Kappa completed the work that had taken them most of a decade. They had now
laid out on paper the basic structure and functions of both the national
organization and its constituent groups. It remained for the leaders of

the society to keep the society going according to the letter and spirit of its newly adopted laws.

Time would be required for the realization of Professor Nash's hope that a reorganized society would "accustom the public . . . to look upon her as a living force and listen reverently and obediently to her teaching." For the moment, the public paid little or no attention to what the society was doing. Readers of *Harper's Magazine* might remember George William Curtis's description of the 1881 affair at Harvard, but Curtis had been preoccupied with the centennial observance and particularly with Wendell Phillips's oration; he did not even mention the convention of delegates from other chapters. Subscribers to the *Harvard University Bulletin* could read Justin Winsor's minutes of the 1881 meetings, but that was hardly a periodical with mass circulation. No such periodical reported or commented on the establishment of the United Chapters. The organization was yet to become a subject of national news.

7

Keeping Up
with the Colleges

The United Chapters emerged in the midst of a revolution in American higher education. Between the Civil War and the First World War, the colleges and universities grew not only in number but, as never before, in total enrollment and in diversity of size and type. It became increasingly difficult to draw a line between those qualified and those unqualified for Phi Beta Kappa charters, and to draw that line was the main responsibility of the new organization. The Senate and the National Council had to develop appropriate standards and procedures. While assisting in the chartering process, the secretary had to maintain a central office and do the chores necessary for holding the organization together from day to day. The successive senators, councillors, and secretaries managed to keep up fairly well with the bewildering growth and change in academe. During the thirty years from 1886 to 1916, they saw to the establishment of sixty-four new chapters—bringing the total to more than four times what it had been when the United Chapters began.

The secretariat developed less rapidly than the society as a whole and less rapidly than the burdens it was expected to bear. Only the devotion and self-sacrifice of the first three secretaries—who also had to serve as treasurers—enabled the central office to perform the duties that the Senate and the Council imposed upon it. And for twenty-one years the secretaries served without pay.

From 1883 to 1889, the secretary was Adolf Werner, a professor at the College of the City of New York. The ΦBK office was Professor Werner's office at the college. From 1889 to 1901, the secretary was Eben B. Parsons, a Williams College graduate of the class of 1859. Parsons was a parson, having filled the pulpit at a Presbyterian church and having received a D.D. degree. He had been editor of the *Williams Quarterly*, alumni necrologist for the college, and compiler of a monumental catalog of his Phi Beta Kappa chapter. He was described as a "short, white-haired gentleman, with a courteous, almost apologetic bearing." While secretary of the United Chapters, he made his living as the Williams registrar and secretary of the faculty. The society's headquarters were in Williamstown, Massachusetts.

In 1901 Oscar M. Voorhees succeeded Parsons. Voorhees, Rutgers '88, was also a minister of the gospel, an ordainee of the Reformed church, and he continued to depend on his ministerial pay. But no such self-effacing character as Parsons, Voorhees was a man of great self-assurance and considerable personal force. With Voorhees as secretary, the society's head office was the study in his parsonage. The parsonage was located first in one and then in another small New Jersey town and after 1909 in New York City.

The secretary had to edit the society's official publications, which presumably were essential for maintaining its unity and continuity and for futhering its avowed aims, particularly the promotion of American scholarship. Elaborate plans for the direct promotion of such scholarship turned out to be premature. In 1889 the Council resolved that it was "eminently desirable to secure in connection with the quadri-centennial of the discovery of America a proper representation of the life of the American people as manifested by their progress in Science and Literature." Therefore a committee should plan a series of historical monographs and should offer prizes for the best general essays on the subject. "This proposal aroused enthusiasm," the minutes stated, but nothing came of it.[1]

While Parsons was secretary, however, in 1900 the society succeeded in bringing out an impressive volume consisting of a "handbook" and a general catalog. This was the kind of work for which Parsons was by temperament and experience ideally suited. He included in the volume not only a directory of all the society's living members but also the text of its constitution and bylaws, a list of its officers and senators, a sketch of its history, a miscellany of its "customs and statistics," and a summary of responses to a questionnaire concerning the current practices of its various chapters. He hired printers in North Adams, a few miles from Williamstown, to manufacture the book.[2]

Under Voorhees the publication of Council proceedings became regularized. In 1904 he launched a series of bulletins which he called "Phi Beta Kappa Publications" and in which he reported not only Senate and Council actions but also "Chapter Activities." In 1910, with the Council's authorization, he began to edit an official quarterly, the *Phi Beta Kappa Key*. This was intended to keep members informed of the affairs of both the United Chapters and the individual chapters. Soon it concentrated rather heavily on historical articles, reflecting Voorhees's interest in the society's history. From the outset, the *Key* was expected to stimulate the morale of members, especially the newly elected students. "By it they will be made to realize," the president of the United Chapters, Edwin A. Grosvenor, declared, "that they are essential parts . . . of a national scholarly brotherhood . . . a mighty educational force."

At the same time that the Council authorized the official quarterly, it also recommended that the officers arrange for the publication of a volume of the best Phi Beta Kappa addresses—a project that the Phi Beta Kappa Alumni in New York had suggested. The Council expected the Harvard and Yale librarians to assist the secretary in making selections. But the work made little progress until Professor Clark S. Northup of Cornell took over as chief.

Northup and his two co-editors chose twenty-six speeches that various chapters had sponsored during the period 1837–1910, among them Ralph Waldo Emerson's and Wendell Phillips's. "The annual address has now become . . . a custom in most of the eighty-six chapters," the editors noted in their preface to the volume. "It has been thought by many that some of the most represenative of these orations should be reprinted, not only because they are in themselves worthy of thus receiving a new lease of life, but also because such a collection would help to emphasize the aims for which the Society has always stood—the cultivation of friendship, literature, and morality." In other words, the book, would be a good advertisement for the society. In 1915, the Houghton Mifflin Company published *Representative Phi Beta Kappa Orations* in a modest edition of one thousand copies.[3]

Royalties from the book, if any, would eke out the society's income, and every little bit would help. The United Chapters desperately needed money, though it had been operating on a shoestring. It paid no rent for office space and no salary to the secretary until 1904, when he began to receive $100 a year. Other regular expenditures went for clerical services, for correspondence, and for travel. Receipts came mainly from a "franchise fee" for every new charter, from a triennial assessment on each chapter, and from subscriptions to the society's publications. The secretary's financial report for the triennium ending in 1904 showed expenses of $289.97 and receipts of about $310 plus $50 in bank interest. All this left $775.70 in the bank, an encouraging gain over the previous triennial balance. But the secretary's salary alone would add $300 to expenses for every triennium in the foreseeable future.

Though the society managed to stay in the black, the bank balance began to shrink, and by 1913 it seemed dangerously low. That year the Council considered raising the triennial assessment on each chapter from $5 to $10 but decided that this kind of tax was inequitable, since the chapters differed greatly in size of membership. The Council voted, instead, to require each new member to pay $1 to the central office. In return, the member would be registered there and would receive a one-year subscription to the *Phi Beta Kappa Key*. Even so, the circulation of the *Key* increased at a disappointing rate, as not even all the Council delegates subscribed. By 1916, only slightly more than 3,000 copies, including those given to new members, were being issued. The periodical was not a great moneymaker.

By this time the financial cares of the office, in addition to the secretarial duties, had become too much for a single secretary-treasurer. According to the United Chapters constitution, the Council was supposed to elect a treasurer as well as a secretary. In 1913 Voorhees appealed to the Council to do so. "It is not good practice," he pointed out, "to make the secretary responsible for all the funds, including hundreds of subscriptions to *The Key,* charge him with the duty as treasurer of paying himself for his services as secretary, and then require him to render an accounting only once in three years." The Council responded by designating a separate treasurer for the first time.

Voorhees got no such immediate response when at the same session he also appealed to the Council for a new society headquarters. He argued persuasively:

> Phi Beta Kappa as an institution is wedded to the cause of higher education. The anomalous fact is that for twelve years its headquarters have been in a pastor's study, subject to the exigencies and uncertainties of twentieth century "circuit riding." To comport with the dignity of the organization, should not Phi Beta Kappa have a habitation as well as a name, a location having a near relation to one or other of the great educational institutions in which it has taken root? If it is decided to continue the *Phi Beta Kappa Key* much material of value will accumulate. When I became Secretary I received the record book, a bundle of letters relating to applications for charters, and some pamphlets. Now our possessions include a desk, a writing machine, a duplicator, a filing cabinet and a case for books. The archives now consist of books, pamphlets and catalogues, some of considerable value, and surplus copies of *The Key.* These are housed in a wooden dwelling with no special insurance or fire protection. Fortunately no loss has thus far been sustained, but I question the wisdom of continuing under these conditions. Has not the time arrived for seeking a permanent home?

But even if the United Chapters had found a permanent home, the money was not available to pay for it.[4]

With such primitive facilities as they could get, Secretaries Werner, Parsons, and Voorhees had carried on the steadily growing operations of the office. A large part of the work consisted of correspondence regarding charter applications. This became increasingly voluminous as more and more colleges and universities rose to a level where they considered themselves worthy of a Phi Beta Kappa chapter.

During the 1880s, while the United Chapters was getting started, the British scholar-politician James Bryce was gathering material for *The American Commonwealth* (1888), his classic study of life in the United States. "The American universities and colleges are in a state of transition," Bryce observed. "True, nearly everything in America is changing," but nothing more noticeably than the educational institutions. Indeed, the changes were dramatic. To consider only the external aspects, the institutions were becoming more and more diverse in purpose, enrollment, wealth, and prestige, while students were flocking to them in numbers far greater than ever before (though far smaller than in years to come).

New small church-related colleges continued to appear, especially in the Midwest and Far West. But in those same areas the state universities grew in number and size, many of them assisted by the Morrill land-grant act of 1862. Captains of industry endowed private institutions such as Vanderbilt, Johns Hopkins, Cornell, Stanford, and Chicago (Rockefeller). These, along with the state universities and the already established private universities, developed graduate and professional schools. Separate professional, technical, and normal schools, some private and some public, also arose. Women's colleges were founded—Vassar, Wellesley, Smith, and others—while many of the private colleges and nearly all the state universities became coeducational. The scene was quite different from what it had been when the nation's colleges, with few exceptions, conformed as closely as they could to a single pattern, the one set by Harvard and Yale.[5]

How many colleges or universities existed at any given time depended on the definition of a college or university. In 1870, making its first list, the United States Bureau of Education included all the institutions granting degrees and arrived at a total of 369. But in 1905 the Carnegie Foundation for the Advancement of Teaching recognized only those having at least six full-time professors, a complete four-year course in the liberal arts and sciences, and a requirement of four years of high school (or the equivalent) for admission. The much more generous Census Bureau counted 563 "institutions of higher education" in 1870 and 951 in 1910. For 1906 the United States Commissioner of Education reported 622 universities, colleges, and technical schools, "including over 100 class B colleges for women only," and 264 normal schools. It would be safe to say that the number of those colleges and universities eligible for consideration by Phi Beta Kappa more than doubled during the three decades after the founding of the United Chapters.

The number of students grew much more rapidly than the number of institutions, but how many were, strictly speaking, college or university students depended, again, on the definition of a college or university. According to one estimate, the college population was multiplied by forty while the country's population was multiplied by three between 1850 and 1906. According to the census figures,the enrollment in institutions of higher education increased three times as fast as did people of college age (eighteen to twenty-one) between 1870 and 1910.

Students attended colleges and universities that were growing larger and larger. In 1850 nearly half of the students—42 percent of them—had been in institutions with an enrollment of fewer than 150, and none in institutions with as many as 400. Yale, with 386, was the largest in the land. In 1904 nearly 80 percent were in institutions with more than 400 students, and nearly 10 percent were in those with more than 4,000. (No college or university yet enrolled as many as 5,000.)

In the early 1900s the majority of students in the Northeast and the South were still attending private colleges or universities. But a larger majority in the Midwest and Far West were going to public institutions, which on the average had considerably larger enrollments than private ones. So large was the majority in the Midwest and the Far West that it meant a majority for the state institutions in the country as a whole.[6]

The small denominational colleges were losing out in their struggle with the state universities. Bryce in the 1880s had predicted this result. Already, he noted, the sectarian colleges were competing desperately for students and resources and were often lobbying to cut state appropriations. "But as the graduates of the State universities became numerous in the legislatures and influential generally, and it is more and more clearly seen that the small colleges cannot, for want of funds, provide the various appliances—libraries, museums, laboratories, and so forth—which universities need, the balance seems likely to incline in favour of the State universities." During the decade 1885–1895, there was little increase in the enrollment of eight reputable church-related colleges in the Midwest, while the number of students trebled in eight state universities taken together (Michigan, Illinois, Wisconsin, Minnesota, Iowa, Kansas, Nebraska, California). By 1895 there were more Presbyterian students in seventeen state universities than in thirty-seven of the best Presbyterian colleges.

Some colleges and universities later qualifying for Phi Beta Kappa charters had little strength or reputation at the start of the United Chapters. In 1880 President F. A. P. Barnard of Columbia, the founder of two Phi Beta Kappa chapters, wondered sarcastically how England, with a population of twenty-three million, could get along with only four degree-granting institutions while Ohio, with three million, boasted thirty-seven of them. At that time the total income of all the Ohio institutions was less than that of Harvard alone.

"The most fully equipped would seem to be the State University at Columbus, with a faculty of 26 teachers; but of its students 141 are in the preparatory department, only 34 in the classical, and 29 in the scientific," Bryce noted. "Oberlin, Wooster, and Marietta (all denominational) have larger totals of students, and are probably quite as efficient, but in these colleges also the majority of students are to be found in the Preparatory Department." (Before high schools were common, it was customary for a college to operate its own academy or preparatory school.)

Marietta, of course, already had a Phi Beta Kappa chapter, and Ohio State, Oberlin, and four other Ohio institutions were to have chapters within twenty years or so after Bryce wrote.[7]

By the early 1900s some of the state universities compared favorably in income with even the most richly endowed private institutions—these depended on tuition fees and investment yields; the state universities had the benefit of tax revenues. Thus, as of 1905, the University of Illinois owned only 40 percent as much property as Johns Hopkins and less than 10 percent as much as Harvard or Columbia. Yet Illinois, with $956,166 for the year, enjoyed an income larger than that of Johns Hopkins, Yale, Chicago, or Stanford. Its income was only slightly smaller than that of Columbia and was 67 percent of that of Harvard, which topped the list.

In the 1880s it seemed to Bryce more difficult to generalize about colleges and universities in the United Sates than in Germany or England. "Not only is the distance between the best and the worst greater . . . but the gradations from the best down to the worst are so imperceptible that one can nowhere draw a line and say that here the true university stops and the pretentious school begins." There was a "vast distance between the standard of a university like Johns Hopkins at the one end of the scale, and that of the colleges of Arkansas at the other."

Even a degree from one of the least distinguished colleges seemed to have "a certain social value," but a degree from one of the favored few had much more. While a lesser institution attracted students only from its immediate vicinity, the more prestigious ones, particularly Harvard and Yale, were beginning to draw from far and wide, as the well-to-do sent their sons away "for the sake of general culture, or of the social advantages" to be gained.

It was perhaps easier to estimate the social value of a degree than to ascertain its real intellectual worth. As late as 1910, the standard of work at leading institutions remained extremely low. Even at Yale and Harvard, a senior could prepare for all his classes by studying an hour or less a day. At Harvard the free-elective system, allowing students to choose their courses at will, made it especially difficult to tell how well educated any graduate might be. "What does an A.B. from Harvard mean in intellectual discipline and development?" the president of the Carnegie Foundation for the Advancement of Teaching, Henry S. Pritchett, asked in 1908. He answered his own question: "Sometimes four years of real work under good men, sometimes three years of disconnected courses (partly snap) passed with the aid of a widow [private tutor]. Perhaps in no institution is the value of the A.B. degree so indeterminate."

The Carnegie Foundation declined to serve as an accrediting agency. The Regents of the State of New York set minimum standards for finances, library, staff, etc., but only for colleges in that state. In 1910 the United States Bureau of Education, together with the American Association of Universities, drew up a list of colleges whose graduates could be presumed to be fitted for graduate work, but so many colleges objected to the list that it was never published. By the early 1900s the North Central Association was beginning to accredit colleges as well as secondary schools in the one region. Nationwide accrediting was yet to come, and when it did, it would only screen out the worst; it would not help in selecting the best.[8]

Meanwhile, in evaluating institutions for Phi Beta Kappa charters, the secretaries, senators, and councillors of the United Chapters were entirely on their own.

The Senate and the Council were slow to develop precise and consistent standards or systematic procedures by which to qualify institutions for charters. The applications at first had to have the endorsement of one chapter and after 1900 the endorsement of five chapters already in existence. Thus there was a certain amount of room for the exercise of personal influence and favoritism. Applicants naturally sought the backing of chapters and individuals presumed to be influential with the Senate or the Council. Chapters sending delegates to the Council had disproportionate power at times because other chapters did not bother to send delegates. At the 1895 Council, for example, only twenty-six of the thirty-four chapters then existing were represented.

At first the Council did not even have a specific requirement that an institution must offer the A.B. degree to be eligible. In 1895, when discussing an application from the Massachusetts Institute of Technology (MIT), which at that time awarded no such degree, the Council delegates had difficulty in reaching an agreement. By a close vote they finally denied the application and referred the following resolution to the various chapters for their approval: "That it is inexpedient to grant a charter of Phi Beta Kappa to any institution which does not grant in regular course the degree of Bachelor of Arts." This became official policy. The principle was soon extended to exclude all except A.B. programs within an institution. An application from the University of Illinois failed in 1901 because it proposed to include the school of library science. (Illinois received a charter in 1907, MIT in 1971.)

Not until the early 1900s did the United Chapters provide an application form. Then, to deal more efficiently with the growing number of applications, it prepared a printed folder for obtaining "Information required of Institutions Petitioning for a Charter" under six headings: "Historical," "Organization," "Equipment," "Financial," "Staff of Instructors," and "Enrollment." Among the questions asked were these: When and under what auspices was the institution founded? How many graduates have there been, and what were their degrees? Is the institution under either church or state control? Is there a preparatory department? What are the buildings, their value, and the library, laboratory, and museum facilities? What are the endowment, the endowment income, the tuition, and the tuition income? How many teaching faculty are there, and what are their ranks and degrees? What is their teaching load—their "average number of recitations a week"? How many students are enrolled in each of the four classes, and what degrees are they candidates for? With this tabulated information before them, the senators could better compare the merits of numerous and competing applicants.[9]

A few institutions did not really have to compete. They were invited to apply. This was the case with Princeton, the last of the colonial colleges to become a Phi Beta Kappa host. At one time there had been a number of Greek-letter social fraternities at Princeton, but the last of them disappeared in 1882, and after that the faculty remained strongly opposed to fraternities of any kind. The Phi Beta Kappa members of the faculty made no effort to secure a charter until they heard from Secretary Parsons in 1895. Then at his suggestion they started a movement that led to the establishment of a Princeton chapter in 1899.[10]

Meanwhile, in 1898 Parsons wrote to Stanford, one of the newest and least ivied of all the universities, having admitted its first students only seven years before. In response, a few of the society members at Stanford met informally, but they failed to find or to arouse much interest in seeking a charter. "Indeed," as one of them later wrote, "grave doubts were expressed as to the advisability of transplanting to this freer western soil an institution which, like Phi Beta Kappa, drew distinctions between graduates of the same class and seemed to create an aristocracy based on marks and other artificial estimates and comparisons."

Not until Parsons appealed to them a second time did the Stanford members—and not quite all of them even then—deign to sign a petition to the United Chapters. The Senate immediately approved it and stood ready to present it to the National Council later the same year. "Previous to that time will you put on file with me the endorsement of some existing chapter," Parsons advised. "It is only a form, but we are called [upon] to observe it." Thus, unenthusiastic though they were, the Stanford people seemed about to get as nearly an instantaneous charter as the United Chapters could possibly conjure up.

Only the merest of coincidences prevented their getting it. At Parsons's suggestion, they asked Cornell for the endorsement that he considered routine. The Cornell chapter, for reasons of its own, had just decided not to endorse any applications for the time being. "The feeble interest in a Stanford chapter was unable to survive this rebuff," according to one of the Phi Beta Kappa professors, "and no further application for endorsement was made." A triennium passed before the Stanfordians renewed their effort, and they did not obtain their charter until

1904. They continued to think of themselves as fundamentally different from Phi Beta Kappas of the effete East. As their poet at their first annual exercises recited in regard to their "venerable sires" of Virginia and New England,

> Grave were the scholars of that band,
>> Those cloistered souls who dwelt in books;
> We are not like them—we demand
>> More ready hands, more open looks.[11]

Most others who sought a charter did it eagerly and sometimes quite systematically, as those at Oberlin did. Even more hostile than Princeton to fraternities, Oberlin had never permitted them since its founding by Congregationalists in 1833. Finally, in 1904 the faculty decided that the college should provide "some distinction for proficiency in scholarship comparable to the honors to be won in other forms of college activity." The faculty appointed a committee to "consider and report upon the various forms of distinction available." After an investigation, the committee recommended that the college seek a chapter of Phi Beta Kappa.

The society's members on the Oberlin faculty were dismayed to learn that they would have to get the endorsements of five chapters and would have to wait three years for the next National Council to meet. By 1907 they were ready with their application and with not five but seven endorsements. They realized that they had scored something of a "triumph" when the Senate unanimously and the Council almost unanimously approved.

"Of the fifteen applications before the Senate this year seven were unsuccessful; two had not secured the necessary endorsements, five were deferred to the meeting three years hence," the leader of the Oberlin campaign recorded. Oberlin's was one of only three applications that the Council granted at the session during which it first received them. This "fortunate outcome" was due mainly to "the fact that Oberlin's equipment and standards met the requirements of the society so completely that there could be no question of her entire fitness to receive the honor. As the president of the United Chapters remarked, 'Oberlin should have had a chapter long ago.'"[12]

Denison University, another church-related school in Ohio, pursued a charter with equal determination and, necessarily, greater persistence. Baptists had founded Denison in 1831 as the Granville Literary and Theological Institution, and most of its early presidents and professors were graduates of the outstanding Baptist university, Brown. To that alma mater the president turned in the 1870s for a Phi Beta Kappa charter, and the Brown chapter voted to approve his application but let him know he must apply to the Ohio Alpha. He tried the Marietta chapter, which gave its consent and referred him to the Western Reserve chapter, which went into eclipse before acting on the petition. That president soon left Denison, and no one else resumed the effort until the twentieth century.

One of the leaders in the revived movement was Emory W. Hunt, a prominent divine who in 1901 took over the presidency of Denison and later the presidency of the Northern Baptist Convention. The other and more effective Denison advocate was Francis W. Shepardson, an alumnus of the class of 1882 and an assistant professor of history at the University of Chicago. Shepardson was a successful Phi Beta Kappa politician. Elected to the chapter at Brown while a graduate student there, he became secretary of the Illinois Beta at Chicago, repeatedly represented it as a delegate to the National Council, and served for twenty-four years (1913-1937) in the Senate.

According to his own account, Shepardson had begun to think of a Phi Beta Kappa chapter for Denison while still an undergraduate there. "The thought was insistent that if Marietta, Western Reserve and Kenyon were entitled to such recognition of their scholarship, Denison had just as good a claim." Soon after Hunt became president of the university, Shepardson with Hunt's cooperation undertook a lobbying campaign. He afterward recalled:

> President Hunt and I began to work, individually and in conjunction. We met difficulties. Denison was not known by enough members of the Senate, the first body in Phi Beta Kappa whose favoring vote was needed. We canvassed the names. We sought the help of those who might endorse the application for a charter. Some of them failed us at a critical moment. We had our case considered by the Senate at a meeting held in New York City on March 8, 1907. The only members who were present who knew anything in particular about Denison were President [Charles F.] Thwing of Western Reserve and Col. John J. McCook of Kenyon.

So, with few friends of Denison on hand, the Senate deferred the case to its 1910 meeting. "During the three years since 1907 special effort has been made both by President Hunt and myself to bring Denison to the favorable attention of the Senators," Shepardson went on. The "happy result" was that in 1910 the Senate gave its approval. Then the National Council, with Shepardson present and active as a delegate, did the same.

As secretary of the University of Chicago chapter, Shepardson knew that others besides him used personal pressure to try to get charters for colleges they favored. "We are placed in a rather embarrassing position here," he wrote from Chicago, "because everyone of our doctors and masters who teaches in an institution anywhere thinks we should endorse petitions freely." He insisted that his own chapter was discriminating—"we are quite anxious to make our endorsements amount to something"—though, in fact, he and his colleagues recommended some institutions that had little or no chance of qualifying.

While advocating chapters for little-known American colleges, Shepardson vigorously objected to Phi Beta Kappa's going abroad and establishing a chapter at the University of Oxford. Quite a few American Rhodes Scholars at Oxford (thirty-four of the eighty-one chosen in 1904 and 1905) were Phi Beta Kappa members. In 1907 the National Council encouraged them to form an association. They proceeded to do so and then applied for a charter that would convert their association into a chapter and enable them to elect members from the Oxford student body. The Senate in 1910 commended the Rhodes Scholars for forming an association but refrained from approving a charter for them.

For a couple of years, the issue remained alive. Secretary Voorhees was particularly enthusiastic about the idea. He recalled what the 1786 orator had said to the Massachusetts Alpha: "Our institution is now confined to a few seminaries in this country. But what obstacle exists to its further extension, or even a communication of it to foreign universities?" Voorhees reported in 1912 that American students at Oxford again were preparing an application.

"I am very sorry to note the scheme for ΦBK in Oxford," Shepardson wrote Voorhees. "I am certain that Cecil Rhodes never wanted any Americanization of students. He wanted our students to become members of a great Anglo-Saxon brotherhood." Other Phi Beta Kappa leaders objected on narrower grounds: it did not seem right to allow students to set up a chapter at a place where their tenure was so impermanent. Neither the Senate nor the Council ever voted to authorize an Oxford chapter.[13]

While the Council was chartering new branches, the number of them also increased through the revival of old ones. In 1883, the year the Council was created, a junior at Yale proposed to revive the chapter there. He found that the charter had not lapsed, for the chapter had never been formally disbanded. Phi Beta Kappa members from the faculty and the community held a reorganization meeting in time to elect students from the class of 1884 and later from all the classes since 1871, when the chapter had ceased to function. In 1886 the Connecticut Alpha ratified the United Chapters constitution and was accepted by the Council.

The Ohio Alpha at Western Reserve after repeated periods of inactivity recovered in 1884 and joined the United Chapters in 1889. The primordial Virginia Alpha had been inactive since 1861, and the College of William and Mary had been struggling to survive ever since the Civil War. Friends of the college, preparing to celebrate its bicentennial in 1893, received a boost when Congress appropriated $60,000 to help restore the main building, which United States troops had burned during the Peninsular campaign. At the bicentennial observance the concluding event was the reorganization of Phi Beta Kappa.

TABLE 1   New Charters by Council,
1883–1916

Council	Year	Chapters	New Charters
First	1883	23	0
Second	1886	23	3
Third	1889	26	4
Fourth	1892	30	3
Fifth	1895	34	6
Sixth	1898	40	10
Seventh	1901	50	3
Eighth	1904	53	10
Ninth	1907	63	8
Tenth	1910	71	6
Eleventh	1913	77	8
Twelfth	1916	86	3

The Alabama Alpha also had been inactive since 1861. In 1886 an Alabama member informed the Senate that the chapter hoped to reorganize; he asked for advice. The Senate ruled that the University of Alabama was not yet in a condition to "support and carry on an efficient Chapter." More than a quarter of a century later, in 1912, the Senate voted unanimously to restore the Alabama Alpha.

From one triennium to another, the Council varied in its generosity with charters (see Table 1). Of the charters granted during the period, more than twice as many went to private institutions (forty-three) as to public ones (twenty-one). Almost as big a majority went to colleges and universities in areas that previously had been underrepresented or not represented at all. A dozen chapters, besides the revived ones in Virginia and Alabama, appeared in the South; twenty-one, in addition to Western Reserve's, in the Midwest; and nine in the Far West, from the Plains to the Pacific. The East, too, became better represented. Pennsylvania, which had had no chapters before 1886, acquired eight in the twenty years thereafter. No longer was Phi Beta Kappa confined almost exclusively to New England and New York. The National Council had become truly national.[14]

8

Classicism or Eclecticism

There were internal as well as external aspects of the late-nineteenth-century revolution in higher education. Not only did colleges and universities grow in number, size, and variety; they also changed drastically in both the social and the intellectual life they offered their students. Phi Beta Kappa was challenged to respond to the new developments and particularly to those concerning the curriculum. With the spread of the elective system, course offerings multiplied and the classics lost their preeminence. These changes were accompanied by endless controversy over the meaning of the "humanities" or the "liberal arts" and over their due place in the educational program.

By the 1880s, when the United Chapters emerged, the transformation of academe was already well under way. As James Bryce then observed in regard to the content and methods of instruction,

Thirty years ago it would have been comparatively easy to describe these, for nearly all the universities and colleges prescribed a regular four years' curriculum to a student, chiefly consisting of classics and mathematics, and leading up to a B.A. degree. A youth had little or no option what he would study, for everybody was expected to take certain classes in each year, and received his degree upon having satisfactorily performed what was in each class required of him. . . . Instruction was mainly, indeed in the small colleges wholly, catechetical. Nowadays the simple uniformity of this traditional system has vanished in the greater universities of the Eastern and Middle States, and in most of the State universities of the West. There are still regular classes, a certain number of which every student must attend, but he is allowed to choose for himself between a variety of courses or curricula, by following any one of which he may obtain a degree. The freedom of choice is greater in some universities, less in others; in some, choice is permitted from the first, in some only after two years. In Harvard this freedom seems to have reached its maximum. . . .

A parallel change has passed upon the methods of teaching. Lecturing with few or no questions to the class interposed is becoming the rule in the larger universities, those especially which adopt the elective system, while what are called "recitations" . . . remain the rule in the more conservative majority of institutions, and are practically universal in Western colleges.[1]

Harvard led the way. True, a few other institutions had earlier allowed students some freedom to choose among courses. "Each student shall be permitted to attend such classes as he may select," William and Mary announced in 1844, "provided, in the opinion of the Faculty, he be competent to pursue the studies of such class with profit; and further, provided he attend at least three departments, unless the Faculty shall allow him to attend a less number." Harvard at that time was just beginning to introduce the elective system and during the next few decades, extended it only gradually, not offering it to freshmen until the 1880s. But Harvard finally carried it to greater extremes than William and Mary or any other predecessor, and Harvard drew much more attention and had much more influence than any of the rest.

Harvard was also ahead with other innovations. By the 1880s the college was holding examinations twice a year, at the end of January and the end of June, and these were nearly all written, not oral as in the old days. The college calendar was reformed to eliminate the long winter vacation (which had been scheduled to enable impecunious scholars to earn their tuition by teaching school) and to extend the academic year from September through June, with ten days' vacation at Christmastime and a week in April.

Harvard's grading system, too, had been revised. "Formerly . . . the students were ranked for the year on a scale of 100," William Roscoe Thayer, a young faculty member, explained at the end of the 1880s, "but no two instructors agreed in their use of it." Some were hard and some were soft graders. A "simple and more rational scheme" was adopted in 1886. "In each of their courses students are now divided into five groups, called A, B, C, D and E: E being composed of those who have not passed." To graduate, a student must "have stood above the group D in at least one-fourth of his college work." For honors, a student's ranking was now calculated "in terms of A, B, C, etc., instead of in per cents." (How this would lessen the inequity between hard and soft graders, Thayer did not explain.)

The most startling innovation at Harvard, except for the introduction of free electives, was the abolition of the requirement that students attend classes. This was going a little too far to suit the Overseers; in 1889 they suggested that at least a careful record of absences should be kept, and this began to be done. "But even such restrictions as these must sooner or later be abandoned," Thayer predicted, "when the idea of what a University should be triumphs—not a reform school, nor a seminary, nor a substitute for paternal superintendence, but a treasury of learning from which every properly qualified person may draw in proportion to his ability."[2]

By 1910, Harvard had begun to retreat a bit. That year, the college discontinued the free-elective system and put into effect a "concentration and distribution" plan. Course offerings were divided among four fields; a student must concentrate in one of the four but must also choose courses from each of the other three. This plan in one form or another was coming to prevail throughout the country. Some institutions still prescribed a great deal of the work and confined electives mainly, if not entirely, to the last two years. Denison University, for example, explained its requirements in 1907 as follows:

> . . . we seek to control the principle of student election of studies by a system of well-defined groups, so drawn as to prevent an undue narrowing of intellectual interests on the one hand, or the incoherent scattering of the seeker after easy courses on the other. Candidates for the A.B. degree are required to present both Latin and Greek for entrance, and to pursue each of these languages at least two years in college. All science groups require the fundamental college courses in Mathematics, Physics, and Chemistry . . . and all groups whatever must contain Freshman Mathematics, two years' work in English, and a year's work in the Department of Philosophy.

Two years of Latin, Greek, and English, plus one of mathematics and philosophy, together with major and minor requirements, would doubtless take up more than half of a student's total time in college.

As of 1907, only a small minority of colleges still required Latin and Greek (or either of the two) for graduation, and an even smaller minority required them for admission. The "point system" of admission was beginning to prevail. A college typically listed twenty to thirty high-school subjects, assigned a certain number of points to each, and expected the applicant to present a minimum point total. Greek and Latin would count in the calculation but were no longer necessary for entrance into most colleges and universities.[3]

        A great debate accompanied the curricular changes of the period, and Phi Beta Kappa provided participants and a forum for much of the discussion. But the organization did not speak with a single voice.

At one extreme stood the die-hard defenders of the classics, and at the other extreme, the advocates of free or almost free electives. In between, some favored a certain amount of choice so long as Greek and Latin continued to be required; others opposed complete freedom but also opposed a Greek or Latin requirement.

Standing out as the foremost champion of the elective system was Charles W. Eliot, for forty years (1869–1909) the president of Harvard. The presidents of Yale and Princeton denounced his ideas, and many of his own faculty members resisted them as best they could. In the minds of those who clung to the traditional curriculum, Eliot figured as an extremely reprehensible if not a downright satanic influence.

This was the man whom the National Council elected as Phi Beta Kappa's very first national president. At that time (1883) the Eliot reforms were approaching their ultimate at Harvard and were becoming increasingly bitter topics of controversy throughout the academic world. In electing Eliot, of course, the Phi Beta Kappa delegates were not necessarily giving an official endorsement of his policies. But they were certainly not indicating disapproval of them.

Siding with Eliot, the 1883 orator at Harvard treated the chapter to an eloquent denunciation of "A College Fetich," by which he meant the insistence on Latin and Greek as entrance requirements and as college studies. The orator was Charles Francis Adams, Jr., Harvard '56, John Quincy Adams's grandson, a prominent lawyer, historian, and railroad authority. Adams said he had forgotten all he ever knew of Greek. It had been of utterly no use to him or to any of his Adams forebears. Along with Latin, it possessed only a snob value as the mark of an "educated man."

"I am no believer in that narrow, scientific, and technological training which now and again we hear extolled," Adams went on. "A practical, and too often a mere vulgar, money-making utility seems to be its natural outcome." He saw a "broadened culture" as the university's true aim. "I would broaden it. No longer content with classic sources, I would have the university seek fresh inspiration at the fountains of living thought."

The forced study of Latin and Greek had been a kind of mental treadmill for Adams. "I am told that I ignore the severe intellectual training I got in learning the Greek grammar and in subsequently applying its rules ... that ... even my slight contact with the Greek masterpiece[s] has left me with a subtile but unmistakable residuum, impalpable, perhaps, but still there, and very precious." To Adams, all this was "pure, unadulterated nonsense."

Adams's "A College Fetich" was published as an article in the *Independent,* a national magazine, and as a separate pamphlet. It provoked the indignation of Daniel Henry Chamberlain, a former carpetbag governor of South Carolina and an alumnus of both Yale College and the Harvard Law School. Chamberlain replied to Adams in *"Not* 'a College Fetich,'" an address he delivered before the Amherst chapter and then the Vermont chapter in 1884. Later he resumed the argument in the pages of the *New Englander and Yale Review,* deploring the Harvard system and glorifying Yale's more restrictive curriculum.[4]

Defenders of the comparatively old-fashioned curriculum justified it on the basis of its faithfulness to the "humanities," and the "liberal arts."

Matthew Arnold, the famous English poet and critic, contributed to the discussion in a lecture on "Literature and Science," which he repeated twenty-nine times during his American tour of 1883–1884. Arnold said the aim of education ought to be "to know the best which has been thought and said in the world." He gave the most important place to the humanities in the traditional sense—that is, to the Greek and Roman classics—but he also made room for "the best" of more recent belles lettres and even "the best" of contemporary scientific works, saying, "I certainly include what in modern times has been thought and said by the great observers of nature." "The best" of both books about humanity and books about nature were quite acceptable to Arnold. With him the subject matter was not so much the criterion as was the quality of the thinking and writing.[5]

A Harvard professor emeritus of Christian morals, Andrew P. Peabody, described "A Liberal Education" in an 1886 issue of the *New Englander and Yale Review.* Peabody wrote:

> *Liberal* literally denotes *belonging to a free man,* and, in its literal sense, a liberal education is that which befits a free man or woman as distinguished from a slave. [It is] an education which makes one a free citizen of the republic of letters, cognizant of his position as a learner and a knower, capable of pursuing his reading, study or research in any one of the several directions which the scholarship of his time may take. Herein lies the distinction between a liberal and a special education. He who is trained as a specialist may find himself not only ignorant but incapable of learning, in various departments that lie open to the general scholar. The liberally educated man knows how to become a specialist in any department, and if he has but little knowledge of it, he has the rudiments from which knowledge may be successfully formed or developed.

Specifically, what should undergraduates study in order to make themselves free citizens of the republic of letters, with the potential of mastering any and all fields of learning? Peabody listed the elements of a liberal education as the following: (1) the "speaking and writing of English correctly"; (2) classical languages, literature, and history; (3) French, German, Italian; (4) "the thorough analysis and philosophical treatment of some one section or epoch of modern history; (5) mathematics through trigonometry; (6) astronomy, physics, chemistry; (7) "the various branches of natural history, miscalled science"; (8) philosophy; and (9) "the most momentous of all . . . Holy Scriptures."

Would Peabody "force all candidates to go through precisely the same curriculum?" His answer: "By no means." He had been, he said, an early advocate of the elective system. "But I would have the choice made only among departments of which the rudiments or fundamental principles have already been acquired." It would seem, though, that once a student had acquired all of Peabody's elements, he or she would have little time left for electives.[6]

The young Harvard instructor Thayer expressed a spirit rather different from that of the retired Harvard professor Peabody. Thayer rejoiced in Eliot's accomplishment, writing in 1890:

> The Classics and Mathematics, before which, as before Gog and Magog, educators fell down and worshipped, declaring them to be the only true agents of culture, have gradually been placed in their proper position—not degraded nor laid on the shelf, but prohibited from excluding proper reverence for Science, History and the Modern Languages, which are now recognized as being important means to culture. And the work done in Greek and Latin and Mathematics, being no longer obligatory, is more earnest than in the days of compulsion, and productive of more good. The old superstition that the degree of A.B. will be unintelligible, unless all who receive it have the same courses, still befogs the eyes of some conservatives; but experience will certainly dissipate this.

Thayer was overoptimistic. Eliot's admirers eventually lost a good deal of ground to his critics. One of the critics, Abbott Lawrence Lowell, succeeded him as Harvard's president in 1909. That year the Princeton president, Woodrow Wilson, delivered the Phi Beta Kappa address in Cambridge. Speaking on "The Spirit of Learning," Wilson "pointed out the error of Eliot's views without mention of him," as one of the anti-Eliot members of the chapter recorded. Charles Francis Adams, that foe of the classics fetich, now believed that Eliot had gone much too far, and he congratulated Wilson as the leading advocate "of the smaller college and of the immediate contact of the more mature with the less mature mind." Lowell indicated to Wilson that he considered himself, Wilson, and Arthur T. Hadley—the presidents of Harvard, Princeton, and Yale—as three like-minded champions of the liberal arts.[7]

While the dispute over the curriculum raged, Phi Beta Kappa remained unsure about the subjects students should study in order to be eligible for membership. In 1895 the National Council put to the chapters the question whether membership ought not to be restricted to those taking the A.B. degree. Some chapters favored this, but others opposed it. "When the attempt was made to preempt for classical students alone the privileges of Phi Beta Kappa in the face of a growing demand for modern languages and science, the Kenyon chapter voted against the proposition," the chapter's historian related. "It admitted candidates for the degree of Bachelor of Philosophy [to] membership, and finally even those unrooted moderns, Bachelors of Science."

"Uniformity has been one of the watchwords of the National Council," Secretary Eben B. Parsons understood, and as a step toward it, at the Council's direction he sent the chapters a questionnaire, the results of which he published in his catalog and handbook of 1900. Some of the questions had to do with election standards and procedures, among them the curricular qualifications. The responses showed that uniformity—with respect to the place of the humanities as well as other matters—was still a long way off. Parsons summarized a part of his findings thus:

> Good moral character and high scholarship are the recognized foundation qualities for membership eligibility. The question at issue has related to the courses of study in which the high scholarship has been obtained. For many years the election was limited to the students pursuing the course leading to the degree of Bachelor of Arts. When the colleges began to give other degrees the question arose regarding Phi Beta Kappa. After prolonged discussion and a reference back to the chapters for their decision the National Council voted that it is not expedient to grant a charter to any institution that does not have the Bachelor of Arts course, but that it is also not expedient to confine the membership to students in that course. At present a very large per cent, probably ninety per cent, of the membership is from the Arts course. The requirements for the Arts degree are not the same in all colleges, though quite generally Latin and Greek, or Latin and a modern language, are required.
>
> The following statements from the chapters indicate the present situation. "No distinction is made between classical and scientific courses" (Amherst); "Members are elected from the A B and the Ph B courses, though eighty-five per cent are from the former" (Brown); "Men are taken from the B A, B L, and B S courses" (Dartmouth); "Men are taken mostly though not altogether from the A B course" (Kenyon); "In earlier days at least nine-tenths have been taken from the A B course, though now the proportion may be about two-thirds" (Marietta); "We take from both the B A and B S courses, though the proportion from the latter is small" (Middlebury).

Apparently, a fairly large number of students were being elected to Phi Beta Kappa without benefit of classical learning. Even those graduating with an A.B. degree, as Parsons pointed out, were not necessarily acquainted with both Latin and Greek—or with either of the two.

Troubled by these findings, the National Council appointed a committee to consider standards of eligibility, especially as these related to the liberal arts. In 1904 the committee reported that humanistic studies were losing ground but that chapter practices were so diverse as to make any agreement unlikely. In 1916 the Council did agree that students in a teacher-training program would not be eligible unless the program had standards "fully equivalent of the course or courses whose members are eligible to Phi Beta Kappa."

Some of the chapters meanwhile revised their standards in such a way as to give at least a little emphasis to the liberal arts. Brown limited membership to candidates for the A.B. or the Ph.B. degree and specified that at least one third of their work must be in the humanities. These were defined as courses in language, literature, philosophy, history, fine arts, political and social science, and political economy. The Michigan chapter excluded students who received more than half of their credits in the natural sciences.

In choosing electives, ΦBK students seem to have conformed to the trends of the time. At the University of Michigan, for example, they took comparatively little in ancient languages or mathematics during the years 1905–1916. They increasingly preferred modern languages, political economy, sociology, history, and philosophy.[8]

If the chapters sometimes showed little regard for the humanities when electing undergraduates, they showed still less regard for them when electing graduate students, alumni, and honorary members. The Johns Hopkins chapter, at its start in 1896, elected a number of the university's doctors of philosophy. It continued to consider a Ph.D. or an M.D. candidate eligible if he had received his bachelor's degree from an institution without Phi Beta Kappa. When in 1914 Johns Hopkins inquired whether other chapters elected graduate or medical students, Secretary Oscar M. Voorhees replied that the University of Chicago did so, but as to how many others, there was a "lack of information."[9]

Usually (as in earlier times) a new chapter began by electing a number of alumni, persons who presumably would already have been members if the chapter had had a longer history. The William and Mary chapter, after its revival in 1893, elected no students at or before graduation but only alumni and others who had been out long enough to gain some sort of distinction. Until 1905, professors automatically became members at CCNY. Most chapters continued to elect honorary members, though some tried to limit the practice. At Dartmouth after 1899, all nominations for honorary membership had to be submitted in writing and referred to a screening committee; no one would be eligible who had failed of election at a college with a chapter. The United Chapters in 1913 recognized three categories of members—student (" members in course"), alumni, and honorary—and recommended that honorary elections be carefully limited.[10]

Honorary membership did not mean the same thing from one chapter to another, and neither did student membership. The chapters differed in the exclusiveness, the method, and the timing of elections, as Parsons discovered in his 1900 survey. With respect to the numbers elected, he observed:

> When the old chapters constituted themselves the United Chapters they reserved some rights and privileges. One of these vested rights was the power to elect one-third of each class to Phi Beta Kappa membership. But since in the new order they have voted that the new chapters shall take only one-fourth of each class, the old chapters, in the interest of uniformity and fairness and good scholarship, have nearly all agreed to restrict themselves to a smaller fractional proportion, so that now only six of the old chapters elect one-third of each class. Yale, Cornell, and the College of the City of

New York elect about one-eighth of each class. Dartmouth and Nebraska take one-sixth. Columbia, Syracuse and Williams take one-fifth. Changes are now in progress that will bring other chapters into these restricted ranks. The prevailing limit outside the great University chapters is one-fourth of each class, the limit fixed for the new chapters by the National Council.

Among the older chapters, Brown and Harvard soon reduced their proportion to approximately one eighth, and in 1913 the National Council set this fraction as the maximum for new chapters. (The fraction applied only to the number of students taking courses that would make them eligible; students in technical programs were not counted.)

"The basis of election is the college system of marks," Parsons found in 1900. But, he noted, the marking systems differed, and so did the election procedures. Some chapters obtained a list of the highest ranking scholars, as many of them as would fill the quota, and these were automatically elected. Other chapters obtained a longer list and chose from it enough names to fill the quota. Undergraduate members did the choosing in six institutions, students and faculty members together in thirteen others, and faculty (sometimes with members from the community) in all the rest, which numbered thirty-one at that time. Some chapters still required a unanimous vote, others as little as a three-fourths majority.

The trend was away from the unanimity requirement, which could occasion unfairness and recrimination, as at CCNY. "In the days before 1905, when a single adverse vote could exclude any candidate for membership," the CCNY chapter's historian recounted, "it was an all too frequent occurrence for a professor or a recent graduate to vote against some individual whom he disliked, or for members who merely felt that too many candidates were being proposed to reject arbitrarily a portion of the list submitted." In 1905 the chapter made three adverse votes necessary for exclusion, and in 1916 five adverse votes."[11]

If the voting process allowed room for prejudice, it also made possible the consideration of qualities other than mere grade-getting ability, and this was the presumed justification for it. The Harvard chapter developed what its members saw as a much improved method of picking out the "best men." As the student newspaper, the *Harvard Crimson,* reported in 1910:

> For many years election to Phi Beta Kappa was based entirely on marks received in college courses. It was in a way automatic: if a man won high enough marks he was assured of election. But with the rapid growth of the College, and especially with the introduction of the elective system, it became evident that some new method of election must be found. The method that was finally selected has several times been modified, and is now believed to secure the election of the best men.
>
> Toward the middle of each year, the College office prepares a list of the highest twelve men in the Junior class, and the highest forty-four men

(exclusive of those already elected) in the Senior class. From these lists the eight men elected in the previous year, known as the "Junior Eight," proceed to elect twenty-two Seniors and eight Juniors, and their choice, subject to the approval of a committee of graduates, of which President Lowell is at present the chairman, constitutes the election.

In making the choice, several considerations besides an aggregate of high grades receive attention. For example, a man who has taken principally easy courses, or a man whose work has fallen off from year to year, may not win election, when another man with less A's to his credit may be elected because he has shown ability in difficult courses and has constantly improved. The narrow specialist and the man who has not specialized at all are equally apt to fail of election. Moreover, the electors take into account success in winning prizes for essays, in debating, and in other intellectual activities. In no case are purely personal grounds—matters of likes and dislikes—considered at all; but a continuous effort is made to recognize real ability and intellectual command, as opposed to mere "grinding."

The proportion of students elected at Harvard was now much smaller than it had been in earlier years. Elections were averaging only about 8 percent of each graduating class, a much smaller percentage than at most colleges.

Electors at the University of Michigan could not take grades into account at all when a chapter of Phi Beta Kappa began its existence there in 1907. The university at that time gave no grades except "passed" and "failed." As a contemporary explained, "Prizes, badges, robes, honors, distinctions, and all such extraneous bribes to scholarship are contrary to the historic ideals of the University of Michigan." Indeed, President James B. Angell had tried to keep Phi Beta Kappa away because he thought it inimical to the spirit of his institution.

While Harvard elected some juniors, and Yale a few sophomores, the majority of chapters—62 percent of them according to Parson's figures in 1900—elected only seniors, and these at the end of their senior year. "In most colleges therefore Phi Beta Kappa is a graduate society," Parsons pointed out. "The trend is away from the original idea of the literary helpfulness and social comradeship of the Virginians."[12]

Phi Beta Kappa leaders wished to see greater uniformity in activities as well as in elections, and they wished to see a larger number of activities. In 1895 the National Council set up a committee to study the society's rituals, customs, and traditions. It also "earnestly recommended to the active members of the several chapters that they hold one or more stated meetings in each month of the college year . . . for the reading of papers, oral discussions, or other appropriate literary exercises in which the members generally may participate." Of course, undergraduate members could not be expected to participate in such exercises in the majority of chapters, since the majority did not have any undergraduate members.

Members elected in course could usually take part in at least one activity—their own initiation. "The form by which the new members are received or initiated differs widely throughout the Fraternity," Parsons said in his 1900 report. "In some chapters the fact of a man's election is reported to him, the admission fee is collected and he helps pay for a dinner to celebrate the event. In other chapters there is an impressive welcoming service."

The service typically began with the candidates answering in unison, "I do," to the following question: "Do you agree to support the Constitution of Phi Beta Kappa and to generously aid in every honorable way to enlarge its usefulness?" (A few chapters still demanded a pledge of secrecy and a promise to help worthy brothers in distress.) Then someone introduced the members-elect, and the presiding officer briefly explained to them the society's history, aims, motto, key, and grip. Having done so, he welcomed them as a group, with the admonition: "Let your new relation urge you to increasing diligence in the pursuit of knowledge and the cultivation of virtue." Finally, the old members greeted the new ones individually, giving them—with the once-secret grip—the right hand of fellowship.

Even this bland ceremony was too much for some chapters, which quit holding any initiation. One of these, the Gamma of New York at CCNY, had followed faithfully for many years the form it had received from the Beta of New York at NYU in 1867. Then in 1905 the chapter began to add new members to its roll with no get-together of any kind. In 1913 the National Council authorized the printing and distribution of a revised ritual in the hope of both stimulating and standardizing initiation ceremonies.

Some "literary" meetings were held by most of the chapters that elected members a year or more before graduation. These meetings generally were infrequent, and they did not necessarily involve students as active participants. The Harvard chapter was a notable exception to the rule of little or no undergraduate activity. In 1820 its student members had ceased to meet regularly for debates or for exercises of any sort. By 1910 they were dining every week in the room the college provided and generous alumni furnished in Memorial Hall. They arranged a banquet for the reception of new members, put on a "spread" on class day, and played an annual baseball game with the brothers from Yale. All this was in addition to the famous oration, poem, and dinner that the chapter continued to sponsor at commencement time.

Most of the chapters provided no regular activities for undergraduates and not many for alumni. Some signs indicated that undergraduate and even graduate activities were decreasing. The Kenyon chapter transformed itself into a kind of faculty club when in the 1896–1897 academic year its members invited colleagues to join in occasional meetings for the presentation of papers. The Brown chapter, which had presented an oration and a poem every other year since 1853, discontinued them after 1912.[13]

Even though Phi Beta Kappa carried on few if any student activities, it continued to exert a definite influence upon student life and thought. This influence, direct and indirect, appeared to be increasing.

The social life of students was changing along with their intellectual life. They were becoming less and less like members of a single collegiate family. As Bryce observed in the 1880s:

> Formerly all or nearly all the students were lodged in buildings called dormitories—which, however, were not merely sleeping places, but contained sitting-rooms jointly tenanted by two or more students—and meals were taken in common. This is still the practice in the smaller colleges, and remains firmly rooted in Yale, Harvard, and Princeton. In the new State universities, and in nearly all universities planted in large cities, the great bulk of the students board with private families, or (more rarely) live in lodgings or hotels, and an increasing number have begun to do so even in places which, like Harvard and Brown University (Rhode Island) and Cornell, have some dormitories. The dormitory plan works well in comparatively small establishments, especially when, as is the case with the smaller denominational colleges, they are almost like large families, and are permeated by a religious spirit. But in the larger universities the tendency is now towards letting the students reside where they please.

So far as the majority of students were concerned, the institution was ceasing to stand *in loco parentis.*

The fraternity gained importance (while the college itself lost it) as a social center. Typically, a fraternity chapter now had "a sort of club house, with several meeting and reading rooms, and sometimes also with bedrooms for the members," as Bryce noted. It was "an object of ambition to be elected a member," and in some institutions as many as a third or a half of the students belonged to a fraternity. During 1879–1898, the number of fraternity chapters grew from 462 to 781, and total membership climbed from about 63,000 to more than 100,000.

Honor societies, which were beginning to appear as independents on campuses here and there, were also gaining in importance. "The oldest and most famous," Bryce wrote, "is called the ΦBK, which is said to mean Φιλοσοφία Βίου Κυβερνήτης and exists in nearly all the leading universities in most of the States." Bryce's informants had misled him a bit, since live chapters at the time numbered only twenty-three, but the exaggeration itself was perhaps a tribute to Phi Beta Kappa's fame. A local society, taking Phi Beta Kappa as its model, sometimes aspired to form the nucleus for a new Phi Beta Kappa chapter. At Vanderbilt, for example, Alpha Theta Phi was organized in 1894 "to stimulate and increase a desire for sound scholarship" but, more particularly, to demonstrate the maintenance of Phi Beta Kappa standards and thus prepare the way for a chapter, a charter for which Vanderbilt received in 1901.[14]

Phi Beta Kappa also began to inspire honor societies of more than local significance. In 1886 a group of men at Cornell founded the Society of the Sigma Xi to recognize and encourage scholarship in the sciences as Phi Beta Kappa did in the humanities. The Sigma Xi medal, too, consisted of an old-fashioned watch key, one on which the Greek letters Sigma and Xi were superimposed. By 1916 the society boasted twenty-nine chapters, slightly less than a third as many as Phi Beta Kappa then had.

Intercollegiate athletics were, by now, competing with scholarship for the attention of students. "Of late years," Bryce learned as early as the 1880s, "the passion for base-ball, foot-ball, rowing, and athletic exercises generally, has become very strong in the universities . . . where fashionable youth congregates, and the student who excels in these seems to be as much a hero among his comrades as a member of the University Eight or Eleven is at Cambridge or Oxford." The new American version of football in particular gained popularity. More than fifty thousand people watched the Yale-Princeton game on Thanksgiving Day in 1893. More and more professors complained about overemphasis of the sport, use of "ringers" (nonstudents) on teams, payments sub rosa to players, and tactics so rough as to maim and kill. Statistics for the 1905 season showed 18 dead and 159 seriously injured. "Death and injuries are not the strongest argument against football," President Eliot of Harvard declared; "that cheating and brutality are profitable is the main evil."[15]

Phi Beta Kappa took no official stand against the football mania of that period, though many members individually disapproved of it. Some, however, had a good word to say for athletics, as did Francis A. Walker, president of the Massachusetts Institute of Technology, when he addressed the Harvard chapter in 1893. "The oration was an able and soldierly defense of athleticism," a visitor from Oxford remarked. "There were some in high authority at Harvard who thought that in a university, where athleticism seems running mad, such a defence was altogether out of place. They maintained, moreover, that the subject was ill-suited for a learned society."

In competition with athletics and other campus attractions, the ΦBK key held its own quite well in the minds of students. It even gained a bit of ground, though not everywhere. "At Cornell the scientific honor society, the Sigma Xi, is more highly esteemed than its venerable rival on the literary side, the Phi Beta Kappa," averred the author of a 1910 book on American universities. "Phi Beta Kappas are in fact not so popular as the Kappa Beta Phis, a group of young men whose ideals and mode of life are quite the reverse of those of the former society."

Yet at about the same time, the author of another book on higher education stated in regard to Phi Beta Kappa: "Its membership and pin are about the only general [as distinct from local] recognition of high rank given in the colleges." The *Denisonian* reported that at Denison there formerly had been "no reward for excellence in the classroom equal to that received for work on the athletic field," but now a ΦBK key meant more to a graduate than a football letter did. Phi Beta Kappa, the *Harvard Crimson* explained, "aims to do for the scholar what the 'H' and the Varsity Club do for the athlete." It seemed to be succeeding in this aim at Harvard.

Certainly it was succeeding at Yale. "The attitude of the students in general has undergone a marked change in the last few years," the Yale chapter reported in 1907. "Of late the students had slighted Phi Beta Kappa, considering it suitable for grinds only." A few years after that, in a poll of the classes of 1907, 1911, and 1914, "membership in Phi Beta Kappa was voted by each class to be the highest undergraduate honor, exceeding even editorship of the college paper, or the winning of the football 'Y.'"[16]

9

Money and the American Scholar

Between World War I and World War II, the leaders of Phi Beta Kappa came to view the society more and more definitely as a champion of the liberal arts. Its basic aim, they increasingly agreed, was to inculcate the values of a liberal education—not only among its members and prospective members but also among college graduates in general and even among the public at large. This meant expanding the operations of the United Chapters to include the publication of a magazine of broad intellectual interest. It also meant enlarging the society's expenses and searching continually for money with which to meet them.

Most urgently the United Chapters needed a new headquarters, one more suitable to the society's dignity and functions than a pastor's study. The society's secretary, Oscar M. Voorhees, thought he had an answer to his prayers in 1916 when he received from an anonymous source an offer of $25,000 toward an endowment, the income from which was to be used for maintaining a central office. There were strings attached to the offer: the society had to raise an additional $75,000, and it had to do so by the end of 1917. The Senate readily accepted these conditions.

At first, the Senate hoped that Albert Shaw, editor of the influential *American Review of Reviews*, would head a fund drive. Shaw declined, pleading overwork, but he gave some advice about fund raising. While he "appreciated the offer of a contingent gift of $25,000," he said, he had "never thought that the general plans of the United Chapters ought to be made to turn hastily around that generous tender." Let Phi Beta Kappa take its time. The society should begin with an educational campaign "which would make everybody feel that the fund was not only desirable but necessary." Meanwhile, "there ought to be made available a revised and perfected catalogue of all living members." Eventually a quote of contributions might be imposed on each chapter.

Voorhees protested to Shaw that the society was not being hurried but had been considering the need for new quarters for the past three years. "And three years hence, at latest, if this plan does not go through, the Council may be under the necessity of hasty action, for a new Secretary will have to be chosen, and provision made for a workshop for him." Voorhees was having to neglect his "primary duties" as a pastor, and he doubted if he could continue under the present arrangement beyond the triennium.[1]

Soon there ceased to be any point in looking for a public-relations expert to direct a money-raising campaign. Once the United States had declared war on Germany on April 6, 1917, Phi Beta Kappa members along with the rest of Americans were preoccupied with the war effort. The society was eager to do its part. When the Balfour commission arrived to arrange for close cooperation between the United States and Great Britain, the William and Mary chapter elected Lord Balfour and his colleagues to honorary membership. The ΦBK Senate congratulated American institutions of higher learning on their wartime role. "The colleges are realizing that the war is a struggle for maintaining the ideals which they and the Phi Beta Kappa embody," the Senate declared in November 1917. "The popular idea that the colleges and their graduates are unpatriotic is dispelled by the academic and military history of the recent months."

It was a time to sacrifice for one's country but not for one's fraternity. Phi Beta Kappa could hardly put on a public fund drive in competition with the government's Liberty Loan campaigns. So the Senate decided to ask each chapter to contribute a couple of dollars per member—which on the basis of nearly forty thousand members would amount to more than the required $75,000. Few chapters responded, however, and the deadline of December 31, 1917, passed without the society's having come close to realizing the hoped-for $100,000 endowment.

After the war the society would confront greater responsibilities and greater needs than ever. That, at least, was the theme of several speeches at the 1919 Council. "We have not the wisdom to forecast the coming changes in higher education, but we are sure they will be great and far-reaching," recapitulated Edward A. Birge, president of the United Chapters and of the University of Wisconsin. "In the face of such a situation the responsibility of Phi Beta Kappa is increased; the scope and significance of its duties are greatly enlarged."

Immediately enlarged were the duties of Secretary Voorhees and his assistants, who crowded into his study to work on a new catalog, one that would serve in part as a fund-raising tool. Voorhees was beginning to wonder whether they could ever finish the job without better facilities, when he heard again from the nameless angel who had made the conditional offer of $25,000. This time the man said he would provide office space rent-free for five years.

The space consisted of a fourteenth-floor apartment in a brand-new building at 145 West 55th Street in New York City. On May 1, 1921, Voorhees and his co-workers moved with their books, records, correspondence, equipment, and supplies from the cramped confines of the Bronx parsonage to the new quarters, which by comparison seemed "quite palatial." Of all the apartments in the building, this was the only one not used as a residence or a studio. Here the staff could work, visitors could drop in, and senators could meet in a "delightful atmosphere" entirely free from the "hurry and bustle" of a regular office building.

Soon after moving in, Phi Beta Kappa received another gift from the anonymous donor—the provision of a salary for a full-time secretary for a period of five years. The Council elected Voorhees to the full-time secretaryship, and he resigned his pastorate to accept the new position in 1922. No longer would he have to divide his time between his church and his fraternity.[2]

"In the last century election to Phi Beta Kappa meant that the student had reached success along lines of study universally accepted as standard in higher education," said President Birge in 1922. "Today the tendency towards vocational training is so great that there is urgent need for an active coherent organization, both within the colleges and outside of them, of the forces that stand for liberal education. Can a better center be found for such an organization than Phi Beta Kappa?"

Birge was appealing to the chapters in advance of their sending delegates to the 1922 Council. In four years, he reminded them, the society along with the country would arrive at its sesquicentennial. "The Fraternity has reached an age and grown to a greatness which forbid it to refuse the challenge that the present day brings to all lovers of learning." Hence the Senate had referred to the Council the question of raising a sesquicentennial fund, the purposes of which would be the following: (1) to maintain the central office, temporarily financed by a gift; (2) to construct at William and Mary a building memorializing the society's origin; and (3) to "make possible a larger active work among the colleges in behalf of liberal education."

When the Council met in 1922, the delegates authorized the endowment drive, with $100,000 to be allotted to William and Mary for the memorial building. To facilitate the drive, the society in 1924 obtained a charter from the University of the State of New York for the Phi Beta Kappa Foundation, which the charter described as an "educational institution for the creation of funds or endowments, for the erection of memorial buildings . . . for the promotion of sound learning by the establishment of fellowships, scholarships, foundations, funds, endowments, prizes or by any other method." The 1925 Council set the goal of the campaign at $1,000,000 and ordered the hiring of a fund-raising firm to complete the drive, if possible, by the 150th anniversary date, December 5, 1926.[3]

The campaign ran into unexpected resistance. It had to compete with the postwar fund drives that a number of institutions were carrying on. "There is an intensive campaign taking place not only in this state but where ever Wisconsin alumni are found to obtain funds for the Wisconsin Union building," protested the chapter secretary at Birge's own institution. "Another request for money at this time would meet with little success." The secretary at Oberlin similarly wrote: "Oberlin has urged her alumni for her own campaign to the limit and they are not a rich constituency."[4]

A few chapters objected to the drive on such grounds as to challenge the authority of the United Chapters. At Harvard a committee of aging alumni—men of the classes of 1869, 1875, and 1881—concluded that Harvardians were "not likely to be much interested in that portion of the plans" having to do with a "considerable endowment," with "prizes for excellence in teaching," or with "subsidies for encouraging scholarship in schools." The committee believed, however, that "Harvard members, and probably Yale members also, would take a lively interest in the William and Mary Memorial, since the Harvard and Yale Chapters, and these Chapters only, received their charters directly from the parent Chapter at William and Mary." Therefore the committee "drew up a circular letter stressing this part of the plan and calling on their fellow members to make a generous response."

Voorhees was troubled when he received a report of the committee's action. "It seems to me," he replied to the chairman, "that the real question is whether Phi Beta Kappa is to have a harmonious development in accordance with the official decisions reached by the National Council or whether chapters are to take their own course regardless of what the National Council may do." The Council "feels justified in asking every loyal member of the Society to give aid and encouragement," but "any chapter that proposes to help only one part of the campaign makes the total work all the more difficult."

To Voorhees, the response from Princeton was still more disturbing than the one from Harvard, for it brought up in even starker form the issue of a chapter's status and powers. The secretary of the Princeton chapter presented a kind of constitutional argument for noncooperation. He wrote Voorhees:

> The officers of Princeton Chapter of Phi Beta Kappa have a dual loyalty and obligation. They are members of Phi Beta Kappa, committed to . . . its endowment campaign. . . . Notwithstanding the consistent opposition of the Chapter to this policy while it was being argued at Triennial meetings it

is bound to play its part . . . as enthusiastically as it can. . . . But we are also members of Princeton University, and the Chapter is really a Chapter of Princeton University, and not merely located in this particular vicinity. Now the University has felt so deeply pledged to its own Alumni *not to solicit them for its own present Endowment need*. . . . Consequently the officers of the Chapter feel *bound* not to *press* in any public meeting in the halls of the University itself the matter of the Phi Beta Kappa drive. . . .

You see, therefore, that we must take a course not wholly to your liking or to that of officers of United Chapters Phi Beta Kappa who have no other business but that of administering general policies and national campaigns, remote from the actual production of scholarship in the various colleges and universities of the country. Our membership in the University prevents our carrying out a policy with the vigor and uncritical enthusiasm that you think ought to operate.

This was a rather insolent letter, with its gratuitous reference to United Chapters officers "who have no other business but that of administering general policies and national campaigns, remote from the actual production of scholarship." And it showed an abysmal ignorance of the history and nature of Phi Beta Kappa. It must have sorely tried the patience and charity even of an ex-minister of the gospel like Voorhees.

Still, he managed a measured reply. "From one point of view, the conditions are a little different from what you seem to indicate," he politely informed the Princetonian. "The theory of Phi Beta Kappa has been from the first that it is an independent organization. The charters are given to members who are connected with the faculty of the institutions and they form an organization independent in all regards." Hence the society has a perfect right to solicit its own members, and its doing so is of no concern to nonmembers among the faculty, staff, students, or alumni.[5]

The William and Mary chapter had no compunction about soliciting its own members or the members of any other chapter. For years, friends of the college had hoped to obtain from the state historical library in Richmond the ΦBK records of 1776–1781 and "to erect in Williamsburg a suitable building in which to house and preserve them." By 1919 the idea was "to establish at William and Mary a permanent Phi Beta Kappa memorial of some sort." At the Senate meeting the next year, the college president outlined a plan for a building to honor the fifty "founders" who had belonged to the original Alpha. He asked for approval and for permission to seek contributions. The Senate gave its commendation and best wishes, though pointing out that it could not obligate the society's members.

The Council did obligate the various chapters, however, when it included the William and Mary project among the aims of the sesquicentennial endowment drive. Thereafter the Virginia advocates of the project broadened their conception of it so as to envisage "not a mere William and Mary building but a national shrine." "It should be a Hall of

Fame for the entire society, in which memorials to the great men of any chapter could be placed," a leading advocate explained to Phi Beta Kappa President Charles F. Thwing. "It seems to me that this idea might be used effectively in the campaign as a means of disarming the criticism that it is for a mere local purpose."

Memorial Hall on the William and Mary campus was ready in time for dedication ceremonies and sesquicentennial observances on November 27, 1926. On hand to listen to speeches, toasts, and poetry readings and to partake of a banquet were delegates from Phi Beta Kappa chapters and associations, from other honorary, professional, and social fraternities, and from various learned societies.[6]

The end result of the endowment drive and the interest generated by it was the construction of a new building but also the restoration of old buildings in Williamsburg. The town had decayed steadily since 1780, when it ceased to be the Virginia capital. William A. R. Goodwin had seen to the restoration of Bruton Parish Church while its rector from 1903 to 1908. Returning to Williamsburg as a William and Mary professor in 1923, Goodwin was shocked to find how badly the town had deteriorated during his absence. He obtained money to restore the George Wythe house and aspired to do the same with other run-down or ruined structures.

Goodwin also interested himself in the Memorial Hall project, and in 1924 he spoke in support of it to a Phi Beta Kappa gathering in New York City. There he met John D. Rockefeller, Jr., Brown '97, Phi Beta Kappa senator, and chairman of the local fund-raising committee. Later Goodwin tried to interest Rockefeller in his Williamsburg plans, and he finally succeeded while Rockefeller was in town for the dedication of Memorial Hall. The eventual result was Colonial Williamsburg, the greatest of American historical tourist attractions.

For Phi Beta Kappa, one of the reconstructed buildings in Colonial Williamsburg had a special historical significance—the Raleigh Tavern. Here, according to their own manuscript minutes, the brothers of the original fraternity held three of their four anniversary celebrations. And here, according to unsubstantiated belief, the five founding brothers also held their very first meeting on December 5, 1776. At the 152nd anniversary of the Virginia Alpha on December 5, 1928, the poet of the occasion retold the cherished story thus:

> There gathered the youth of the town
>   Into the Tavern of Raleigh,
> To add to its fame and renown,—
>   There unto the Hall of Apollo,
> That noble band of youth came
>   To found the Phi Beta Kappa.
> And make immortal its name.

Year after year, guides conducting tourists to the Apollo Room assured them solemnly that this was the birthplace of Phi Beta Kappa.[7]

Rockefeller made available not only what was thought to be the society's first home but also what was presumed to be its final home. Its office in the New York apartment building would cease to be rent-free in 1927, according to the anonymous donor's grant. Before that time, the donor died and his identity was revealed. He was Francis Phelps Dodge, a graduate of Yale and heir to a fortune who, bed-ridden with arthritis, had devoted himself to giving away his wealth in various philanthropies. He had encouraged the sesquicentennial endowment drive by promising—and making—a generous contribution to it. Rockefeller, who also had contributed heavily to the campaign, now provided $16,000 for the purchase of stock in the corporation owning the apartment house. By paying an annual assessment of not more than 6 percent on this stock, the society would be able to keep the rooms that served as its headquarters.

Even with the generosity of such philanthropists as Rockefeller and Dodge, the sesquicentennial campaign fell far short of its $1,000,000 goal. In 1928 the trustees of the Phi Beta Kappa Foundation reported to the New York Board of Regents that nearly $350,000 had been raised, including the $100,000 for Memorial Hall and the $16,000 for office space. There remained some $35,000 in unpaid pledges, however, and once these were realized the United Chapters would have a nest egg of more than $250,000, which was the equivalent of perhaps ten times that much in the dollars of the 1980s.[8]

As it had done off and on from its very beginning, the United Chapters continued to look for new means of advancing the cause of humane learning. It paid for the printing of one thousand copies of *Representative Phi Beta Kappa Orations*, Second Series, which appeared in 1927 under the imprint of what the society called, after its early missionary, "The Elisha Parmele Press." In cooperation with the American Classical League, the society encouraged high schools to celebrate, on October 15, 1930, the two-thousandth anniversary of the Roman poet Virgil's birth. Under the society's auspices, 337 classicists spoke at 422 schools in 46 states.

At the invitation of the American Association for the Advancement of Science, the United Chapters began in 1935 to sponsor an annual lecture at the Association's annual meeting. Each lecture was to be delivered by someone other than a scientist and was to deal with a subject other than a scientific one. The intention was to illustrate the basic unity of all knowledge, to recognize science as an essential part of liberal culture, and to encourage scientists to take an increasing interest in the humanities.

But the society's most important venture of the period was the founding of a scholarly and literary magazine. This magazine was the brain-child of William A. Shimer, Harvard A.B. (1918) and Ph.D., who in 1930 came to the United Chapters office from the Ohio State University, where he had been an assistant professor of English. Shimer was to serve as assistant secretary of the United Chapters for a year to learn pro-cedures and then was to succeed Voorhees as secretary. After thirty years' service, Voorhees was retiring from the secretaryship in 1931 to devote himself to the duties of the society's historian.

Shimer had not been assistant secretary very long when he broached the idea of an expanded and elevated Phi Beta Kappa quarterly. Voor-hees endorsed the undertaking and expressed optimism about its ability to pay for itself after an initial subsidy. But at least one of the senators had misgivings. "If friends can be found in New York who will finance the journal for several years during its period of necessary struggle, well and good," Francis W. Shepardson said. "I do not see how, under any possibility, it would be proper to take the money of the Phi Beta Kappa Foundation to use in a venture of this nature." Shimer replied that it was "difficult to understand how any project could be more directly in keep-ing with the purpose of the Foundation." The proposed journal would encourage scholarship by providing a medium for "scholarly articles of a general nature." Anyhow, it was "quite possible that very little or perhaps no demand" would "need to be made upon the Foundation's funds."

The Senate approved and so did the Council. In planning the maga-zine, Shimer sought advice from various writers and publishers. He appointed a board of editors, one of whom was John Erskine, professor of English at Columbia University and prolific author of both scholarly works and best-selling novels, among the latter *The Private Life of Helen of Troy* (1925). Erskine suggested that the magazine adopt the title of Emer-son's famous Phi Beta Kappa address of 1837, *The American Scholar*. The new quarterly was expected to include news of Phi Beta Kappa activities as well as articles of general interest, thus taking the place of the old quarterly, the *Phi Beta Kappa Key*, which ceased publication in 1931. Each contributor of $50 or more to the endowment fund had received a lifetime subscription to the *Key;* this was now transferred to the *Scholar*. The first issue appeared in January 1932, announcing its main objective as "the promotion in America of liberal scholarship."[9]

Shimer as editor, along with the staff he had assembled to help put out the magazine, anxiously awaited reactions. "The only comment I heard on the first number of *The American Scholar* was that there was too much in it about Phi Beta Kappa," Erskine reported. "This criticism came from experienced newspaper and magazine people." Ada L. Comstock, presi-dent of Radcliffe College and member of the editorial board, said the "total effect" on her was "too much like a Phi Beta Kappa meeting." Another member thought the magazine lacked broad appeal because of its attempt to combine scholarly literature with fraternity news.

Still another member of the board, Dean Christian Gauss of Princeton, considered the first issue "too professorial and academic." He proposed that, to avoid this evil, he and his colleagues "plan out the contents of the numbers and decide upon subjects which are of particular interest to educated men and then ask some competent Phi Beta Kappa scholar to give us an article upon it." This proposal became the practice, except that authorship was not confined to Phi Beta Kappa scholars. The editorial board consisted not of figureheads, as such boards often do, but of men and women who took an active part in both the proposal and the appraisal of manuscripts.[10]

In planning the contents, Shimer and his advisers tried to steer a course between such magazines as *Harper's* and *Atlantic* on the one hand and the professional and technical journals on the other. To request an article on a specific topic could lead to embarrassment when the resulting manuscript proved unacceptable. One invited and rejected author wondered sarcastically whether Shimer had "really wanted an article or simply a subscription to 'American Scholar.'" This author thought well of his work and said he anticipated "no difficulty in selling it to a magazine of general circulation, such as 'Atlantic Monthly,' 'Scribner's,' or 'Harper's.'" I can appreciate your feeling," Shimer responded, ". . . for I have had many manuscripts of my own returned." But "we are trying to avoid the use of articles which would be accepted by the magazines you mention."

Shimer and his assistants were eager to mollify readers as well as writers. When one subscriber declined to renew, saying he did not like the magazine, Shimer's assistant in charge of publicity, Angela Melville, wrote to ask the man why he did not like it. "We are sincere in asking this question," she said; " . . . remember that we are, in these early months of the magazine's life, struggling with many difficulties, one of the greatest of which is lack of funds to pay large honoraria to authors." The ex-subscriber answered that he was "not cultured enough to appreciate" the magazine, despite his "having received a key from a New England college." He read *Liberty, Time,* and the *Readers' Digest,* but he was "a traveling salesman—not a college professor," so how could he be expected to read the *American Scholar?* "Be honest and admit that you aren't publishing it for the likes of me," he challenged. "You are exactly the sort of person who ought to be reading THE AMERICAN SCHOLAR," Melville wrote back, "because you are interested in what's happening in the world today."[11]

During its first three years, the magazine gave increasing space to articles of general interest, some of them by well-known writers, but it continued also to publish essays on comparatively recondite themes. Subjects and writers ranged from "Private Property or Capitalism" by Herbert Agar, "Speculation in American and English Stock Markets" by Ralph C. Hawtrey, and "The Churches in Germany" by Reinhold Niebuhr to "Louis Agassiz" by Helen Ann Warren, "Shelley: Values and Imagination" by Bennett Weaver, and "Medicine from Toads" by Ed-

ward Podolsky. Of 135 articles in the first twelve numbers, only 73 were
written by ΦBK members.

Reporting to the 1934 Council, a committee on the *American Scholar*
congratulated the editor and the editorial board for having produced a
"splendid magazine," one that enabled the society "not only to demonstrate in an attractive manner the type of liberal or cultural scholarship
symbolized by the ΦBK key but also to stimulate such scholarship and
such publications in America." The committee believed, however, that
"the magazine should not be merged with a news sheet." The Council,
which three years earlier had reluctantly authorized the publication,
now enthusiastically endorsed it.

Following the committee's recommendation (and the editorial board's
long-standing preference), the Council provided for removing Phi Beta
Kappa news from the *American Scholar* and publishing such news separately. In 1934 the United Chapters issued *ΦBK Annals,* a nicely printed
eighty-page pamphlet recording the society's activities for the year, a
copy of which was sent gratis to each of the now approximately seventy
thousand members. The *Annals* included a job register for victims of the
Great Depression, many of the members being completely unemployed
or "devoting their scholarly talents to the shining of shoes." But, after
one issue, the *Annals* itself succumbed to the Depression, having proved
too expensive to keep up. It was succeeded by the *Key Reporter,* a quarterly house organ that began in 1935 and soon took the form of a rather
modest leaflet. This continued to list members in search of employment.[12]

As for the *American Scholar,* Shimer worked hard to boost its circulation and make it prosper in spite of the hard times. He obtained endorsements from celebrities, such as the Virginia novelist Helen Glasgow, in his effort to "convince educated Americans without costly
advertising that THE AMERICAN SCHOLAR is a good magazine."

Growing rather desperate by 1937, Shimer considered "the desirability of finding a new name for the magazine and of changing its
format, perhaps to that of the *Readers' Digest* size, with a view to news-
stand sales." Erskine vigorously objected to altering either the name or
the format. "And if you put yourself in the hands of promotional agencies," he warned, "you need not bother about your editorial policy because they will decide it. That is the great bane of magazine publishing in
our country—the sales force dictate the editorial policy to an unfortu-
nate extent."

Shimer next got an inspiration for massive free advertising. A popular
radio program, with an estimated four million listeners, consisted of a
weekly spelling bee. For this program, which was unsponsored, Shimer
thought of "one team to be composed of five widely known members of
the Editorial Board and the other team of five young women recently
elected to ΦBK in eastern colleges." The broadcast would "help us to

overcome the popular tendency to make the names of the Society and the magazine suggestive of an academic, high-brow, and stodgy attitude toward life." Erskine dismissed this as "one of the silliest ideas" he had ever heard of, and the rest of the board was unenthusiastic at most, but Shimer went ahead and scheduled the spelling match. Then he suddenly called it off, having learned that the program had been made commercial and had been "sold to the Energine Cleaning Fluid Company."[13]

During the years 1938–1940, Shimer succeeded in arranging numerous broadcasts of a more dignified sort, which gave publicity to both the *American Scholar* and Phi Beta Kappa. One of these broadcasts, on the serious-minded *America's Town Meeting of the Air,* presented a discussion of the question "What Do We Need for 1940: Philosophy or Religion?" Such programs may not have improved the high-brow and stodgy image that worried Shimer. Still, many chapters, associations, and individual members tuned in, and most of them seemed to be favorably impressed.

The *American Scholar* had started remarkably well for having been launched on the ebb tide of the Depression. At the end of its third year, it could boast 4481 subscribers, 478 of them nonmembers and 233 of them libraries. By the end of the tenth year, however, the total had increased to only 6,000, of whom about 5,000 belonged to Phi Beta Kappa—hardly more than 5 percent of the society's membership. According to different estimates, the magazine would need as many as 8,000 or possibly 10,000 subscribers if it were to "carry itself."

At their 1941 meeting the senators reconsidered the problem of the *American Scholar.* One of them suggested adding to its appeal to Phi Beta Kappa members by publishing society news in it (as had been done in the beginning). Shimer explained that

> all materials concerning the Society had been excluded . . . for several reasons: (1) to keep the general public from thinking of the publication as a house organ, (2) to overcome the assumption by some that because they consider Phi Beta Kappa as a Society is dry and stodgy, the American Scholar articles would therefore not be of general interest, (3) to publish news of the Society in the form of an inexpensive magazine such as THE KEY REPORTER, which can be distributed to all members.

Another senator thought perhaps the society ought to discontinue the magazine if unable to afford such a "luxury." One of his colleagues replied:

> We are spending a good deal of time trying to find ways in which Phi Beta Kappa can further the cause of liberal education. Yet here we are worrying about a little deficit in one phase of our activity. My idea of THE AMERICAN SCHOLAR is that it is one of the very few things that our Society is contributing to the cause of liberal education in this country. If the organization as a whole can keep in the black, I do not think we should be

concerned about a deficit in one phase of the work. I think the magazine has brought more good will and esteem for our Society than anything we have done.

The senators disposed of the problem by referring it to a committee.[14]

The society was broadening its aims. "Important as has been its influence in obtaining for properly qualified students adequate recognition of honors quality in academic work while they are still undergraduates," Shimer declared in 1934, "a ΦBK function of even greater importance has been the stimulation and maintenance of active scholarly and cultural interests in its members after graduation." This meant taking on additional activities and expenses at a time when its financial resources were shrinking on account of the Depression.

At the same time, members showed an increasing interest in the society, if attendance at Council meetings was any measure. Previously the number of chapters sending representatives had fluctuated between 48 and 91 percent of the total. At the 1934 meeting in Cincinnati, 97 percent of the chapters sent representatives, and these numbered 236 as compared with the largest previous attendance of 172 in 1931. To maintain a good attendance in the future, the 1934 Council committed the United Chapters to paying the expenses of one delegate from each chapter.

This added to the financial burden of the society, and the burden grew as the society began to schedule its meetings in comparatively faraway places. The object was to emphasize its *national* character by recognizing the wide geographical distribution of its members, but one of the results was to increase the traveling expenses of most of the delegates. The first eight meetings (1883–1904) had been held in Saratoga Springs, New York. Except for the 1907 meeting in Williamsburg, Virginia, all the rest had been in the Northeast or in Ohio. In 1937 the Council met for the first time in the Deep South, in Atlanta, and in 1940 for the first time in the Far West, in San Francisco.

To pay the delegates' expenses, the United Chapters collected 50 cents per initiate from each chapter. Altogether, the central office received a registration fee of $2 for every new member, though no annual dues from existing members. A few of the chapters had funds from which to meet these charges, but most of the chapters imposed initiation fees, which ranged from $2 to $30. And in most cases the student being honored had to buy his own key. The United Chapters held a monopoly on the keys and sold them at a profit. It lost this source of income by absorbing the increase in cost when the New Deal administration reduced the gold content of the dollar and thus tripled the price of gold.

The $2 registration fee remained the central office's main support except for the endowment yield. This yield fell off as dividends and interest declined, especially the interest on the society's real-estate mortgages. To manage its finances more efficiently, the society in 1934 recast the relationship between the United Chapters and the Phi Beta Kappa Foundation. Hitherto, each of the two agencies had been responsible for certain activities and expenses. Henceforth, the Foundation would concentrate on managing the investments and would provide grants to the United Chapters, which would make all the detailed expenditures. The treasurer of both the society and the Foundation would be a financial expert, beginning with the deputy governor of the Federal Reserve Bank of New York, who would serve without pay.

The financial statement for the 1931–1934 triennium showed $38,123.25 of endowment income spent, including $3,646.73 for the *American Scholar* in general and $4,353.27 for its authors' honoraria. But salaries ($10,315), office expenses and equipment ($1,342.32), rent ($1,440), and "general administration" ($2,913.56) were listed as separate items. Some portion of each of these expenditures could properly be assigned also to the subsidization of the *American Scholar*. The conclusion, then, would be that the publication was costing more than one fourth of the society's endowment income. For the 1934–1937 triennium, the budget allotted somewhat less—$6,000 instead of $8,000—to the support of the magazine.

If the money going to the magazine's support had been used for the United Chapters' operating expenses, the collections from new members could have been reduced or eliminated. That is to say, these young initiates were helping to subsidize the magazine. Some of their money was going indirectly into the pockets of the authors who received honoraria. Shimer did not put the matter exactly that way, but he pointed out that for "a substantial proportion of its financial support" the society depended on its new members. "It should be borne in mind," he admonished, "that all too frequently the most capable, the most serious-minded, the hardest-working of the undergraduates—in other words, the very group from which the membership of Phi Beta Kappa is drawn— are the hardest pressed financially, and the least able to pay the entrance fees . . . and in addition purchase their own keys."[15]

Shimer did not propose to eliminate the subsidy to the *American Scholar,* his pet project. Instead, he wanted to spend more money on it. He considered the existing office too small—that fourteenth-floor apartment which Voorhees had seen a dozen years earlier as ideal for the United Chapters' permanent home. "It is not Mr. Shimer's idea, necessarily, that we should erect headquarters but that we should have adequate space for the editing of THE AMERICAN SCHOLAR and for the national office of the Society," his assistant Angela Melville explained privately. "We certainly have not got sufficient space now and we know that our work is greatly handicapped for that reason."

But Shimer also hoped to relieve the "youngest members" of the burden they bore in maintaining the headquarters. "He believes that this would be better done by funds contributed by Phi Beta Kappa members through the Phi Beta Kappa Foundation and that the outward symbol of the honor which membership in the Society implies would mean more if, instead of an order blank at the time of his initiation, a new member were presented in a dignified way with a Key." In other words, the thing to do was to augment the Foundation's funds and not reduce its grants to the *American Scholar.* "The Phi Beta Kappa Foundation needs $1,500,000 more endowment," Shimer announced.

So in 1933 the United Chapters started another money-raising effort. "We are not going after funds for the Foundation in any public way," Melville informed Bruce Barton of the prominent advertising firm Batten, Barton, Durstine & Osborn, "but by quietly working through our own membership and the use of a dignified leaflet setting forth the work and aims of the Foundation, we believe that a gradual increase of funds will come to the enterprise so as to enable it to develop the work already begun and to advance some of the hopes and ideals for future undertakings."[16]

Though the effort got some free assistance from Barton, it ran into resistance on the part of local chapters, as earlier campaigns had done. Apparently no time was ever appropriate for asking college faculty to contribute—and certainly not during the Depression, when the colleges themselves were finding it very hard to raise money. In the circumstance, $1,500,000 seemed like an enormous sum. "Such a plan in times like these is a grandiose scheme," a Kenyon College professor complained in the educators' journal *School and Society.* "But aside from this practical consideration, the question arises as to why Phi Beta Kappa should become another educational bureaucracy as it inevitably would with the resources afforded for publicity and propaganda by such a large endowment." In reply, Shimer protested that the United Chapters was relying on "steady growth" through small contributions, not on a big fund drive, and had nothing like bureaucracy in mind. Nevertheless, the contributions lagged, and the fund-raising effort soon petered out.[17]

A few years later, in 1938, the United Chapters tried again. In two successive issues the *Key Reporter* appealed for a $1 contribution from each of the approximately eighty thousand members. Again there was little response.

By this time, many Americans, and especially intellectuals, were seriously concerned about the spread of totalitarianism and the threat of a second world war. Naziism in Germany and militarism in Japan appeared to imperil freedom throughout the world. Some Americans thought they endangered the security as well as the democracy of the United States. President Franklin D. Roosevelt talked of the need to prepare for *defense,* and he and other spokesmen for the administration emphasized the word more and more after the Germans invaded Poland on September 1, 1939.

Here was a cue for the Phi Beta Kappa fund-raisers. Early in 1939 they started a drive for the Phi Beta Kappa Defense Fund for the Humanities and Intellectual Freedom. "To the defense!" was the slogan; $300,000 the comparatively modest goal. "The new world-wide struggle for freedom cannot hope to salvage civilization, except as it seeks the chart of its course in the ΦBK motto, 'The love of wisdom is the helmsman of life.'" So the president of the United Chapters, Frank Pierrepont Graves, proclaimed while commending the proposed fund to the society's membership. "But under the economic and political pressures of today even educators are casting the humanities overboard, leaving the helm of life at the mercy of this or that particular skill or technique—a proceeding dangerous in uncharted seas."

The "sober, scholarly Phi Beta Kappa Society" was "emerging from its ivied sanctuary into the thick of world affairs, launching a campaign to scotch 'isms' and preserve intellectual freedom in the U.S.," *Time* magazine reported after a great kick-off banquet in New York City. "In that cause Phi Beta Kappa brought together an odd collection of conservatives, liberals, jurists, admen, politicians who had nothing outwardly in common but their little gold keys." Among the personages taking part were Chief Justice Charles Evans Hughes and Associate Justice Owen J. Roberts, two of the seven key owners on the Supreme Court. Other fund-raising dinners were held in 1940 and 1941, but despite its illustrious sponsorship, the campaign netted only $80,000 of the targeted $300,000.

Meanwhile, at its December 1939 meeting, the Senate considered three other plans for improving the society's finances. One proposal was to ask each member for dues of $1 a year, but Shimer explained that the billing would probably cost as much as would be collected. The senators voted down that plan but approved the other two. The first of these called for the establishment of "sustaining or contributing memberships," and the second for the establishment of "a group of Phi Beta Kappa Associates, paying $100 each for ten years." In the long run, the contributing memberships were to become the society's largest source of income, but for the time being, the Associates idea seemed the more promising.

In February 1940 fourteen men met at the Harvard Club in New York City and agreed to form an organization of two hundred Phi Beta Kappa Associates who would constitute a "living endowment" for the society, each of them pledging $100 a year for a decade (at the end of which they could renew their pledge or be replaced). At the going rate of interest, the resulting $20,000 per annum would equal the return on an investment of $500,000. The Associates were well on the way toward enlisting their twentieth member by the time the Japanese attacked Pearl Harbor on December 7, 1941, and not long after that, the Associates reached their objective, despite the country's absorption in World War II.[18]

10

Applications
and Qualifications

Between the wars, as colleges and universities continued to grow in size and number if not also in quality, more and more of them considered themselves worthy of a Phi Beta Kappa chapter. Which of them should get one? This was for the existing chapters to decide through their delegates to the National Council. To cope with the increase in applications, the United Chapters repeatedly refined its standards and altered its procedures for the qualifications of colleges and universities. It even assumed the right to suspend charters at institutions falling below the standards. While limiting the multiplication of chapters, the society encouraged the growth of ΦBK alumni associations.

Between 1920 and 1940, the number of institutions of higher education—universities, liberal-arts colleges, professional and technical schools, teachers colleges, and normal schools— increased by 64 percent, from 1,041 to 1,708. The count of undergraduates attending those institutions went up by 150 percent, from 582,000 to 1,388,000, in spite of a Depression-induced decline between 1930 and 1936. These undergraduates represented slightly more than 8 percent of the eighteen- to twenty-year-olds in 1920 and nearly 16 percent in 1940. During the same period, the junior colleges were multiplied by almost nine, going from 52 to 456, and their enrollment by more than eighteen, from 8,000 to 150,000.

Presumably these figures tended both to reflect and to realize the American dream of rising in the world. "In this last decade, higher education has become such a fetich in America that all the youth of the country, rich or poor, from the cities and the farms, fit or unfit, are seeking the roads that lead to the universities," a 1928 guidebook to

colleges declared with some exaggeration. "To each one, or to his parents, a college degree is a stamp of social superiority, its lack, a social
stigma. Each one believes that it is a magic key to happiness, success, and
riches."[1]

But the perceived value of the degree varied considerably from one
institution to another, as had been the case in previous decades. If anything, the disparity between the most and the least prestigious institutions was widening. Certainly the student bodies differed a great deal in
range of social backgrounds as well as in distribution of native talent. As
a rule, the state institutions, even the largest and wealthiest, were compelled to accept nearly all of the state's high-school graduates who applied for admission, regardless of their academic records. But some of
the private colleges and universities were becoming more and more
selective. They sorted out the better prospects with the aid of entrance
examinations and intelligence tests, which the army had made familiar
by using them on recruits during World War I.

The academic rating of the institutions themselves was somewhat
more difficult than the academic rating of their prospective students. By
the 1930s, the regional accrediting agencies were well established, but
they were coming under criticism for basing accreditation too much on
quantitative and too little on qualitative data. They could get only a
rough indication of the quality of instruction by counting such things as
campus buildings, laboratory facilities, library holdings, and faculty with
doctoral degrees. Accreditation meant adequacy at best, not necessarily
excellence, and many so-called colleges lacked even that credential. As
Frank Aydelotte, a Phi Beta Kappa senator and an authority on higher
education, remarked in 1942, "there are hundreds of institutions calling
themselves colleges and universities, licensed by their state Boards of
Education to confer degrees, which enjoy no recognition from regularly
organized educational associations."

One agency went beyond the regional accrediting agencies in rating
liberal arts colleges and university colleges of arts and science. This was
the Association of American Universities (AAU), dating from 1900,
through which the universities with large graduate and professional
programs cooperated to appraise the colleges' preparation of undergraduates for advanced study. It was a mark of distinction for a college
to appear on the AAU's approved list.[2]

Another list was still shorter and more select. It consisted of those
institutions that had qualified not only for the AAU's approval but also
for a ΦBK charter.

As early as 1916, Secretary Oscar M.
Voorhees had urged the Council to tighten up the chartering process.
The "number of hopeful institutions is well nigh legion," Voorhees then
pointed out. "The Secretary has no authority to prevent these hopeful

people from pushing their cases before the chapters." Still, he "must stand between assertive representatives of fairly worthy institutions and the Senate and Council," and it was "not pleasant." Hence he called attention to the "desirability of providing an arrangement for discouraging applications from institutions that were quite evidently not in the Phi Beta Kappa class."

The 1916 Council authorized the Senate to act, and the Senate appointed a Committee on Applications, to which all inquiries regarding charters were to be referred. As the secretary reported at the end of the triennium, "the new committee took a strongly conservative attitude and gave no inquirers positive encouragement." This, from his point of view, was all to the good, for inquiries continued to pile up. "The fact is flattering, perhaps, to Phi Beta Kappa, but not always as to the judgment of inquirers respecting the institutions they represent." If all the inquiries had led to applications and all the applications had led to charters, the total number of chapters would have doubled in a few years. Instead, only four new chapters eventuated from the 1916–1919 triennium.

The 1919 Council left to the Senate the question of devising a different chartering system, one that would spare the Senate committee the task of discouraging applications. The Senate came up with a plan for putting the responsibility on the chapters themselves. They would be organized in regional groups and would nominate the colleges they considered most deserving within their respective regions.

In recommending this proposal to the 1922 Council, President Edward A. Birge characterized it as "an attempt to restore the conditions of 1882" so far as possible. The existing procedure, he said, had worked well in the beginning, when there were only twenty-three chapters and they were confined to the Northeast, except for three in Ohio. "These institutions were neighbors and comparatively well known to each other." Presumably, they could easily and wisely handle the few applications that came from other neighbors. Now that there were ninety-three chapters scattered across the country, the old neighborliness was gone, but much of it could supposedly be recovered within the proposed geographical groupings.

The new bylaw, as adopted by the 1922 Council, contained the following provisions: (1) the United States would be divided into five districts, and the chapters within each would constitute a "district unit or conference"; (2) each conference would have the right to nominate as many as three institutions every triennium from the district; (3) from those thus nominated, the Senate would select the ones it deemed "most worthy of immediate consideration," would "encourage from them applications," and would transmit these to the Council for approval or disapproval.[3]

The five districts were to be the North Atlantic, South Atlantic, North Central, South Central, and Western. According to the 1920 census, the population of these districts was 50, 14, 34, 19, and 9 million respectively. As of 1922, the count of existing chapters in each was 43, 8, 29, 5, and 7.

The districts varied not only in the number of colleges possessing charters but also in the number entitled to consideration for charters—that is, the number having the Association of American Universities' endorsement. "A study of the Association's list reveals the names of seventy-five institutions otherwise apparently eligible that have no Phi Beta Kappa charters," Voorhees reported in 1927. "Of these, eighteen are in the North Central District; ten in the South Atlantic; thirty-one in the North Central; and nine in the Western District." Some of these, Voorhees conceded, were "fully equal" to or even "evidently stronger" than some of the institutions already sheltering chapters. "This makes it clear that our requirements have advanced."

Voorhees did not appear to be troubled by the illogicalities and inequities of the new system. In the North Central district, dozens of chapters had to agree in making the choice, and they could recommend no more than three institutions each triennium, even though their district contained thirty-one that deserved consideration. In the South Central district, a mere handful had to agree, and they were entitled to recommend as many as three institutions though the district contained only seven that were eligible! Besides, it was unrealistic to think of each district as a kind of neighborhood. Two institutions more than a thousand miles apart, such as Colorado College and Whitman College, both members of the Western conference, could hardly be considered neighbors.

At colleges already sheltering chapters, the members generally favored at least the principle of restriction, but at colleges without chapters the Phi Beta Kappa faculty sometimes took a quite different view. Such faculty, "working hard to maintain high educational standards," were handicapped because they could not "offer the recognition that Phi Beta Kappa affords," as one of them protested in 1924. The multiplication of chapters, he argued, would be to the advantage of the society as a whole. It would bring additional revenue, make possible a heightened influence on the public school system, and enhance "the position of the Society in American thought and life." The William and Mary founders of the society had resolved that its principles "should be extended to the wise and virtuous of every degree and of whatever country."

> It was the fond hope of many of us that we might see in the very near future chapters of our Society established in every College of Arts and Sciences in the country, where the training of the faculty, the material equipment, the character of the student body, and the educational history of the institution justified it. But this would mean that charters must be granted at a far greater rate than at present.[4]

Such appeals did nothing to change the existing policy. Some of the chapters, however, soon became dissatisfied with the district system, which caused them a certain amount of difficulty and confusion from the very beginning. The Oberlin chapter, for example, sent its recommendations in 1923 to the Western Reserve chapter (which, as the senior Alpha in the district, was to take charge) and then received requests from three more colleges for recommendations. "Are we supposed, after having made our selections, to refrain from further recommendation," the Oberlin committee chairman inquired of Voorhees, "or would it be acceptable for us to add to the list or send in the names of other colleges that we regard as suitable material for chapters?" Voorhees simply instructed him to have the chapter arrange in order of preference the six institutions that had already received the most votes in the district.[5]

The greatest cause of dissatisfaction was the geographical bias of the district system. During the first two triennia of its operation (1922–1928), it yielded four charters for the South Atlantic, three each for the North Central and South Central, and five for the Western districts, but none at all for the North Atlantic, which contained the second largest number of institutions eligible for charters. In response to complaints from the North Atlantic district, the 1928 Council split the two most populous districts, the North Atlantic and the North Central, thus producing a total of seven.

This reduced the inequalities but did not bring uniformity in district standards, nor did it eliminate what Voorhees referred to as "unseemly efforts to influence the voting of chapters." The Brown chapter suggested that a "special board or commission, technically expert, should be constituted to investigate and report on applying institutions." Soon the New England chapters created such a commission. These and other indications of discontent prompted the Senate to set up a Committee on Criteria and Methods to study the problem. The committee proposed discontinuing the district system and using something like the New England commission for the country as a whole.

Accordingly, the 1931 Council set up a standing Committee on Qualifications to investigate and select institutions to recommend for charters. The committee would consist of six members who would serve staggered six-year terms, three members retiring at the end of each triennium. At the 1931 Council meeting, some delegates wanted to get rid of the district arrangements entirely, but others insisted that these should continue to have a role in the chartering process, and so they were retained for advisory purposes. The procedure would be for the committee to request a preliminary statement from each of the inquiring colleges, to study and visit those whose statements indicated the greatest promise, and to send the list of selected colleges to the district conferences for their comment.

When the new Committee on Qualifications reported to the 1934 Council, some of the delegates were ready for revolt. They had "a fear of centralization of control in ΦBK, a fear which had been developing for many years," as Voorhees's successor William A. Shimer observed. Now it seemed to them that a central committee was taking still more power away from the chapters and their representatives on the Council. "Shall we be a rubber stamp?" one delegate demanded, then went on to declare: "We must retain some degree of chapter responsibility."

The committee's chairman, David A. Robertson, addressed the Council in defense of the new procedure. He had beside him a stack of folders eight feet high. The committee, he explained, had sifted through all those data, had investigated thirty-seven colleges and universities, had paid experienced examiners to visit every institution, and had given the chapters in each district an opportunity to express their opinions. Pointing to the pile of data, he invited the delegates to make their own selection. He said:

> Now, if you are going to do it, the material is here. Are you ready to go through this material as you honestly must if you are to undertake it at all? Are you going to sit for two days as we did on your behalf and go through those data or are you going to trust this Committee and the Senate? There is the issue perfectly flat. Is the work going to be done on behalf of the Council by persons that have been perfectly honest in devoting their thought and their time to this enterprise or is each one here going to go through that material? It is there and we submit it to you. Go to it.

None of the delegates went to it. The disgruntled ones soon quieted down. "By the end of the third day," Shimer was happy to note, "complete harmony reigned, and the delegates were able to feel that the society's motto, philosophy the guide of life, had been exemplified."

By 1942, when the war interrupted the work, the Committee on Qualifications had considered a total of about 250 inquiring institutions, had investigated 86 of them, and had recommended 20, all of which the Senate and the Council had approved for charters. That was not nearly enough to keep up with the increase in eligible institutions as measured by the Association of American Universities' lengthening list of those it endorsed. The number of institutions without ΦBK chapters on that list—which had come to 75 by the mid-1920s—reached as many as 125 during the 1930s.[6]

Perhaps the United Chapters ought to have the power of disciplining chapters and revoking as well as granting charters. This proposition gained support as the society confronted threats to academic and intellectual freedom—threats that conceivably could eventuate in an institution's loss of scholarly character.

A serious threat, as most members saw it, came from the proposal and the passage of state laws against the teaching of evolution. Oklahoma and Florida in 1923 and Tennessee in 1925 adopted such laws, while several other states seemed on the verge of doing so. According to the Tennessee legislation, which applied to all state-supported schools, it would be "unlawful for any teacher to teach any theory that denies the Story of Divine Creation of man as taught in the Bible."

The American Civil Liberties Union (ACLU), then only five years old, promptly challenged the Tennessee law. At the prompting of the ACLU, the young high-school teacher John T. Scopes stood trial for using a biology text that presented the Darwinian theory. The trial made headlines throughout the country, especially when the famous criminal lawyer Clarence Darrow examined the anti-evolution leader and three-time presidential candidate William Jennings Bryan as a Bible expert. To some Northern intellectuals, it seemed that Darrow made a monkey out of Bryan and that, despite Scopes's conviction, the anti-evolution movement was dying of ridicule. In fact, the anti-evolutionists prevailed in most of the South, and biology texts used in the North as well as the South were rewritten to conform to the fundamentalist beliefs and regulations of the Southern states.[7]

State colleges and universities—and private ones with fundamentalist denominational ties—appeared to be in danger of repression like the public schools. This heightened the apprehensions of delegates to the 1925 Council. They adopted a resolution denouncing "the present tendency to suppress freedom of thought and speech in our colleges" and declaring it to be "the sense of this convention that no college that gives evidence of denying this freedom will be considered worthy of a chapter in Phi Beta Kappa."

This resolution did not please every member of the society. "It places the society in definite opposition to all colleges maintained by Christian churches," one member complained in a letter to the *New York Times.* He argued that Phi Beta Kappa was violating civil liberty rather than upholding it. "One of the essential elements in civil liberty is the right of voluntary association," he wrote, "—the right of persons who have come to have any view on any subject whatever to associate themselves for the propagation of their view and to educate their children accordingly." But another correspondent replied that Phi Beta Kappa ought to be entitled to the same right that the critic asserted for groups in general. "Surely he ought not to grant the 'right' to be intolerant to every organization except the Phi Beta Kappa."[8]

While the Council's freedom-of-thought resolution went too far for some members, it did not go far enough for others. In 1926 the Johns Hopkins chapter unanimously adopted the following resolution:

> It is the sense of the Alpha of Maryland Chapter of Phi Beta Kappa that the Senate, in cooperation with the American Association of University Professors, should recommend to the National Council that a new chapter be

not established, and should recommend the withdrawal of an established chapter, at such institutions that [*sic*] have violated, in the opinion of the Senate and the American Association of University Professors, the principle of Academic Freedom.

There was a certain irony in the Johns Hopkins chapter's taking this advanced position—to revoke the Phi Beta Kappa charter at any anti-evolution college or university—since the Johns Hopkins University had provided the only "scientist" to testify at the Scopes trial in support of the Tennessee law. This was the physician Howard A. Kelly, who believed that a "Christian must stand very literally with the Word regarding the creation of man."

The Johns Hopkins chapter's resolution inspired a sympathetic but skeptical editorial in the *Baltimore Sun*, the newspaper of the great debunker H. L. Mencken, who had reported the Scopes trial with sardonic glee. The chapter's gesture was "in accord with the tradition of Phi Beta Kappa," the *Sun* said, but it would have little or no effect. The politicians and "political parsons" campaigning against evolution had strong popular backing in states like Tennessee. But Phi Beta Kappa "can ask for the support only of the comparatively few men who understand the value of learning or who actually believe in the old-fashioned American idea of freedom." And to remove Phi Beta Kappa would only lessen the influence of men of this kind.

The Johns Hopkins resolution aroused still less enthusiasm on the part of Secretary Voorhees. The Senate, he explained to the chapter secretary, had already "gone on record" that it would not recommend any institution where academic freedom was denied. The withdrawal of charters already granted was a quite different matter. "The question of the authority for this has been talked over annually but it has never been felt that any chapter had so failed in its duties as to make the question a practical one." Voorhees continued: "I confess that I regret that your resolution should have tied up the Phi Beta Kappa Senate with the American Association of University Professors." The Senate had always "insisted on maintaining its own independence of action" and no doubt would give "scant consideration" to a proposal for joint action with the AAUP. Voorhees might have added—but did not—that the AAUP was a little too much like the ACLU, which had a reputation as a defender of labor unions and radical groups. In taking a stand for intellectual freedom, Phi Beta Kappa seemingly needed to keep its distance from less-reputable advocates of the same high cause.[9]

For some time, the disciplinary power of the United Chapters remained unclear. An institution with a chapter might cease to maintain standards that would qualify the institution for a charter, yet the Council apparently could do nothing to disqualify the institution. Accordingly, the Committee on Qualifications in 1934 made the following proposals:

"that, since chapters may deteriorate and in time bring discredit to the society, constitutional provisions should be made for the revocation as well as for the granting of charters"; but "that individual chapters should continue to be allowed, in the determination of their policies and the conduct of their affairs, the maximum of freedom compatible with the purposes and well-being of the ΦBK society at large."

Already the chapter at the College of the City of New York was carrying freedom to an extreme that threatened the well-being of that chapter if not the well-being of the society as a whole. At CCNY the faculty and undergraduates constituted only a small proportion of the active membership; alumni residing in and around the city dominated the meetings. These alumni took a keen interest in the college and did not hesitate to give it advice. They requested the reinstatement of some expelled students, called for an inquiry into alleged pro-Nazi student organizations, protested against certain American Legion activities on the campus, objected to the dismissal of an instructor, and condemned the methods of a state legislative committee investigating the college.

Finally, in 1934 the activist majority demanded the removal of the college president. This was too much for some of the faculty belonging to the chapter. The chapter's secretary and its treasurer resigned, and a number of the members quit going to meetings. The dissidents talked of organizing a new chapter or a Phi Beta Kappa group of some other kind. In their opinion at least, the old chapter had suffered a serious deterioration.

Taking up the CCNY affair, the Committee on Qualifications in 1935 gave the activists an implied rebuke. The committee declared that it "would not be proper for any Chapter to adopt resolutions or to take other action on questions not directly related to the method of selecting members." In other words, a chapter had no business doing anything or discussing anything except elections to membership! This declaration seemed to put the United Chapters on the side of restricting rather than protecting academic freedom.

The CCNY chapter responded: "we do not think the purpose of our Society can well be served by any rule that will prevent us from adopting any resolution or taking any action for the promotion of scholarship and the safeguarding of intellectual freedom not only in our country at large but even in the very institution from which we draw our members." Convinced of its error, the Committee on Qualifications amended its statement about what the chapters should confine themselves to. It struck out the phrase "the method of selecting members" and substituted for it "the promotion of scholarship."

A new constitution adopted in 1937 specified the disciplinary powers of the United Chapters. According to the revised document, the Council can prevent a chapter from electing members and can amend or suspend its charter if the chapter "has disregarded any of the provisions of its charter" or if "there has been a serious deterioration in the sheltering institution." This provision remained a dead letter from the start. The Council never took any of the newly authorized disciplinary steps against any chapter, no matter what the chapter or its sheltering institution might have done or have failed to do.[10]

                          While discouraging the rapid multi-plication of chapters, the Council undertook to encourage the quick establishment of Phi Beta Kappa alumni associations. Voorhees in partic-ular liked to see these multiply, not only because they might further the society's scholarly aims but also because they could assist in its money-raising efforts. During the 1925–1928 triennium, when the sesquicen-tennial fund drive was at its height, the number of associations nearly doubled, reaching a total of sixty, of which four were located overseas—in Rome, Paris, London, and Peking. The associations varied in size and in activity, but most of them sponsored occasional lectures, if nothing more.[11]

Successful though the organizing effort was in most parts of the United States, it made slow headway in the Chicago area, as the chapters at Northwestern University and the University of Chicago at first re-fused to cooperate with one another or with the United Chapters. In 1925 Voorhees appealed to Francis W. Shepardson, a University of Chi-cago history professor and the Phi Beta Kappa vice president. "I think the time has come when our Society should speak out with some authori-ty," Voorhees wrote him. Shepardson, in turn, appealed to the president of Northwestern, Walter Dill Scott. "After a good deal of conference," he told Scott, "we thought the most feasible plan of awakening interest was to have the Northwestern and Chicago chapters invite to their annual address in June all members of the society from whatever chapter, divid-ing the territory geographically, possibly by the Chicago River." Scott replied: "I know the local chapter will be anxious to cooperate." But nothing happened at that time.

Seven years later, in 1932, the Northwestern chapter was still holding back. Its secretary protested to Secretary Shimer:

> I must confess I have serious doubts as to the usefulness of a local associa-tion. There are already so many addresses, conferences, and other cultural opportunities in Evanston and Chicago that people who have the inclina-tion to patronize them are compelled to choose which they will attend and it is becoming really difficult to secure a good-sized audience even for very well-known speakers. It was the opinion of our executive committee last year that, if and when we could get a mailing list of all members of Phi Beta Kappa in this neighborhood, we should invite them to be present to hear our annual oration and any other addresses for which the local chapter is

responsible. I do not see what else a local association could do along this line. Of course there are other possible activities in which an association might engage but some of them, at least, would involve the raising of funds, which would be very difficult in Evanston just now.

Sending a list of Phi Beta Kappa members resident in Evanston, Shimer's assistant Angela Melville replied: "Mr. Shimer's idea of the work of Associations is, I think, that they may, besides organizing for their own stimulation and interest, actually do a great deal to advance scholarship in their communities, working through local high schools, colleges without chapters, etc., by recognizing in some way outstanding scholarship in individuals."

Finally, in 1935 the United Chapters succeeded in bringing about an association for the Chicago area. Of the 3,000 key holders living there, 430 joined the association and 247 attended its first regular meeting, to partake in a banquet and hear an address on "The Responsibility of the Educated Man."[12]

By that time, the number of associations had grown to approximately eighty, some of them in such faraway places as Rome, Peking, Shanghai, Tokyo, Manila, Tabriz (Iran), and Beirut. There was no longer an active one, however, in either Paris or London. The secretary of the London association confessed in 1936 that he had found himself "unable to carry on the duties" and was "afraid the Association must be regarded as nonexistent." Then two years later in response to an appeal from Shimer, an official of the English-Speaking Union agreed to arrange a December 5 dinner, at which the London branch could be reorganized.

The president of the oldest and most active of all such groups, the Phi Beta Kappa Alumni in New York, characterized the typical association of the 1930s as "a means of encouraging scholarship and culture outside of college walls, a way to apply scholarship to the saving of America." The New York group met three times a year for a dinner and a speech, the speaker usually being a prominent figure. Most of the associations met only once a year, for a lecture by someone less well known. Topics— similar to those of the annual orations of the chapters—ranged from "Character in the Making" to "What Shakespeare Means to Me," "How the 'New Deal' Affects ΦBK," and "Whither America?" A few of the associations carried on the kind of scholarship-rewarding activity that Shimer had in mind. They recognized high-school honor students by inviting them to meetings, sending them letters of congratulation, giving them books or other prizes, or urging their schools to form honor societies.

As of 1934, an association with as many as twenty-five members could send to the Triennial Council a delegate with all privileges except voting. "A constitutional committee," Shimer reported at that time, "is now considering a proposal to give such delegates a vote, to grant charters to the Associations, and otherwise to dignify further their status in the society."

In the revised constitution of 1937, the Council received authority to charter associations, and those thus accredited were to have full representation in the Council, though not full voting rights. Association delegates could not yet vote on all the questions that chapter delegates could—not on such matters as the election of senators, the adoption of constitutional amendments, or the chartering of new chapters.[13]

11

Merit, Marks, and Membership

While the United Chapters selected institutions, the separate chapters selected individuals, and during the interwar period there was as much disagreement about the selection process for individuals as for institutions. Chapters differed in regard to such questions as these: Should they elect students other than undergraduates majoring in the liberal arts (and what did the liberal arts consist of)? Should they give memberships quite freely to alumni, faculty, and friends of the college? Should they base undergraduate elections on course grades, or should they take into account other evidence of scholarly attainment and promise? These questions were not new, but they became increasingly urgent during the 1920s and 1930s. The United Chapters made only limited progress toward answering them and bringing about uniformity and improvement in chapter practices.

Phi Beta Kappa friends of the liberal arts had to contend with a curriculum that grew more and more complex and gave a larger and larger place to practical and vocational as opposed to cultural and liberal studies. The classics continued to lose ground; by 1923 all the well-known New England colleges except Amherst and Williams had dropped Greek and Latin requirements for admission. By 1933 the great majority of institutions throughout the country allowed at least some credits in vocational courses to count toward the Bachelor of Arts degree. No longer did the B.A. itself necessarily mean four years of exposure to the arts and sciences.

Colleges experimented with various plans for reforming the curriculum to provide a more truly liberal education, particularly for the abler students. Some of the plans aimed to individualize instruction; others, to integrate it. On the one hand, colleges introduced tutorial and preceptorial systems, seminars, individual studies, and honors programs. (Regarding honors work, Frank Aydelotte declared in 1942: "It might be called the most important educational development of the period between the two world wars.") On the other hand, colleges introduced broad "survey" courses, interdepartmental majors, comprehensive examinations, and "great books" programs.[1]

The "great books" idea was especially controversial. Its leading advocate, President Robert M. Hutchins of the University of Chicago, maintained that the best liberal-arts curriculum consisted of the most seminal writings of all time. Chancellor Harry Woodburn Chase of New York University, one of the founding members of the *American Scholar*'s editorial board, condemned Hutchins's book *The Higher Learning in America* (1936) in the pages of the Phi Beta Kappa magazine. Hutchins, according to Chase, was reverting to the long-discredited notion of eighteenth- and nineteenth-century educators and even to those of medieval schoolmen. Hutchins seemed to think that subject matter was less important than the learning process, the subjects of study being merely setting-up exercises for the mind. His list of great books, supposedly a universal curriculum, was reminiscent of another universal curriculum, the trivium and quadrivium of the Middle Ages.

The educational philosopher John Dewey, in *Experience and Education* (1938), disagreed with "traditionalists" such as Hutchins. Dewey said "the new philosophy of education is committed to some kind of empirical and experimental philosophy." But his long career of promoting "progressive" education hardly commended his ideas to promoters of the liberal arts.[2]

There being no consensus on the nature or even the desirability of a liberal education, it is not surprising that Phi Beta Kappa was slow to agree officially on the kind of courses that would make a student eligible for membership. By the time of World War I, some chapters were even electing graduate and medical students. In 1914 the secretary of the Johns Hopkins chapter, the first to elect them, inquired of Secretary Oscar M. Voorhees of the United Chapters how many other chapters did so. He asked further: "What is your own feeling as to the propriety of such elections? And what do you think of . . . electing graduate students who come from institutions which have Phi Beta Kappa Chapters of their own?"

Voorhees could muster only a vague and tentative reply. "Your inquiry," he wrote, "raises a question that has never received general discussion by either Senate or Council but which will soon have to be faced and some decisions reached." He thought, but was not sure, that the University of Chicago and the University of Virginia and possibly a few others also chose from the graduate and medical schools. "While there are many physicians and surgeons who are members of Phi Beta Kappa the election of men on the basis of work in medical schools seems at first thought contrary to the history and genius of Phi Beta Kappa"—and it

TABLE 2   Degrees of New Members in Course,
1933–1934

Degree Received	Initiates Receiving	Chapters Electing
B.A.	1493	86
Ph.B.	18	2
Litt.B.	2	2
B.S.	196	37
E.E.	3	1
B.Ed.	1	1
LL.B.	8	3
M.A.	5	2
M.S.	2	2
M.E.	1	1
Ph.D.	35	10
M.D.	10	2
LL.D.	1	1
Not reported	1190	33

certainly was contrary to the idea of recognizing cultural rather than vocational studies—"yet where these schools are on a basis quite similar to others in the graduate department of the university I cannot see that they should be excluded if the others are included." The practice of electing graduates of colleges with Phi Beta Kappa chapters, however, had "serious drawbacks" but then again, perhaps it would "seem unfair" to discriminate against such students.[3]

The number of chapters electing from graduate schools increased: 28 out of a total of 119 were doing so by 1933. Of the 119, only 43 confined their elections to candidates for the B.A. degree. For the academic year 1933–1934, approximately 16 percent of the degrees reported were *not* B.A.s, and a few were in such distant fields as pedagogy and engineering (see Table 2).

In 1934 Secretary William A. Shimer announced that the society was considering proposals to discourage the election of graduate students and to put increased emphasis on "the liberal or cultural motive in scholarship." The Committee on Criteria and Methods proceeded to recommend that "only students whose work had been at least 75 percent definitely liberal in character should be eligible for membership in course." The committee defined liberal studies as those "designed primarily for a knowledge or understanding or appreciation of the natural and social world in which we live as contrasted with training intended primarily to develop skill or professional techniques, such as most courses in departments or schools of home economics, business administration, or applied arts, or those for the preparation of teachers."

The essence of these recommendations found a place in the revised constitution of 1937. This provided that members should be "chosen principally from undergraduates in colleges of liberal arts and sciences," but it did not forbid the election of graduate students, and a few chapters continued to elect them, though not in large numbers.[4]

At one time it had been fairly common for new chapters, as soon as they were organized, to elect a large number of alumni and honorary members. After World War I, the United Chapters began an effort to restrict alumni and honorary along with graduate-student elections. The central office ran into special difficulties in dealing with certain chapters.

The William and Mary chapter presented unusual problems. Indeed, this chapter was unique in having elected *only* alumni and honorary members since its revival in 1893. A student had to wait at least a year or two after graduation, and ordinarily much longer, before the chapter would consider him. Some men of each graduating class went on to do graduate work at the University of Virginia, which obtained a chapter in 1908. "As we usually send a pretty good class of men there," a William and Mary spokesman complained in 1912, "they are frequently invited to join the University Chapter, and that causes us to lose them." He thought the Charlottesville people ought to hold off until the Williamsburg people had had time to make up their minds!

In the 1920s William and Mary, having become coeducational, grew "from a faculty of seven and a student body of 150–200 to a faculty of sixty and a student body of 1200." The chapter now undertook to revise its election policies and bring them into line with those of the society as a whole. Voorhees urged its members to make alumni ineligible until fifteen years after their graduation. Earlier, in the case of Bates College, he had recommended a ten-year wait, but he no longer thought that enough time. "The time is rather short for the average college graduate to find himself." The best model was Brown's. "At the Brown chapter Alumni are nominated from the fifteen, twenty, and twenty-five year classes only."

The William and Mary members had long been proud of their alumni and honorary elections. These were based on nothing but true merit, the *Richmond Times-Dispatch* had once remarked in an editorial congratulating the chapter. When "charges are rife that colleges and universities barter and sell their LL.D.s for endowments and buildings or to curry favor with the men of wealth and power," the editorial went on, "it is refreshing altogether to remember that the Phi Beta Kappa Society has always been uncontaminated, and recognized naught but clear and pure scholarship or noble service."

Such being the William and Mary chapter's reputation, it was especially shocking to all concerned when this very chapter was accused of "bargain and sale" in the election of an honorary member. The accusation came in 1928 after the election of Otto H. Kahn, a New York banker associated with Kuhn, Loeb and Company and a patron of the opera and the arts. The informant, a New Yorker whom Voorhees did not name in protesting to the president of the college, J. A. C. Chandler, had heard that Kahn had "made a substantial contribution towards certain financial requirements of William and Mary." Hence—and this was particularly galling to Voorhees—the financier had been "unable to make any contribution to the Phi Beta Kappa Endowment Fund."

"It really hurts me very much that any Phi Beta Kappa member should feel that the William and Mary Chapter would be guilty of a bargain and sale," President Chandler replied. "Nobody has ever thought of asking Mr. Kahn for any money," the president assured Voorhees. "However, he said when he was here to be initiated that he hoped some day he could do something for art here, which, of course, would be very gratifying." Since then, Kahn had offered a little financial help. He had been kind enough to "lend some money to some poor students" and pay the interest on a note the college owed. That was all.[5]

The Princeton chapter asserted its independence from the United Chapters in regard to elections, as it did in regard to solicitations for Phi Beta Kappa endowment funds. Voorhees realized there was something amiss about the chapter when he was gathering information for the 1923 general catalog. Here were Princeton graduates listed as members in course from the classes of 1896, 1897, and 1898, but Princeton did not have a chapter until 1899! The secretary explained to Voorhees that when the Phi Beta Kappa men on the faculty applied for a charter in 1895, they formed a "provisional" chapter and invited the rest of the faculty to join. Subsequently, they elected about a tenth of each graduating class to "provisional membership, such membership to be regarded as complete when the charter should be granted." Voorhees now insisted that those prematurely elected should be listed as alumni members and their keys dated 1899. "I don't really want an alumni member's key," one of those early members objected, and Voorhees finally "yielded to his insistence that he was a full member from the time of his election."[6]

Several years later, in 1927, Voorhees had another tiff with the Princeton chapter. At this time the chapter took in fifty-four alumni at once, including a graduate of the class of 1863. Voorhees, after consulting with ΦBK President Charles F. Thwing, reminded the Princeton secretary that the Senate had recently adopted a report defining alumni membership and confining it to those whose postgraduate achievements entitled them to the honor. "The report also definitely advises chapters of the propriety of selecting members in Anniversary years, and plainly does not encourage such wholesale elections as your action indicates." Voorhees and Thwing requested that, at the very least, the chapter bring in for the face-to-face ceremony the constitution required those of the fifty-four alumni who had been initiated *in absentia*. Voorhees regretted the necessity, but he was "under obligation to carry out the regulations of the Society and to report any irregularities to the Senate, which has final responsibility for seeing that Phi Beta Kappa is kept on a high plane."

The "reply of the Executive Committee acting for the Princeton Chapter" was unyielding and unapologetic. According to the committee, a local chapter bylaw, as amended in 1915, authorized the election of alumni who had ranked among the highest tenth of their classes. By 1926, a total of eighty had been "duly elected." It then seemed that, "in equity," the same honor ought to go to all the rest of those still living whose record entitled them to it. The recent fifty-four made "a notable array of names," the names of men who had "actually distinguished themselves." What constituted initiation for them was up to the Princeton chapter to decide. "And, after all, the local Chapter is the best judge of what keeps 'Phi Beta Kappa on a high plane.'"

Voorhees could do no more than send a fresh copy of the constitution and bylaws of the United Chapters and point out that the Princetonians had violated these in revising their own constitution and bylaws. "The Constitution was given to the chapter by the National Council and cannot be amended except by that body."[7]

The Johns Hopkins chapter was guilty of a less serious infraction of the rules. In 1928 its secretary reported the election of three faculty men whom he was listing as alumni, since all the faculty were considered "ipso facto" Johns Hopkins alumni, whether or not they were Johns Hopkins graduates. If the United Chapters should prefer to list the three men as honorary members, well and good, though they were hardly persons of "national prominence." "The practice of Johns Hopkins in the matter," the secretary concluded, "has gone on since 1895 unchallenged and though less ancient than that of Harvard and Yale has still some prescription of long use."

No longer did the practice go unchallenged. Voorhees promptly mailed a copy of the current Phi Beta Kappa manual, showing that the constitution provided for the election of only three categories of members: student ("in course"), alumni, and honorary. Unless professors were graduates of the university, they could not be called alumni members, and if they were "not sufficiently distinguished to be called Honorary Members," they should not be elected at all. Once elected, members of the three categories enjoyed equal rights and privileges in a particular chapter—whether as original or as "associate" members of it—and in the United Chapters. "Many of those elected to Honorary membership have given remarkably helpful service, and a few of them have been members of the Phi Beta Kappa Senate."[8]

While some chapters chose too many alumni and honorary members, the Oberlin chapter chose too few—too few to satisfy even Voorhees. In 1925 he hoped to get an election for William E. Barton, a graduate of Oberlin's theological seminary, a well-known author of books on Abraham Lincoln, and the father of the advertising genius Bruce Barton. The younger Barton, while helping with the sesquicentennial fund drive, intimated to Voorhees that he would like to see his father thus honored. So Voorhees, eager to please such an influential member of the society, appealed to Henry Churchill King, president of Oberlin College and supposedly a "close friend . . . for many years" of the elder Barton. But King could not sway the Oberlin chapter, if, indeed, he even tried to do so. The next year William E. Barton was elected to honorary membership at his son's alma mater, Amherst College. (It was two years after this when Voorhees relayed to the William and Mary president the charge of "bargain and sale" in the election of Otto H. Kahn.)

One of the Oberlin professors also thought the Oberlin chapter too stingy with honorary elections—at least in his own case. This man, of German extraction, had opposed the United States going to war against Germany in 1917. In 1932 he complained to Secretary Shimer that his colleagues, while electing others on the faculty with "key standing" but with diplomas from institutions lacking Phi Beta Kappa, had turned him down. "Now during and after the world war . . . they used election to the society to punish pacifists and non-conformists and reward the violent race haters." To Shimer's discreet inquiry, the chapter secretary responded that the chapter had voted in 1912 to make honorary elections "very conservatively" and, since then, had elected only a half-dozen faculty members, three of them graduates of Oberlin, the other three graduates of foreign universities, among whom he mentioned the distinguished Hungarian political scientist Oscar Jaszi. "How anybody can charge that he has been discriminated against . . . I cannot understand."[9]

Oberlin's conservative policy was in line with the United Chapters' long-standing objective—a careful regulation and a drastic reduction of alumni and honorary elections. In the 1920s and 1930s, the society made considerable progress toward achieving this. For the 1923 general catalog, Voorhees tabulated the memberships from 1776 to 1922 and found the totals to be the following: honorary, 2,081; alumni, 11,459; in course, 42,221. That is, the honorary and alumni together amounted to 26 percent of the aggregate. During the 1920s, these together came to less than 11 percent. "In the last triennium," Shimer reported in 1934, "an average of 4 alumni members and 1 honorary member per chapter was elected."

Shimer added: "The by-laws of 91 [chapters] do not fix a limit, 6 chapters specify a small number annually, 6 specify 2, 8 specify 3, and 5 specify from 5 to 10." The revised constitution of 1937 did not set a maximum but did provide that the number of honorary and alumni members should be "strictly limited by the chapter by-laws" and that the only persons eligible should be those who "by contributions in the fields of humane sciences and letters or by works of pure literature have given clear evidence of the possession of distinguished scholarly capacities."[10]

If it was hard to pick the most deserving alumni and honorary members, it was even harder to be sure of choosing the most meritorious members in course. All kinds of questions arose in regard to the election process. Some of these occurred to a William and Mary alumnus when in 1926 his chapter undertook to revise its procedures. He sent letters to more than a dozen chapters to inquire about their practices. In the letter to Wellesley College he asked:

> Is the selection by a committee or by the chapter at large? How many votes are required to blackball? Do you elect at the end of the senior year or earlier? Do you take into account grades of the entire college course, thus excluding all who start in the higher classes, or do you consider only the grades of the last year or two? Do you wait until the final examinations to make up your list, or do you act upon grades of and up to the preceding half session? If the latter, and a student who is reported favorably fails on her final tests, how do you handle such a case? What consideration do you give to other characteristics than mere scholarship, and how do you find out about such characteristics? What proportion or number of your graduating class do you elect to membership?

These questions might be reduced to the following: *Who* does the electing? *When* are elections held? *How many* students are elected? *What* are the criteria for election? This last question—bringing up the issue of grades alone versus grades and other standards—was the most troublesome.

According to the official policy of the United Chapters, as Voorhees expressed it in 1927, "the responsibility for the elections should rest with the members of the faculty who are members of Phi Beta Kappa." At some places, however, the responsibility rested largely with the undergraduate members. At Harvard, for example, the dean's office provided them with a list of the highest-ranking students, from which they were to choose about half. A 1933 report of the United Chapters revealed a variety of electors. "The electors are the faculty in 71 chapters, active members in 14, members in course in 13, a committee in 9, and all members in 8."

Undergraduates could elect or help elect new members only at colleges where elections were held before the end of the senior year. Voorhees thought elections should be held during the junior year for part of the quota and early in the senior year for at least some of the rest, as was done at Harvard. "The question of Junior membership has the authority of years behind it and the direct approval of the Council," he said in 1922. During the 1930s about half of the chapters did elect juniors, but as a rule only a very few of them, and the seniors were not initiated until quite near commencement time.

There was, meanwhile, a tendency for chapters to decrease the proportion of the class elected. In the revised United Chapters constitution of 1937, the maximum was reduced from 25 to 15 percent. Many of the chapters already limited the number to less than 15 percent in their bylaws and did not reach their own maximums in actual practice. But wide variations in bylaws and in practices continued.[11]

Though such inconsistencies gave rise to some concern, this was nothing in comparison with the controversy that raged over the selection of members on the basis of grades. The practice seemed deplorable to at least a few ΦBK faculty, such as the Johns Hopkins professor who in 1927 refused to contribute to the endowment fund because he did not believe in "rewards of merit" but only in "activities that make more keen the interest in the intellectual life for its own sake." When Voorhees asked him for suggestions, the professor elaborated:

> The intangible "prize" in the form of honor recognition of success in scholarship as indicated by grades seems to me on the whole undignified and unworthy and . . . it persuades a considerable number of students capable of better things to work in part at least with unworthy ideals, stimulated not wholly by genuine interest in the subjects studied but in part by selfish considerations of personal prestige. . . .
>
> I believe that Phi Beta Kappa should recast her methods and reform her ideals, getting away wholly from the "prize" idea and endeavoring rather to stimulate true scholarly interest. At present she is something of a menace to real scholarship. She should be a definite aid, standing for the real thing rather than the empty honor.
>
> I speak not without some first-hand knowledge for I have been one of the founders of three chapters, Johns Hopkins University, Goucher College, and Oberlin College and have attended one of the tri-annual conventions. I have watched with a good bit of interest the effect of Phi Beta Kappa in Goucher and in Oberlin. I think at Hopkins it has had very little influence. I have deplored some features of its influence in Goucher and at Oberlin.[12]

Taking much the same view, a few students declined election to the society. Two of them made news in the *New York Times* during the 1920s. The first Vassar student ever to reject the key did so because she never had "believed in rewarding high or low marks to students." The first at Dartmouth said he "did not consider it an honor." He explained that "the present system of marks in college does not show the true ability of a student."

The *New York Herald-Tribune*, in an editorial on February 22, 1928, commended the Dartmouth student. Seniors may well hold Phi Beta Kappa memberships in "contempt," the editorial writer said, since "boys become members or not automatically as a result of scholastic standing mathematically established by the faculty." Voorhees wrote a letter to the paper in reply. "There is no requirement that all those who receive high rank must be automatically elected," he declared. "Any chapter that applies the automatic rule is untrue to the tradition of the fraternity, for it existed more than half a century—nearly seventy-five years in fact—without ever taking marks into consideration." Indeed, there were no marks to be taken into consideration in those early years. "That Phi Beta Kappa elections are not held in 'contempt,'" Voorhees added, "is evident from the fact that, while over 2,500 memberships are offered each year, the number of declensions during my quarter-century of service average less than one a year." But he did not succeed in putting an end to the controversy.

Some Phi Beta Kappa leaders themselves were quite critical in the 1930s. The society has done nothing to "enrich the intellectual life of the college world," charged Edward Ellery, acting president of Union College and former president of the New York State Association of Phi Beta Kappa. It merely "seconds the motion" of professors who have graded the student, he said, and it makes no distinction between easy and difficult courses. He contended that the society ought to "blacklist" courses that "require little or no intellectual effort."

"About half of the chapters have insisted on grades as the sole basis, for, they say, any other criteria would be subjective and open to wire-pulling," Shimer informed the *New York Times* in 1934. "Other chapters, fortunately, are willing to take a certain amount of hot water in order to distinguish more accurately the truly liberal scholar from the grind, the faithful lesson learner, and the follower of snap courses." Some of these other chapters were basing elections not only on grades but also on "one or more of the following considerations: general promise, character of courses, cultural qualities, breadth of interest, extracurricular activities, leadership, personality, interest in the welfare of the university."[13]

Here and there a college or university tried to do something about the discrepancy in grading between individual professors and between entire departments. Oberlin experimented from 1930 to 1934 with a ranking system, which required easy as well as difficult graders to list their students in order from best to worst instead of giving most of them A's and B's (or D's and F's). The University of Chicago introduced comprehensive exams to be graded by readers other than the persons teaching the courses. Under the old system, the Chicago dean reported, "not a few" students were elected to Phi Beta Kappa because they had taken "pipe" courses. Under the new system the initiates proved to be "more worthy members."

Harvard had begun as early as 1916 to give special emphasis to independent work and to base honors at graduation not only on course grades but also on the results of broad written examinations, an oral exam, and an honors thesis. Until 1934, however, the Harvard chapter persisted in electing largely on the basis of course grades. As a result, some students were elected to Phi Beta Kappa but failed to win honors, while others graduated with honors but failed to make Phi Beta Kappa. The reform of 1934 increased the number elected at the end of the senior year (while decreasing the number elected earlier) and confined the choice to candidates for the degree *magna* or *summa cum laude*. Even more important, the new system required that in the earlier elections the course grades be supplemented by the following: tutors' reports regarding intellectual curiosity, originality, and scholarly promise; performance on divisional and comprehensive examinations; and data regarding prizes won and literary or other academic achievements of an extracurricular sort.[14]

Honors work at some colleges seemed to lead away from election to Phi Beta Kappa rather than toward it. Often the honors work was graded like any other course, and the grade was averaged with the rest of the grades. Since the honors work was more difficult, it tended to reduce the average and hence the prospects for a ΦBK key. At Dickinson College, for example, as an expert on honors programs observed in 1942, "some students avoid honors work because it inevitably lessens the amount of time they can spend on their regular courses and thus diminishes their chances of qualifying for Phi Beta Kappa."

At Yale a group of undergraduate members in 1936 tried unsuccessfully to bring about a reform comparable to the one at Harvard. They proposed to take into account not only grade averages but also such factors as "the difficulty of courses taken, the quality of preparation for college as judged by the preparatory school attended, and the improvement or decline in grades since freshman year." In support of this proposal, the editor of the *Yale News* wrote: "Every undergraduate knows that marks are really not much of a criterion either of conscientious scholarship or of intellectual ability." But opponents maintained that only an automatic and arbitrary system of election could be free from "politics," and the chapter overwhelmingly voted down the proposed change. "My own views are that it seems wise to stick to the scholarship marks as the only fair path to membership in the scholarship society," the dean of Yale College commented. "Over a long period the general scholarship averages do represent, we find, the value of the college work done by a student."

At the University of Pennsylvania, an "electoral board" interviewed candidates before certifying them for election. Shimer eagerly inquired about the Pennsylvania procedure, wondering whether the board took into consideration "other criteria" such as "breadth of interest." "If so," he wrote the chapter secretary, "it seems to me that such a method may avoid both the evils of depending solely upon grade averages and favoritism and prejudice which may influence decisions when nominations are debated in open Chapter meetings."[15]

Still, the dilemma persisted. On the one hand, it could certainly be unfair to depend on grades when there were gross discrepancies in grading. On the other hand, it could be at least equally unfair to discriminate on the basis of "personality," "leadership," or extracurricular activities, to say nothing of the prestige of the preparatory school attended.

There was not only the frequent charge that the society honored grade-grabbing more than true scholarship. "Another criticism of Phi Beta Kappa is to the effect that the society is dead," Shimer noted in 1934, "—that it elects the student and gives him a dinner and that that is the end of his active relation to the society." Sure enough, most of the chapters did little or nothing except choose members and put on an initiation ceremony, a dinner, and an address once a year. Shimer conceded: "The society's purpose, the encouragement of scholarship and cultural interests among college students and graduates, is effected mainly by the election of students to membership."[16]

A few of the chapters did more than that. For example, in the 1920s the Denison chapter accumulated a small fund for assisting needy Phi Beta Kappa seniors and for subsidizing the publication of professors' books. By 1940 the Berkeley chapter was sponsoring an honor students' association and operating an advisory bureau, which aided faculty advisers, urged freshmen to take courses other than "snaps," and supplied speakers for high-school assemblies. Meanwhile the Harvard chapter offered a tutoring service for deficient students and published the best honors theses of the graduating seniors.[17]

12

Great Men—or Grinds?

Presumably the 1920s were years of frivolity on the college campus—years of the coed flapper, the college boy with hip flask and raccoon coat, and the prevalence of a rah-rah spirit. The 1930s, with the onset of the Depression, took a more serious turn. Through good times and bad, Phi Beta Kappa more than held its own in competition with other campus attractions. It became a household word, the significance of which was a subject of fairly widespread discussion. Psychologists and others tried to analyze the ΦBK student, and they often concluded that he (or she, but she was commonly overlooked in this connection) had a much better than average chance for success in life. Still, the reputation of the society and its membership remained a matter of some dispute.

To some, the collegiate atmosphere seemed hostile to such an organization as Phi Beta Kappa. "College life," the president of Reed College lamented in 1917, was considered more important than books or classes. Fraternity leaders and football heroes were supposedly the most promising candidates for later success. Hence the "gentleman's grade," the C or its equivalent, was the most respectable, and the terms "grind" and "greasy grind" indicated the contempt of the "good fellow" for the serious student.

After the war, scholarship seemed to lose what little respectability it had left. New stadiums appeared, seating seventy thousand and more. When a committee of the American Association of University Professors condemned the commercialization of college athletics, the professors were merely showing their envy of coaches with superior earning power, the popular and patriotic magazine *Liberty* charged in its issue of July 17, 1926. "The problem is not the elimination or restriction of football," the magazine said, "but how long it will be before red-blooded colleges demand the elimination or restriction of those afflicted with this inferiority complex."

Considering the apparent *Zeitgeist,* one would expect the scholarly undergraduate to have been disparaged and discouraged. That is precisely what a journalist reported in 1925 after visiting nearly all of the best-known colleges in the country. "I know of no college where high scholarship in and of itself . . . commands great social prestige," he wrote. Most graduates who have made Phi Beta Kappa "look back upon their college life with memories of poverty [and] isolation."[1]

The writer might have drawn a different conclusion if he had actually interviewed a fair sampling of either graduates or undergraduates. Contemporary questionnaires given to seniors showed that Phi Beta Kappa consistently enjoyed a great deal of respect on the campus. "The varsity letter has come, in the development of athletics, to compete with the old Greek-lettered key," the *New York Times* observed in 1925. "It is significant, however, of the appeal that this scholarly symbol still makes that in Princeton University 150 seniors expressed a preference for the Phi Beta Kappa honor as against 114 for the varsity 'P.'" Year after year a majority of seniors rated the Phi Beta Kappa key as the highest college honor not only at Princeton but also at Yale, Rutgers, Amherst, Gettysburg, and other institutions where seniors customarily responded to the same kind of questionnaire.

"This is a far cry from the days when Phi Beta Kappa was scorned by the generality and accepted apologetically by the recipients," the *New York Times* editorialized in 1928. There seemed to be a "changing intellectual atmosphere in the colleges," and the editor wondered why. "Whatever the reasons, the fact would seem to be established that A's and B's in class are no longer regarded as unseemly in a good all-around college man; the 'gentleman's C' has lost prestige." But there was no real evidence that students were changing their attitude toward scholarship or toward Phi Beta Kappa. Even before the war, a majority of the seniors at Yale and elsewhere had voted for the key as the most coveted college honor.[2]

In any event, it was fallacious to suppose that the ablest students were typically friendless, withdrawn, utterly preoccupied with their books. The prospective ΦBK winners participated as much as others—and more than most—in extracurricular activities. A Wellesley senior, elected in 1926, believed that her fellow members had come from the two extremes: the "grinds" on the one hand and the campus leaders on the other. But that was just one student's opinion. At Vassar about half of the students, under the supervision of a student committee, kept a record of all the time they spent on studies, activities, and exercise. The results, published in 1927, showed that the students elected to Phi Beta Kappa were somewhat more active than the average.[3]

A Colgate professor made a study with far more persuasive conclusions. which he published in 1929. He looked into the campus activities of the 416 men who had graduated from Colgate as Phi Beta Kappa members during the previous thirty years. Of the 416, he learned, fully 68 had won a letter in one or more varsity sports (not counting the men who had been on the varsity squad but failed to make a letter or those who had taken part only in intramural competition). That made a ratio of one to six, a higher ratio than that of Phi Beta Kappa members to the entire number of graduates. "Colgate's first intercollegiate football team back in the 90's had eight members who were elected to Phi Beta Kappa." Of the total of 416 ΦBK men, 143 or one third "won positions of great importance in the student body." More than half of the presidents of the students' association were from among these men. More than two thirds of the editors in chief of the campus newspaper—twenty-two of the thirty—had been elected to the Colgate chapter.

As United Chapters Secretary Oscar M. Voorhees pointed out, Phi Beta Kappa and football had been together from the very beginning. When Rutgers and Princeton started the game in 1869, four members of the Rutgers team were also members of the Rutgers chapter (Princeton did not yet have a chapter). When Rutgers played Columbia in 1870, there were six Phi Betes on the Rutgers team and at least one and possibly more on the Columbia team. Voorhees, while boasting of the society's athletic record, sympathized with the protest that some academics in the 1920s were making against the "tyranny of football." He thought "four public games a season should be the limit."

The *Key Reporter* called attention to the fact that during the 1930s the society could claim as members at least two major league baseball players, the second baseman of the New York Giants and the catcher of the Boston Red Sox. But the *Key Reporter* offered its prime example of the symbiosis of athletics and academics when it published a picture of Byron R. ("Whizzer") White, Phi Beta Kappa, University of Colorado, 1938. As the caption noted, White had been an all-conference selection in football, basketball, and baseball; had been the leading national scorer as a halfback; and had won a Rhodes Scholarship. He was deferring his appearance at Oxford so that he could play a season of professional football. (White was to receive an appointment to the Supreme Court in 1962.)

Another famous athlete of the 1930s provided less favorable advertising for Phi Beta Kappa. This was Helen Wills (Moody), then considered the greatest woman tennis player who had ever lived, an eight-time winner at Wimbledon. A graduate of the University of California, Berkeley, Wills publicly confessed to having gone "gunning cold-bloodedly" for a ΦBK key. "I had an almost complete lack of interest in learning for the sake of knowing something," she wrote in the July 1936 issue of *Scribner's Magazine.* "I was, in the truest sense of the word, a cup hunter in the field of scholarship." The *Key Reporter* quoted her confession without comment.[4]

Despite the athletic fame of a few Phi Betes, the stereotype of the "grind" persisted. "The common belief that Phi Beta Kappa men in our colleges are generally social bores has no basis in fact," a psychologist at Colgate announced in 1935 after a year-long investigation. He had asked students to confide which of their classmates they liked and which they disliked the most. From the responses he concluded that there was no correlation between a student's popularity and his grades. Still, the psychologist's composite picture of the Phi Beta Kappa men hardly made them look like the life of the party. "The studious dread parties as a rule, he says, consider flirting and betting wrong, have absolutely no desire to indulge in drinking, and have worried about their souls," the *New York Times* reported. "They are inclined to be intimate with but a few friends and are apparently less interested in the opposite sex than their less studious classmates." Perhaps some of the Phi Betes were pulling the psychologist's leg.

A professor at the College of the City of New York also compared Phi Beta Kappa students with others. He found, in 1936, that at CCNY there was "no difference between the groups in introversion-extroversion or in inferiority feeling." On the average, however, the ΦBK man was younger, taller, healthier, and "better developed physically." The professor hoped that his study would "dispel some of the popular erroneous beliefs concerning ΦBK men" and would lead to other studies that would permit more widely based generalizations.

Other studies tended to suggest that Phi Beta Kappa students were more intelligent (as measured by standard "intelligence" tests) than their fellow students. They were almost uniformly good in all their courses; indeed, they would have to be in order to be elected. A student would find himself out of the running, no matter how brilliant he might be, if his brilliance shone in but a single field. The Phi Betes, during those Depression years, were somewhat more likely than others to be working their way through college.

While the CCNY professor thought members of the society were especially healthy, the famous surgeon George W. Crile considered most of them sick. "So high is the scholastic record among patients with hyperthyroidism and so many individuals of Phi Beta Kappa are to be found among them, that although hyperthyroidism may appear years after graduation, in a certain sense we may say that even Phi Beta Kappa is a disease." So wrote Dr. Crile in *The Phenomena of Life* (1936). "Certainly there is no record of an individual with myxedema (hypothyroidism, or sub-activity of the gland) attaining Phi Beta Kappa rank."[5]

If ΦBK students had been mostly bookworms and drudges, they presumably would have had a rather poor chance of succeeding in later life. Actually, members of the society achieved success at a greater rate than nonmembers during the early twentieth century. So, at least, its defenders maintained, but critics disagreed. One's view depended largely on one's definition of success.

The most convincing evidence of high achievement came from the pages of *Who's Who in America,* the first edition of which appeared in 1900, the same year that the first general catalog of Phi Beta Kappa members appeared. Soon educators here and there got the idea of comparing the listings in the two books. A University of Illinois professor looked into *Who's Who* for graduates of institutions that had had a chapter for at least twenty years. There he found, as he reported in *Popular Science Monthly* for March 1903, fully 5.9 percent of the ΦBK graduates but only 2.1 percent of all the graduates, high, medium, and low. "Our conclusion must be that the Phi Beta Kappa man's chances of success are nearly three times those of his classmates as a whole."

A Wesleyan University professor later discovered a somewhat greater statistical disparity between members and nonmembers among Wesleyan alumni in *Who's Who.* As William T. Foster summarized the findings in his book *Should Students Study?* (1917), the Wesleyan professor "concludes that of the highest-honor graduates (the two or three leading scholars of each class) one out of two will become distinguished; of Phi Beta Kappa men, one out of three; of the rest, one out of ten." Foster also cited similar studies that led to similar results at other colleges.

Each new edition of *Who's Who* seemed to reconfirm the point. One investigator discovered that of the 26,915 persons included in the 1927 edition, nearly 6,000 were members of Phi Beta Kappa. In the *North American Review* he gave national publicity to his conclusion: " . . . the total membership of Phi Beta Kappa is but one in three thousand of our population—that is, three one-hundredths of one percent—but that numerically insignificant minority has furnished many times, probably one hundred times, its quota of our men of fame, of our truly successful men, our leaders of the State, of the bench and the bar, of art and letters, of scientific achievement, of civic affairs in general."

Not all commentators were equally impressed. "Although students of scholarly attainment are likely to become professors and professors have more than an ordinary chance of being entered in 'Who's Who,'" the *New York Times* observed in 1926, "these records are taken to show that there is a connection between scholarship and later distinction." No doubt the volumes did give disproportionately large space to educators, clergymen, social workers, and the like—and disproportionately little to businessmen or to celebrities in sports, entertainment, or the arts.

Such considerations did not deter the United Chapters from endorsing the claim so frequently made. "In general . . . the measure of one's attainment in school and college is the measure of one's success in life," an official publication of the society asserted in 1934. "Perhaps the most interesting confirmation is to be found in a comparative study of ΦBK records and *Who's Who.*"[6]

There were other ways of estimating achievement. One way was to get the opinions of classmates or other contemporaries. Researching in 1908 and 1909 for a series of articles to appear in *The Independent*, a journalist undertook to compare the careers of "society men"—those belonging to the "senior societies" known as Wolf's Head, Scroll and Key, and Skull and Bones, none of which based its selections on scholarship—with the careers of Phi Beta Kappa men at Yale. "I asked seven Yale graduates in classes from 1872 to 1896 to mark in the directory of graduates the names of their classmates who had in some way distinguished themselves since graduation," he reported. His conclusions: "the senior societies and the Phi Beta Kappa, though their standards of judgment are different, are equally successful in picking out the men of superior ability," and "a student belonging to either of these groups has twice the chance of future prominence as one belonging to neither."

In preparing for his 1917 book, *Should Students Study?*, William T. Foster asked three men to select the Harvard graduates of the class of 1894 whom they considered outstandingly successful. They chose twenty-three. Foster then compared the undergraduate records of these twenty-three with the records of twenty-three others he picked out at random. The "successful" group had garnered a total of 196 A's; the others, only 56.

An Indiana University professor took a different approach. He looked into *American Men of Science* for 1921 to find the thousand scientists whose names were starred because their fellow scientists had commended them for especially meritorious work. From the 1923 Phi Beta Kappa catalog, he ascertained that more than half of the "starred scientists"—518 of the 1,000—were members of the society. Later he checked the roll of the National Academy of Science, listing the most outstanding scientists, and found Phi Beta Kappa members even more conspicuously represented. They had gained scientific recognition, he said, despite the fact that specialization, so essential to such recognition, seemed incompatible with "general excellence in all courses taken in college." All this, he concluded, provided "one more bit of evidence of the great value to society of the comparatively small proportion of students who win the golden Key." But only 219 of his 518 key holders among the starred scientists had won it as undergraduates; all the rest were elected as alumni or honorary members.[7]

Whether or not the Phi Beta Kappa people were of unusual value to society, they appeared to be of less than average value in the marketplace. Such, at least, was the conclusion that a psychology instructor at the University of Michigan reached after collecting information on the earnings of the Michigan class of 1912 during the decade following graduation. "College 'Mixers' Earn Most Afterward," the *New York Times* headlined in 1923; "Phi Beta Kappa Men Prosper Least Pecuniarily." The psychologist's figures indicated that those most prominent socially as students and most active in student affairs were averaging more than $10,000 a year; the "M" athletes, $6,400; the ordinary students, $5,800;

and the Phi Beta Kappa members, only $3,000! "The statistics do not prove that the men of the high grades are incapable of earning big money," the psychologist explained. "They merely show that they do not prefer those lines of work which are most lucrative."

The *New York Times* questioned the validity of the Michigan instructor's findings. He had collected income data on only 98 of the 257 men graduating in 1912; his results might have been different if had included all the men of a number of classes and colleges. "He seems to have ignored the possibility that the boys who belonged to the fraternities largely devoted to social activities were those who, after graduation, commanded the influence and help of fathers who could put them in positions where advancement was fast," the *Times* added. "And, anyway, earning capacity . . . is neither the only nor the best test of success."

In reality, the earning capacity of the scholarly types was not so bad, according to Walter S. Gifford, president of the American Telephone and Telegraph Company. Writing in the May 1928 issue of *Harper's Magazine*, Gifford referred approvingly to the frequent appearance of Phi Beta Kappa members in *Who's Who*. "This does not necessarily bear directly upon the relationship between scholarship and business," he pointed out, "because *Who's Who* is not intended as a guide to business distinction." He thought he could show the relationship between scholarship and business, however, by analyzing the salaries of AT&T employees who had been out of college for five years or more. He divided the employees into thirds according to their college grades. "In general," he summarized, "men in the first third of their college classes are more likely to be found in the highest third in their group in salary and those in the lowest third in scholarship to be in the lowest third of salary."

Further evidence of a correlation between scholarship and earning power appeared in John R. Tunis's *Was College Worth While?* (1936). Tunis sampled four groups from the Harvard class of 1911: athletes, Phi Betes, "clubmen," and "average unheard-of men." He concluded: " . . . financially the Phi Beta Kappas lead the field." In the *Key Reporter* a reviewer judged that the study was "statistically flawed," for Tunis had taken twenty-five examples of each category—twenty-five of twenty-seven Phi Beta Kappa members but twenty-five of much larger numbers of each of the other categories. The critic retained his preexisting belief that "membership in the Society correlates fairly well with a preference for a high 'psychic income' rather than a swollen pay-check from a dull or dreary calling."[8]

Whether true or not, it was widely presumed that most Phi Beta Kappa members, no matter how high their intelligence, occupied a rather low position on the economic scale. This supposed fact led pundits to raise a question of eugenics, the pseudoscience that began to attract a good deal of attention soon after World War I. Eugenicists advocated the selective breeding of human beings to produce a superior people. Frank

Pierrepont Graves, president of the University of the State of New York and state commissioner of education, brought up the subject in a Phi Beta Kappa address of 1925. Phi Beta Kappa, Graves said, ought to be "an aristocracy of service rather than an aristocracy of mere brains." There was need for a new kind of test to supplement the intelligence test and measure the moral and spiritual qualities essential to leadership in service, he went on. "Thus far his ideas were well within the horizon of his hearers," the *New York Times* commented.

> But when he proceeded to recommend eugenics as a means of recruiting leadership in service, he may well have given Phi Beta Kappa men what is sometimes called a pause.
>
> Not many of them are likely to doubt that the mental and moral qualities of such leadership are hereditary [since Charles Darwin and Francis Galton have provided evidence that] qualities of leadership are derived in much the same manner as the points of prize stock. Statistics compiled from "Who's Who" give similar results. The army mental tests showed that intelligence quotients in different occupational groups correspond with a precision truly amazing to the mental, and largely even to the moral, qualities required. A subsequent study showed that the intelligence quotient of the children of professional men was 125, of the children of semi-professional and "higher business" men 118, of the children of skilled laborers 107 and of the children of semi-skilled and unskilled laborers 92. Another study showed that high intelligence is "approximately five times as frequent" among children of men in the higher occupations. Members of Phi Beta Kappa can scarcely doubt that, biologically, they are well qualified to assist in the cause of eugenics. What gives them pause is the fact that the problem is not primarily biologic but economic.
>
> With the best will in the world and the best germ cells, they simply cannot afford a family, at least until somewhat too late. If our universities were to be recruited only from the children of graduates and were to get all of them, they would shrink by one-half in each generation. In point of fact they are expanding hugely—which is doubtless why they have, as Dr. Graves says, "too much poor student material." It is not to be doubted that children of high grade intelligence are to be found in the under strata of our life. But when they become members of Phi Beta Kappa they fall inexorably under the working of economic law. To put it briefly, they are obliged to bring up children, if any, on the pay of not even skilled but of unskilled laborers. How long should it take to extinguish all of the more precious biologic strains in a nation?[9]

The "precious biologic strains" of the Phi Beta Kappa stock were not, in fact, about to die out. A large proportion of the society's members did, however, continue to go into school- and college-teaching and other comparatively low-paying "service" occupations. Between the 1840s and the 1890s, the proportion becoming teachers had increased from less than 10 to more than 25 percent, according to a study by the pioneer educational psychologist Edward L. Thorndike. Between the 1900s and the 1930s, the proportion apparently rose still higher. In 1924 the president of the United Chapters, Charles F. Thwing, found most of the hundreds of members living abroad to be engaged in educational or missionary work. His totals were 250 in China, 82 in Japan, 75 in India, 50 in South America, 20 in Africa, 100 in England, 50 in France, 20 in Italy, 15 in Germany, 14 in Turkey, and 21 in Mexico. He saw these people as serving the cause of civilization and world peace.[10]

Though the majority of members may have lived at a rather modest economic level, a surprisingly large number of them attained positions of importance, power, and fame. That, at least, was the message that Secretary Voorhees kept reiterating in the *Phi Beta Kappa Key* during his fund drives of the 1920s. All of the first six Americans to win the Nobel Prize were "wearers of the golden key," he pointed out—so were all four members of the U.S. delegation to the Washington Conference of 1921–1922; twenty-eight of the sixty-three men and women selected by 1923 for New York University's Hall of Fame; eighteen of forty-four secretaries of state; twenty-six of sixty-three justices of the Supreme Court; and with the accession of Calvin Coolidge, eleven of twenty-five presidents of the United States.

But Coolidge did not receive his golden key until after his inauguration as vice president, when he became an honorary member. Quite a few of the other notable personages on Voorhees's lists also held honorary or alumni memberships. This was true of many subsequently prominent public figures as well. Though Theodore Roosevelt earned his key as a Harvard undergraduate, Franklin D. Roosevelt did not. While F. D. R. was president, he related a story to Representative Lester Hill of Alabama, a fellow ΦBK member, and Hill repeated it as follows:

> He [Roosevelt] said that at a certain conference of governors, while he was Governor of New York, he and two other governors were seated in a friendly chat, and one of them made mention of the fact that each of the three governors was wearing a Phi Beta Kappa key. The President said that he immediately remarked that his key was honorary and not awarded for academic attainments. Thereupon, each of the other governors confessed that his key was also honorary, and there was a laugh all around.

According to the *Key Reporter*, which printed this story, it raised the question "why honorary membership should be regarded as less an honor than membership attained as a student." The answer would seem to be that, indeed, many of the honorary members were among the most honorable in the society, but they generally had been elected in recognition of their achievements after graduation. To include them in a list of the society's notables, if the list were intended to illustrate the Phi Beta Kappa student's prospects for success, would look like circular reasoning of a kind unworthy of a Phi Beta Kappa intellect.[11]

Since Phi Beta Kappa proudly re-
flected the glory of its most famous members, the society could hardly
avoid the shame of the most notorious ones. For more than a century,
there had been no disgrace comparable to that brought upon the
Dartmouth chapter by John Henry, whom the chapter expelled after he
turned out to be an unprincipled double agent at the time of the War of
1812. But in the 1920s Nathan F. Leopold, Jr., a member of another
chapter, was to be convicted of a far more heinous crime.

Some people thought James Branch Cabell a big enough embarrassment to the society. Cabell was the author of *Jurgen* (1919), the story of a
young man who traveled through "the heaven of his grandmother and
the hell of his fathers" and other fantastic regions. The book contained
no dirty words, no descriptions of sex acts, and nothing whatever that
would have interested a censor in the 1980s, yet the New York Anti-Vice
Society tried to suppress it in 1920. The next year, the William and Mary
chapter elected Cabell as an honorary member. A nonmember in Washington, D.C., protested to the United Chapters against the election of the
man who had written "the most indecent book ever published in this
country." (A Cabell had belonged to the original society at William and
Mary, and the Cabell family had preserved its records for a couple of
generations after the dissolution of that society.)[12]

Incomparably more serious was the case of Nathan F. Leopold, Jr. He
was elected to the University of Chicago chapter on March 20, 1923. On
May 21, 1924, he and Richard Loeb, both of them then nineteen-year-
old graduate students at the university, kidnaped and killed thirteen-
year-old Bobby Franks.

"The newspapers have given a great deal of publicity to the fact that
two young men who committed an atrocious murder in Chicago are
members of the Phi Beta Kappa," a 1911 Vanderbilt graduate wrote to
Voorhees. "I am impelled to inquire whether there are any rules or
regulations which might be used to purge our fraternity of two members
who have brought discredit upon everyone and everything with which
they are connected."

Voorhees replied that there was no record of Loeb's having been
elected (at Michigan, where he had gone as an undergraduate) but that
Leopold was indeed a member of the Chicago chapter. "The constitution of that chapter provides that good moral character is a qualification
of membership, and that any member having lost this qualification can
be expelled by a ⅘ vote at a regular annual meeting of the chapter."
Voorhees wrote to the chapter's secretary to urge prompt action.

Many years later, in 1961, after Leopold's release from prison, where
he had been sentenced to a life term, he ordered a Phi Beta Kappa key
"to replace one which he gave his girl friend in happier days," as the
chapter secretary informed the United Chapters office. "As it is reported

to me, he wishes to obtain a replacement to present to his wife." The secretary found that Leopold's membership card had been marked "drop" and had been put in a file of deceased members. But there was no record of the chapter's having taken official action on the case.

The advice from the United Chapters was that its files also lacked any record of Leopold's expulsion, that keys should be worn only by members and not by wives or girl friends, that it would be unwise "to drop him by special action of the chapter at this late date," and that "chapters do not always take disciplinary action. Alger Hiss, for example, is still a member of Phi Beta Kappa." That was the Alger Hiss who, while a State Department official between 1936 and 1947, had supplied the Communist courier Whitaker Chambers with state secrets for the Soviet Union—according to Chambers's allegation. Hiss denied the charge, was convicted of perjury for having done so, and was sentenced in 1950 to five years in prison.[13]

$R$are was the instance of conspicuous culpability on the part of a member, and during the 1930s both the self-image and the reputation of the society remained enviable on the whole.

The Princeton chapter boasted in 1930 that the town of Princeton had "the most intelligent population of any city in the United States," since it contained 238 ΦBK members. This gave it a ratio of 1 member to every 42 inhabitants—comparably, 1 to 96 in Arbor, Michigan; 1 to 107 in Ithaca, New York; 1 to 211 in Cambridge, Massachusetts; 1 to 327 in New Haven, Connecticut; 1 to 1,390 in Chicago; 1 to 2,700 in San Francisco; 1 to 3,400 in New Orleans; 1 to 3,600 in New York City; and 1 to 4,700 in Los Angeles. The Oberlin chapter retorted that Oberlin must be an even more intelligent place than Princeton. "The population of Oberlin is 4,275, and the number of members of Phi Beta Kappa, exclusive of transitory student population, is 102, making an average of 1 to 41."

The society received favorable publicity when ninety professors took part in a "mental contest" at the 1931 meeting of the American Psychological Association in Toronto. The Phi Beta Kappa men "made consistently higher scores" than the rest. "In what might be called an intellectual 'world series,'" the *New York Times* reported, "the Phi Beta Kappa team has won the pennant."[14]

With the worsening of the Depression, however, even some members began to wonder whether the society was meeting the needs of the times. In response to an appeal for funds, a retired Teachers College professor complained in *School and Society* in 1934: "The Phi Beta Kappa Foundation could make no greater contribution to culture just now than by admitting that its past glorification of the scholar has contained within itself an unmentioned sanction for social relations, even economic relations, that now are proved to be unsound." By failing to question their own assumptions, scholars were permitting their scholarship to "become automatically an adjunct of the social *status quo*." In sum, "the educational philosophy of this honor society is already senescent."

Replying in the same journal, Secretary William A. Shimer of the United Chapters said the critic seemed to think that "Phi Beta Kappa should summon college teachers to some more or less particular social goal or economic theory, such as Socialism or Fascism." Shimer himself believed that "scholarship for its own sake, so-called, would seem to have more justification than art for art's sake." He reasserted the policy that the critic had condemned as outdated and inadequate: "Phi Beta Kappa gives recognition to scholars. Phi Beta Kappa does not give recognition to economic and social philosophies."[15]

To some extent, newspaper comments no doubt reflected public opinion in regard to the society and its members. From time to time in the late 1930s, the *Key Reporter* reprinted such comments, favorable and not so favorable, under the heading "As Others See Us." Here is a sampling:

Some of the most unassuming and kindest people I have ever seen were Phi Beta Kappas.—Tyler (Tex.) *Courier-Times.*

The highest ideals of scholarship, leadership, character and service are embodied in this society whose name is known in school circles all over the nation.—Wilmington (N.C.) *Star News.*

Phi Beta Kappa . . . whose famous "key" emblem stands the continent over for excellence in study and depth of knowledge.—Manchester (N.H.) *Union.*

A Phi Beta Kappa key is a good thing to have for a souvenir, but if the truth were known there are a good many people who never saw the inside of a college who know more than some of the men and women who have won this symbol.—Bridgeton (Me.) *News.*[16]

13

Sisters in
the Brotherhood

During World War I, the movement
for woman suffrage gained momentum as more and more states changed
their voting laws. Then in 1920 the country adopted both the Eighteenth
Amendment, which prohibited the sale of intoxicating beverages, and the
Nineteenth, which forbade denying the right to vote on account of sex.
These amendments were called "twin victories for feminism," and American women, "already emancipated in everything save politics," were now
said to be emancipated in that regard also. At the very time they were
overcoming old obstacles to equality in politics, however, women were
encountering the threat of new obstacles to equality in Phi Beta Kappa.
Though by 1920 they had escaped the danger of discriminatory action by
the United Chapters, they could not yet compete freely with men for
membership in all of the individual chapters, not even in all of those at
coeducational institutions. Many women remained less than first-class
citizens in the society despite impressive gains that some women made in it
during the next two decades.

By the time of World War I, women
were becoming a menace to Phi Beta Kappa, or so its leaders had begun
to fear. They anxiously discussed what to do about the prospect of the
society's being overrun by hordes of female initiates. Such a possibility
had, of course, been far from the minds of the brethren a century
earlier, when women could not even hope to enter college, to say nothing of aspiring to a ΦBK key. A newspaper editor expressed the prevailing male attitude when he declared in 1828 that the "most acceptable
degree" for a young lady was the "degree of M. R. S."
In that earlier time the Harvard chapter did welcome women to the

audience at the annual oration, apparently considering them arbiters of culture though incapable of scholarship. "The number of the literati and of our fair countrywomen who attend," the *Columbian Centinel* of Boston opined in 1817, "is a flattering proof of the importance of the society and of the refinement of our age." As for the women, they year after year looked eagerly for seats in the crowded Cambridge church where the orator held forth. In 1829 the daughter of Harvard's president recorded that she and other women were waiting at the door to rush in as soon as it was unbolted. "It was really surprising," she wrote, "to see the ladies leap over the tops of the pews."[1]

When the first college classes met at Oberlin in 1834, women were among the students, but they were not admitted to the regular collegiate program until 1837 and did not receive their first B.A. degrees until 1841. One of the female graduates of that year, Caroline Mary Rudd, was elected to Phi Beta Kappa when the Oberlin chapter was organized in 1907, and so was a female graduate of 1844, Emily Francis Fairchild. These two were the earliest women students ever to be elected, though not the first to be elected in course.

The total of women students gradually increased as other denominational colleges and the new state universities of the West followed Oberlin's example, some of the existing Eastern institutions also became coeducational, and separate or coordinate colleges for women were founded. This trend was accompanied by apprehensions that higher learning might prove too rigorous for the weaker sex. In 1883 the president of the University of Michigan, James B. Angell, assured the English social worker Emily Faithfull that, in her words, "none of the ladies had found the curriculum too heavy for their physical endurance." During the academic year 1885–1886, the University of Michigan enrolled 135 women and 461 men. At the same time, Wellesley College had 520 students, all of them women (along with 75 faculty members, 61 of them women). According to federal statistics, the number of female recipients of the bachelor's degree or first professional degree rose from 2,485 in 1880, to 5,237 in 1900, to 16,642 in 1920, and to 76,954 in 1940. Meanwhile, of the total of degree recipients, male and female, the percentage of women remained about the same in 1900 as in 1880, at slightly less than 20 percent, but rose to almost 35 percent in 1920 and to more than 40 percent in 1940.[2]

Before long, the proportion of women students elected to Phi Beta Kappa began to increase much more rapidly than that, but at first, the elections were quite slow, running into considerable resistance. Chapters in men's colleges, such as Kenyon, long objected to the election of women even in coeducational colleges. The Vermont chapter took a bold step when in 1875 it resolved that "all the graduates of this University should be eligible to membership without distinction of sex" and again when the next day it proceeded to elect Ellen Eliza Hamilton and Lida Mason, the first women anywhere to be elected in course. Even so, there was some-

thing a bit grudging about the way the chapter initiated the two "ladies"—separately from the two men who were elected at the same time.
(Some years later Ellen Eliza Hamilton married one of the two men.)

The next year at Wesleyan University the chapter approved by a
majority but not a unanimous vote the committee report that women
were "eligible to membership on the same condition as men." Though
the Cornell chapter admitted women from its founding in 1882, it did so
only after overcoming the objections of some of its male founders. One
of them complained that "rather hastily it was voted to accept the precedent [of Vermont] and let the fair creatures in. Personally I have doubts
about the wisdom of such a course. It seems to me in the first place
absurd to admit women to a *Fraternity*, and, secondly, that the whole
tradition and character of the concern make it exclusively a male affair."
It would seem, indeed, that the society was a brotherhood by definition
as well as by tradition!

The first woman's college to apply for a charter was Barnard, which
Columbia University had established as a coordinate college to provide
courses for women without admitting them to out-and-out coeducation.
In 1895 the Senate and the National Council rejected the Barnard application "on the ground that granting it would be in effect the erection
of two chapters in one institution." The first woman's college to receive a
charter was Vassar, whose chapter organized in 1899.[3]

Already the chapter at Brown—which, like Columbia, had set up an
affiliated women's college (Pembroke)—was discussing the problem of
the "deserving sisters" on the campus. After two years of deliberation,
the brothers at Brown hit upon an ingenious solution. In 1900 they
created a women's section of the existing chapter, with the proviso that
"membership therein should in no wise diminish the number of men
eligible for election," that the women elected "must manifest a scholarship not inferior to the lowest among the men," and that "the number
of women elected from a given class should bear no larger proportion to
the whole number of women in that class" than was the case with the
men elected. The Brown chapter took this action without obtaining a
separate charter for the women's section. A year later the United Chapters followed the Brown precedent to the extent of granting a charter
for a similar section at Columbia to accommodate Barnard students.

Nevertheless, when Radcliffe applied for a charter in 1910, the Senate
objected that Radcliffe did not need one, since the Harvard chapter
could elect students from the related college for women. But the Harvardians refused to do so or to establish a women's section, insisting that
Radcliffe was not an integral part of the university. The brothers requested a separate charter for Radcliffe. At the next Council in 1913, the
question provoked a discussion that ended only when a facsimile of the
Massachusetts Alpha charter of 1779 was produced to show that the
members were to be "gentlemen." The Council then voted for a separate
Radcliffe chapter, and it was organized in 1914. By 1922, a total of

eleven women's colleges had either a section of a chapter or a chapter of their own.[4]

At the same time the United Chapters was launched in 1883, there were only twenty-six women members of the society, twelve of them elected at Vermont, eight at Wesleyan, and five at Cornell. By 1900 the count had increased to 532, of whom 179 had been elected at Vassar (mostly alumnae, the chapter being only a year old), 51 at Cornell, and 353 at nineteen other chapters. Then, in the single year 1913, more women were elected as undergraduates than the entire number of women elected up to 1900. The tally for 1913 (with all but one chapter reporting) was 563 women to 567 men.

Those were the figures for members elected in course. If honorary and alumni members also were counted, the male majority would be considerably larger for the time being, but it would go on shrinking. The elections in 1914 were distributed as follows: at twenty-six men's colleges, 544; at six women's colleges, 177; at fifty-one coeducational colleges, 1,373, of whom 770 were women and only 603 were men. Several of the coeducational chapters had been reduced to electing very few men in course.[5]

This feminizing trend interested the author of *Great American Universities* (1910), who observed:

> In the University of Minnesota this year [1908–9] the Phi Beta Kappa, choosing students from the Senior class for their literary proficiency, elected thirteen women and four men. The Sigma Xi, basing its selection on the ability to carry on scientific research, elected thirty-one men and four women. In the College of Science, Literature, and the Arts, the college proper according to the old idea, the women outnumber the men two to one. In the higher classes and in the purely literary courses, the proportion of women is still greater and tends to increase. There are now over 1400 women in the University of Minnesota, a larger number than in any women's college in the United States.

Apparently literature and Phi Beta Kappa were for women, science and Sigma Xi for men—at least at Minnesota. And after visiting the University of Michigan, the same author remarked: "In coeducational institutions, if it [ΦBK] is awarded strictly according to grades, it is apt to be monopolized by the more diligent sex and goes by the name of 'the woman's club.'"[6]

The woman's club! Phi Beta Kappa appeared to be turning effeminate. This was the crisis that, in the view of some of its leaders, the society faced by the time its Twelfth National Council met in 1916.

R eporting to the assembled delegates in 1916, United Chapters Secretary Oscar M. Voorhees presented statistical tables to show the rapid increase in the number of undergraduate women being elected. Then he said:

> I speak of this now . . . to raise the question as to whether some of our chapters may not fail of the highest reputation by reason of the preponderance of women. I am of the opinion that this condition results from the fact that a larger proportion of women than men, owing to the men's more varied interests, take the courses now accepted as leading to eligibility. The situation warrants careful study. I raise the question whether election on the basis of grades only has not been pushed to the extreme. We need to be reminded that Phi Beta Kappa attained its first extended reputation through men elected on another basis.
>
> The reputation of Phi Beta Kappa to-day rests in part upon the distinctions won and the worthy service rendered by the members who graduated from ten to fifty years ago. Its reputation in the years to come will rest in part upon the work of those whom we are now electing to membership. It will be generally conceded, I think, that a larger share of its reputation must come from its men than from its women members. The way must be kept open to the men, and promising students be encouraged to seek membership. In my opinion there are other marks in undergraduates of the promise of usefulness than mere grades. Is it too much to ask that our chapter officers give this matter careful consideration?

The delegates gave the matter enough consideration to refer it to a Committee on the Proportionate Election of Men and Women, with instructions to report at the next triennial Council in 1919.

The secretary's remarks and the Council's action provoked some rather sharp comment. The *New York Evening Post* could hardly believe that "some means of limiting the proportion of women" was actually being considered, even though women were being elected in rapidly increasing numbers. "The secretary suggests that this situation should be altered by taking other things than scholastic distinction into consideration," the *Evening Post* noted. "What these other things should be he does not specify, but . . . they must be male characteristics." The paper went on to argue against any such policy.

A more devastating response came from Leta S. Hollingworth, a thirty-year-old specialist in child psychology who had just joined the faculty of Teachers College, Columbia University. Hollingworth wrote in *School and Society:*

> That a position so untenable and action so invidious should be taken by the representatives of a body supposedly recruited on the basis of superior intelligence, offers food for ironical reflection. The irony grows when one remembers that less than half a century ago one of the most insistent arguments against admitting women to higher education was that standards of scholarship would be lowered through the inability of the women to keep pace with the men!

Women, Hollingworth continued, do less well after graduation because they have to raise children. They often receive less encouragement while in college because their professors expect them to get married and not have professional careers. "The remarks of our secretary might lead to a discussion of the whole present-day condition of the higher education of women." They will continue to fall short of the highest attainment until they are able to "bring about a change in the attitude of the academic world, as exemplified in our secretary. This women can do by modifying social customs gradually, so that they may both fulfill their intellectual promise and enjoy a normal domestic life, as men have always done."[7]

Such critical comment did not faze the secretary, but he kept on worrying about the effeminization of the society. At the next meeting of the Council in 1919, he had even more alarming statistics to report than before. In 1918 the line had finally been crossed: more women than men had been elected in course—884 to 701. But the committee that had been appointed to consider the matter was less concerned than the secretary.

The chairman (not "chairperson") of the committee was Mary Emma Woolley, president of Mount Holyoke College. Woolley was the first woman to be graduated from Brown University and, elected in 1907, also the first women to serve as a Phi Beta Kappa senator. Though active in the cause of woman suffrage, she was hardly a radical agitator for women's rights. While chairing the committee, she gave the annual address for the Columbia chapter and took as her topic "A Redefinition of Fraternity." She might have redefined the word to mean sisterhood as well as brotherhood, but she did no such thing. Instead, she talked about such ideals as an "international fraternity" for the postwar keeping of world peace.

The Woolley committee had under advisement a suggestion from the Brown chapter for limiting the number of women elected in coeducational institutions. Having solved the problem at Brown by setting up a separate women's section, the Brown brothers proposed that coeducational colleges solve their problem by doing essentially the same thing. Let every chapter be "organized into two sections, one for men, the other for women, each electing a proper ratio of the whole number to be elected, proportional to the number of each sex registered as candidates for literary degrees."

From a questionnaire it sent to the fifty-six chapters in coeducational or partly coeducational institutions, the committee ascertained that more than half of the chapters—twenty-nine of them—elected a larger number of women than of men. The recent percentages of women and men overall were 62.8 and 37.2 respectively. In nineteen of the twenty-nine institutions, more women than men were enrolled in the college of liberal arts. But the committee found that in the affiliated colleges for women—such as those at Brown, Columbia, Hobart, Tulane, Tufts, and

Western Reserve—the elections were "distinct, the same basis for election prevailing in each section." And the committee also discovered that chapters in a number of completely coeducational institutions were following "practically the same plan"; they were "electing a certain percentage of men in the class and an equal percentage of women." Among the chapters doing so were those at Allegheny, Bates, Colby, Denison, Grinnell, Middlebury, Northwestern, and West Virginia. In other words, these and other chapters were already putting into effect the essence of the Brown idea.

So the committee recommended that the Council take no action but leave the question "entirely to the local chapters" for their decision. The committee also urged that election should continue to be "on the basis of scholarship," though expressing sympathy with the wish of one committeeman "that there were some equitable way by which talented young men who at the beginning of their junior year enter practically professional courses, where the grading is more severe, might have a chance to enter the Fraternity." In sum, the committee favored giving women "every possible opportunity on terms of absolute equality with men." The Council accepted the committee's report and adopted its recommendations.[8]

The delegates seemed unaware that the committee was recommending and they themselves were taking a contradictory stand. It did not really make sense to endorse the principle of equal percentages for women and men and at the same time to proclaim "absolute equality" of opportunity for women. The equal-percentages plan actually amounted to a quota system that would restrict women's opportunities.

This point was rather dramatically illustrated by an item that the *New York Herald Tribune* published in 1936 under the headline "After 31 Years of Error, He Finds Wife Is Smarter." The story, with a Chicago dateline, ran as follows:

> For thirty-one years Alderman John Massen, Sr., was secure in the knowledge that when both he and his wife attended Northwestern University he was elected to Phi Beta Kappa, scholastic fraternity, and she was not.
>
> But yesterday Mrs. Massen received from Dr. Walter Dill Scott, university president, the following letter:
>
> "For your personal gratification I want you to know that your husband did not make a better record at Northwestern than you. Your four-year scholastic average [on a basis of 7] was 6.0226, and that of your distinguished husband was 5.9848.
>
> "The fact that he made Phi Beta Kappa and you did not is no justification for his 'lording it over you.' At that time the university awarded Phi Beta Kappa to a certain percentage of men and women."[9]

The efforts to limit female membership reflected a kind of gynephobia—a fear of the scholarly woman, at least—that prevailed in much of academe until World War II, if not beyond.

A presumed expert undertook to analyze and allay this fear in the October 1920 issue of the professional journal *Education*. The expert was J. H. Doyle, Ph.D., a former consulting psychologist at the Culver Military Academy. "That strange question," Doyle began, "is now really being asked: Are women mentally superior to men?" Certainly, as the statistics of Phi Beta Kappa elections demonstrated, college women were getting better grades than college men. But did that necessarily mean that the men were mentally inferior? "Phi Beta Kappa seems to think so down deep within, and so does the general school public, if one is to accept the tone of the press, and particularly the comment that is to be heard in college and university circles."

Doyle disagreed. Women excel not in brains, he believed, but in "interest" and "application." He proceeded to give reasons why they apply themselves more assiduously. First, they are biologically "static" while men are biologically "dynamic," and "book education is relatively a static affair." Second, women are sociologically restricted to fewer roles, to practically nothing but scholarship while in college. "But with men it is far different. The populace still demands athletics, even as in the days of Greece and Rome—yea, even the Phi Beta Kappa girl demands athletics, for her dreams and her admiration are universally for the dynamic hero emerging from some arena of physical contest rather than for the typical man who wins scholarship honors in Phi Beta Kappa!"

Third, while male students are free to carouse and waste their time, female students are not; there is little they can do except study. This is especially true of the most scholarly ones, for they are unlikely to be distracted by dates: "women are social favorites with men in inverse proportion to their Phi Beta Kappa qualities." Fourth, women get better grades because "in our co-educational colleges and universities the faculties are made up mostly of men," and "the appeal to men of the eternal feminine enters to a certain extent into grades given to women." (Did the expert mean to say that male professors found the greatest "eternal feminine" appeal in those same coeds who were by no means the "social favorites" of college boys?)[10]

According to other sources, women were quite aware of the prejudice that Doyle noted, and it caused some of them to deprecate Phi Beta Kappa. When a Vassar student declined election in 1922, it was thought to be a case of "feminine flouting" of the fraternity. "This tendency of women undergraduates and alumnae to speak lightly of Phi Beta Kappa, after they have nearly studied their heads adrift in getting the marks to entitle them to election, has been growing for several years," a male commentator said in the *New York Times*. "It is an open secret that some of the winners of the key have felt that in this age of flappers it is a distinct handicap if matrimony is their goal."[11]

If women Phi Betes did manage to attain the goal of matrimony, the fact occasioned surprise and a bit of amusement, at least in certain male quarters. Such was the response to a 1934 statistical study that analyzed the careers of forty-five women and forty-three men who had graduated from Knox College with ΦBK keys during the years 1917–1927. "An entertaining sidelight on these statistics," the *New York Times* commented, "is the report that thirty of the forty-five women initiates are married"—two thirds of them! That was indeed entertaining, if not downright hilarious.

The humor of the woman Phi Bete's predicament did not escape Virginia Scott Miner, a Phi Bete from Northwestern. In the summer 1942 issue of the *Key Reporter,* she offered the following "Advice to the Admirer of a Potential ΦBK":

> Never compliment her brains
> If you would win the fair—
> Assume she *might* come in from rains
> But drop the matter there.
>
> Praise her grades, her high I.Q.,
> You'll find the whole thing rankles—
> She'd any day prefer that you
> Would compliment her ankles![12]

When he addressed the chapter at the Randolph-Macon Woman's College in 1921, Secretary Voorhees presumed, with regard to the admission of women to Phi Beta Kappa, that "the fathers would have stood aghast if such action had been proposed" to them. Yet he could not be sure, since in their charter for Harvard the "fathers" (those youths!) had declared it "repugnant to the liberal principles of societies that they should be confined to any particular place, men, or description of men." In any event, the more recent representatives of Phi Beta Kappa, when they granted a charter to Vassar, demonstrated that the society was "not a citadel of conservatism but the very apostle of progress." On another occasion Voorhees said: "We often remark that, in the recognition of women, Phi Beta Kappa did not come trailing at the end of the procession. It led the way."

ΦBK Senator Francis W. Shepardson talked in much the same self-satisfied manner when at the Williamsburg sesquicentennial ceremonies in 1926, he introduced a woman to speak for the women of the society. He said he "thought how far our fraternity had gone ahead, both in the admission of women to membership on equal terms with men and, particularly, in providing a place on our Sesquicentennial program for the gifted and gracious president of Mt. Holyoke, Dr. Mary E. Woolley, the first woman to be elected to the Senate of Phi Beta Kappa."

After this somewhat condescending introduction, Senator Woolley proceeded to make a cheerful, grateful, and gently defensive address.

Yes, the fraternity had done a great deal for women, she allowed, but it would be fair to ask what women had done in return. "The real significance of this organization consists in the character of the human beings who have been its representatives." These included great men, to be sure, but they also included great women, and she listed a number of the most famous ones, among them Elsie Singmaster, Ellen Glasgow, Mary Johnston, Dorothy Canfield, Katherine Lee Bates, and Ida M. Tarbell. (Here, by implication, she was rebutting Voorhees's 1916 assertion that the society's reputation depended much more on its men than on its women members.) "No, we are not without *our* roll of honor, we women of Phi Beta Kappa, a roll that we assure you shall grow in numbers and distinction in the years to come," she declared. "I predict that the day will never come when you will regret the inclusion of women in Phi Beta Kappa!"[13]

Women continued to gain distinction and to make progress in the society but only against continuing resistance. Men even objected to authorizing chapters in additional women's colleges. "I am opposed to Agnes Scott being granted a Phi Beta Kappa charter at this time," the William and Mary president confessed in 1924, during the period when applicants needed the support of existing chapters in a given region. "So many of our institutions now are coeducational that if, in addition to the coeducational institutions, we grant charters to such institutions as accept women only, we will soon have many more women than men in Phi Beta Kappa in the South." Nevertheless, the next Council voted a charter for Agnes Scott.[14]

Women had an especially long struggle for recognition at the University of Pennsylvania. They began to attend the medical school in the 1870s, the law school in the 1880s, and the school of education in the 1910s. But they continued to be excluded from the college of liberal arts.

On the advice of her son, a ΦBK graduate of Pennsylvania, the chairman (again, not the "chairperson") of the alumnae association's committee on educational policies wrote to Voorhees in 1921 "to make inquiry as to how Phi Beta Kappa may be opened to women" at the university. "I am told by a member of my committee that the matter rests upon the shoulders of the local chapter, which is violently opposed to the admission of women." This raised a "difficult question," Voorhees immediately replied. "I imagine that the opening of the University of Pennsylvania to women will prove a slow process, and until a regular liberal arts course is made available to them, I do not see how any pressure can be rightly brought to bear upon the chapter to establish a women's section."

United Chapters Secretary William A. Shimer answered in the same spirit when in 1932 another woman raised the question of Pennsylvania's exclusory stand. "I should say," Shimer wrote, "that the University of Pennsylvania does not elect women to membership in their Chapter because the elections are confined to the college of arts and sciences which is not co-educational." The next year the university finally provided a separate undergraduate college for women, and just one year after that, the Council chartered a section of the Pennsylvania Delta for the women's college.[15]

By 1941 a total of twenty-one women's colleges had received ΦBK charters, in addition to the affiliated colleges that had separate sections but not separate charters. The roster of such colleges with the ΦBK founding dates is as follows: Vassar 1899, Barnard (section of the New York Delta at Columbia) 1901, Smith 1904, Wellesley 1904, Mount Holyoke 1905, Goucher 1905, Flora Stone Mather (section of Ohio Alpha at Western Reserve) 1906, Radcliffe 1914, Randolph-Macon Woman's College 1917, Hunter 1920, New Jersey College for Women (section of New Jersey Alpha at Rutgers) 1922, Agnes Scott 1926, Mills 1929, Wheaton 1932, Woman's College of the University of North Carolina (section of North Carolina Alpha at Chapel Hill) 1934, Connecticut College 1935, Florida State College for Women 1935, University of Pennsylvania College of Liberal Arts for Women (section of Pennsylvania Delta) 1935, St. Catherine 1938, Elmira 1940, and Milwaukee-Downer 1941.

Women continued to be excluded from membership in the oldest and most active of the ΦBK alumni associations, the one in New York City. Women were not even allowed to attend any of the New York association's meetings until 1910, when for the first time members were permitted to bring their wives as guests. The presence of the ladies, according to the minutes of the meeting, "was a decided innovation and added greatly to the enjoyment of the occasion." It did not add enough to the enjoyment, however, to persuade the men as yet, to admit women Phi Betes as members. The women had to form a separate association.

The ΦBK alumni association of Philadelphia also confined its membership to men. "Would you be interested in helping organize an association for the alumnae in Philadelphia similar to that here in New York City?" Shimer in 1932 inquired of the woman who had asked him why the chapter at the university was closed to women. "The men and women [in New York] have separate associations and meet jointly once each year."

In 1938, after the establishment of the women's college and the organization of the ΦBK section, Shimer again brought up the idea. He invited the 507 women members living in Philadelphia to meet with him there to discuss the desirability of forming a graduate association. "Several replies raised the question: 'Why segregate the women?'" he informed the secretary of the men's chapter at the university, and he invited him to be present to help with the answer. The chapter secretary wired back: "Strong sentiment in local association against admitting women."[16]

Within the United Chapters, women were gradually getting ahead. Ada L. Comstock, president of Radcliffe College, served on the original board of editors of the *American Scholar*. Six women had held the office of senator by 1940. In that year the Council elected the first woman president—Marjorie Hope Nicolson, dean of Smith College.

A Phi Beta Kappa graduate of Dean Nicolson's college was education editor of the *New York Times*. In 1938 this editor, writing in the *Key Reporter*, hailed the progress that women had made in the academic world. The previous fall, she said, Oberlin had celebrated the "centennial of woman's painless entrance into the college classroom." Some forty years after entering college at Oberlin, with "even less difficulty" women "strolled" into Phi Beta Kappa at Vermont. "Of all the barred doors which woman has forced in her century's struggle to become a citizen of the world," the education editor averred, "none had yielded so easily as those of the college and of ΦBK."[17]

In scholarship and in scholarly honors, women had indeed come a long way, but their progress had hardly been painless or easy.

14

To Quicken the Activities

After the United States entered World War II in 1941, and especially after it extended the draft to include eighteen-year-olds in 1943, enrollments in American colleges and universities fell off precipitously. Then once the war was over, they rose even more rapidly, soon reaching unprecedented heights. The numbers elected to Phi Beta Kappa also increased as never before. Indeed, more were elected during the years 1940–1967 (140,000) than during the entire period of 1776–1940 (113,000). By 1974 the living membership numbered approximately 250,000. The United Chapters strove to keep pace with this remarkable growth by finding adequate facilities and personnel. At the same time, the United Chapters added greatly to its functions by starting important new enterprises for the promotion of scholarship.

The war brought about a change in the society's leadership. Since 1940, when he accepted a deanship at Bucknell University, William A. Shimer had been able to make do as a part-time secretary of the United Chapters because of the wartime slackening of its activities. Then in 1943 he resigned to take command of the Navy's educational (V–12) unit at Emory and Henry. At his departure he was credited with having been mainly responsible for the establishment of the *American Scholar*, the *Key Reporter*, the Committee on Qualifications, and the Phi Beta Kappa Associates. Already an executive assistant, Dorothy E. Blair, had assumed responsibility for much of Phi Beta Kappa's day-to-day business, and she remained in charge until a new executive secretary could succeed Shimer.[1]

To succeed him, Hiram Haydn moved from Greensboro, North Carolina, to New York in 1944. Haydn—B.A. Amherst, M.A. Western Reserve, Ph.D. Columbia—had taught English at a private school in

Cleveland, at Western Reserve, and at the Woman's College of the University of North Carolina. He was not to remain long in the position of Phi Beta Kappa secretary. "The secretaryship baffled me," he afterwards confessed; "the job seemed to require the talents of a public relations man, an antiquarian, and a service station operator."

A task immediately confronting Haydn as secretary was to see to the publication of *The History of Phi Beta Kappa* (which came out in 1945), and to him it was a most unpleasant task. He did not like either the manuscript or its author, Oscar M. Voorhees, who had been Phi Beta Kappa secretary from 1901 to 1931 and official historian since that time. Voorhees had assembled a great deal of information, much of which without him would have been consigned to oblivion, but Haydn thought the work was "composed of equal parts of statistics, roll-calling, glue and sawdust." In cooperation with a Crown Publishers' editor, he undertook to edit it so as to make it publishable. "Dr. Voorhees was mortified," Haydn later recalled. "He fastened me with his bald eagle's peeled eye and demanded, 'Young man, for whom are you working? For Phi Beta Kappa, or for Crown Publishers?' He was unconsciously prophetic."

After a little more than a year in the United Chapters office, Haydn did go to work for Crown as an assistant editor, thus beginning a notable career in book publishing. He resigned as secretary of the United Chapters but remained as editor of the *American Scholar*, a job that he thoroughly enjoyed. In 1945 the Senate separated the editorship from the secretaryship and started a search for another secretary. Again there was an interregnum, this one lasting for about two years, during which a succession of acting secretaries kept the office going.[2]

The last of these acting secretaries was Carl Billman, who was already serving as an assistant secretary. Billman, born in 1913, had earned B.A. an M.A. degrees in history at Harvard and had taught there and at St. Mark's School. Before the end of 1947, the Senate elected him to the secretaryship. He was to hold the position for more than twenty-six years, almost as long as Voorhees had held it. But Billman, while equally devoted to the interests of the society, performed very differently from Voorhees as an executive. Quiet, self-effacing, he remained as inconspicuous as possible at Senate meetings, taking the minutes himself. "He conceives of his job as a service," Haydn explained, "not as the opportunity to exercise power."

During Billman's tenure the United Chapters searched for and finally found a new "permanent home." By the early 1940s the society, having left its elegant apartment to find more space, was occupying rooms at 12 East 44th Street in New York. Then to escape a rent increase, the society moved its headquarters to nearby 5 East 44th but soon found the accommodations there "so small that the twelve members of the staff were constantly tripping over each other and over the Addressograph machines." The staff continued to grow and to crowd the available space (by 1966 there were to be twenty-three full-time members).

In 1948 the society relocated at 415 First Avenue. This was quite a comedown from the "highbrow purlieus" of the previous Fifth Avenue neighborhood, as a *New Yorker* reporter discovered when he visited the office in 1950. "The block includes a junk shop, a couple of bars, some ancient apartment buildings, and two vacant stores, formerly occupied by sellers of automobile parts." At No. 415, a two-story building, the only sign was a huge green one bearing the words "The Home of Mrs. Altman's Strudels, Pastries." The proprietor showed the reporter to a side entrance leading to the Phi Beta Kappa office upstairs.

Billman, greeting the caller, was not surprised he had had difficulty finding the place. "Happens all the time," Billman said. Just recently he had sent out a "posse" to look for "the lady president of a ladies' college" who was twenty minutes late for an appointment. "She was discovered wandering about on the third floor of the building next door, outside a dance studio." Still, Billman was quite satisfied with the quarters. "This loft has been divided into five offices and a storeroom," he pointed out, "and is not only roomier than any other place we've had but cheaper." The main drawback was that drunks kept staggering in—none of them wearing a Phi Beta Kappa key.[3]

These quarters were only temporary, anyhow. Already a committee was engaged in the search for a permanent habitation. Princeton, Philadelphia, and Washington, as well as New York, seemed like possible sites. While waiting for something to turn up, the society in 1951 resorted to yet another stopgap, transferring its staff and its belongings to the Phi Beta Kappa Memorial Hall on the William and Mary campus. On December 29, 1953, a fire destroyed most of this building except for the north wing, which housed the offices of the United Chapters. Most of the society's correspondence had been stored at Columbia University; its records on hand were saved. (By 1957 a completely new Memorial Hall had been built with money from insurance, appropriations of the Virginia legislature, and contributions from Phi Beta Kappa members, including $250,000 from John D. Rockefeller, Jr.)

In 1954 the society bought and began to remodel a mansion at 1811 Q Street, N.W., just off Du Pont Circle, in Washington, D.C. Built in 1912 for a prominent attorney, this house was well located, well constructed, handsome, and commodious; its three floors and basement provided ample space for the society's needs, at least for the time being. Early in 1955 the society moved in. Its records, though surviving the recent fire, were in something of a mess, and for a long time they were to remain without proper organization and care. The price of the Q Street building was $75,000, and the remodeling cost another $20,000. To meet this expense, the Phi Beta Kappa Foundation had been setting aside money for several years, and it now was making an investment in real estate that during the next few decades was to increase tremendously in dollar value.

Meanwhile, the Foundation continued to accumulate funds, partly from the contributions of the two hundred Phi Beta Kappa Associates, each of whom paid $100 a year for ten years and then was replaced by another volunteer. By 1975 the Foundation had received $708,615 from the Associates, had allocated $432,102 of it to operating expenses, and had added the balance, $276,513, to the endowment. By this time the endowment had a book value of approximately $2,000,000.[4]

The society was depending on endowment income, however, for a smaller and smaller proportion of its operating expenses. Leaving out of account the receipts from the *American Scholar* and the *Key Reporter* (neither of which paid for itself), nearly 30 percent of the money for current expenditures had come from invested funds and from Associates' contributions in 1940–1941, but only about 15 percent did so in 1975–1976. The share coming from the registration fees of new members and from the sale of keys and certificates fell off even more. Together, the registration fee and the key royalty had once constituted the society's largest source of revenue, and as late as 1940–1941, they accounted for nearly half of it—more than 43 percent. In 1975–1976 they provided only 13 percent. Even more dramatic than this decline was the rise in the proportion of net income derived from sustaining memberships. These contributed a mere 7 percent of the total in 1940–1941 and 46 percent in 1975–1976.

After Billman's death in 1974, the *Key Reporter* noted that he had directed the society during its period of greatest growth and had contributed "with skill, energy, tact, and gentle patience" to the success of the society's greatly expanded activities. His successor, Kenneth M. Greene, came to the office in 1975 from the presidency of Lasell Junior College in Newton, Massachusetts. With a B.A. from Brown and an M.A. and a Ph.D. from Columbia, Greene had previously taught English at Simmons College. He possessed additional administrative experience as director of the school of education at Simmons and as a member of the executive board of the Association of Independent Colleges and Universities in Massachusetts.

Greene approached his new position "with respect for the dignity, wisdom, and graciousness that Carl Billman imparted to the executive function of the United Chapters," he said. "The objective of Phi Beta Kappa—the advancement of humane learning—has long been the compelling objective of the days I have spent in the classroom, at the administrator's desk, and at the conference table."[5]

Hiram Haydn faced challenges from the time he took over the editorship of the *American Scholar* in 1944. Its circulation, declining because of the war, needed to be improved, yet its standards of excellence and independence must be maintained. Maintaining these standards became more and more difficult as the postwar Red Scare developed into Joe McCarthyism, and the task did not get much easier as McCarthyism subsided into the chronic anti-Communist hysteria of the Cold War.

"I was young, new to New York, and excited by the 'great world,'" Haydn later reminisced about his arrival in 1944 (though he was not entirely new to New York, having lived there long enough to earn a doctorate from Columbia); "I was also and (at thirty-six) belatedly an active liberal." Promptly he plunged into what turned out to be "angry years, rife with tumult and controversy." He was to look back "with some pride at the part the *Scholar* took in the struggle for intellectual and academic freedom."

At the start of his editorship, Haydn chose to keep the editorial policies and procedures basically unchanged. As before, the acceptance of any manuscript would require the affirmative vote of at least two of the twelve members of the editorial board in addition to the vote of the editor himself. Haydn did make some innovations, however, beginning with his very first issue. He introduced three new departments. The "*American Scholar* Forum" presented alternately debates on controversial issues and symposia on topics of current interest. The "Revolving Bookstand" allowed a guest critic, a different one each time, to review at length some recent and notable book. "Under Whatever Sky," conducted by the Columbia University philosopher Irwin Edman, provided commentary on the passing intellectual scene.[6]

One of Haydn's conceptions failed to bear fruit. He made an arrangement with Simon and Schuster to publish an "American Scholar Series" of books. "These books," as he envisaged them, "will all be expansions of articles published in the magazine." In a couple of years he had "three or four vaguish possibilities lined up," but he had "yet to see S. & S. evince any real enthusiasm for any suggestions," and he could only conclude that the series was "slowly dying on its feet."

In one respect, Haydn was somewhat less willing to meddle with the magazine than were the members of his editorial board. He voted along with them to change the magazine's subtitle from "A Quarterly for the Independent Thinker" to "An Independent Quarterly." Then a day later he had second thoughts. "Might it not be that many of the good brethren and sisters of Phi Beta Kappa would resent the caption 'Independent Quarterly' when that quarterly is published by Phi Beta Kappa?" he asked in a letter to the board members. "As most of you know, there has been a minority but strong sentiment in the past that THE AMERICAN SCHOLAR was too independent of Phi Beta Kappa anyway."[7]

Haydn knew that he himself was largely to blame (or to credit) for this sentiment, for he had made the magazine more independent of the society than ever. Originally, all members of the editorial board were required to be members of Phi Beta Kappa. Though some were journalists and men of affairs, most were academics. Haydn thought of "balancing the numbers in academic life with others from 'the other world.'"

Officially, the Senate appointed the board but did so on the recommendation of the editor. He appealed to the senators, pointing out that only
six of the twenty-eight of them had been elected to Phi Beta Kappa as
undergraduates; "all the others had become *honorary* members later!"
His argument must have been persuasive. "My amendment carried, and
before long we didn't even know which of our [board] members were
also Phi Beta Kappas."

As soon as the terms of four members expired in 1944, Haydn nominated and the Senate "warily" approved as replacements the news-
papermen R. L. Duffus and Max Lerner, the diplomat and historian
Sumner Welles, and the singer and actor Paul Robeson. "Wow, four new
members," exclaimed Will D. Howe, who was both a senator and a board
member. "Two Communists and two fellow travelers!" Lerner was Rus-
sian-born (and Jewish), but only Robeson (who was black) could fairly be
considered a Communist or a sympathizer. He had been elected to Phi
Beta Kappa—and also selected for Walter Camp's All-American football
team—while a student at Rutgers.

Haydn received no reply to his letter informing Robeson of the nomi-
nation, so he wrote again. Finally a telegram arrived saying Robeson was
on tour with *Othello* and could not meet with the board for several
months but would gladly accept if his services could be dispensed with
until then. The time came, and he failed to appear. In fact, during the
three years he was a member, he never attended a single meeting of the
board. He did write an editorial and read several manuscripts, but that
was all. He had no real influence on the course of the *American Scholar.*
Hence it is ironic that his appointment should have provoked the outcry
it did.[8]

From Phi Beta Kappa member and others, Haydn kept getting letters
demanding Robeson's removal. In one of the letters, a friend of Will D.
Howe's protested against "adding communist collectivist inclined front-
ers to the editorial staff" and said he could not understand "why the
Board came to elect leftwingers such as Paul Robeson and Max Lerner."
To show that he was unprejudiced—that he was not "affected by racial
considerations"—this correspondent avowed he would "nominate with
enthusiasm" the wealthy investment banker Bernard Baruch if a Jew
were to be chosen, and he would prefer the heavyweight champion Joe
Lewis if a Negro were to be.

Haydn replied that Robeson and Lerner had been appointed, like
other board members, because of their distinguished achievements,
their special knowledge, and their editorial ability. "Left wing, right
wing, all around the block—it is utterly immaterial to me for the purposes of THE AMERICAN SCHOLAR so long as they have the above
qualifications and do not try to intrude doctrinaire political views upon
the magazine." Lerner had been very helpful, Haydn said; Robeson, so
far, had been preoccupied with other interests. "Incidentally, when you

examine the constituency of the rest of the Board, beginning with your friend Dr. Howe, I don't believe that you will find it possible that THE AMERICAN SCHOLAR is fostering any subversive tendencies."

There was a preponderance of "Pinks" on the board, Howe's friend reiterated, and the magazine was "considered quite Pink in spots." The man wondered why "so many literati have so little of what Solomon called 'wisdom' or 'understanding.'" Apparently, "in practical matters about seventy-five per cent of our Ph.D.'s are also d.ph.'s"—that is, "damn phules." To this Haydn responded:

> Speaking for myself, I can add that the "big red scare" leaves me somewhat less than terrified. I am not a Communist; I am not a fellow-traveler; I am simply a quite ordinary American citizen who dislikes to be pushed around by people who decide that association with this or that other individual is dangerous. . . . Whether or not THE AMERICAN SCHOLAR is considered "quite pink in spots" seems to me really unimportant. It has demonstrably no political message or line of any sort.

Haydn's right-wing correspondent insisted on the last word. He retorted: "I would say that this airy view of the red-pink menace to American institutions is a made-in-Moscow cliché for the cloistered—yes, the American scholar!"[9]

Despite the continuing complaints of right-wingers and anti-intellectuals, Haydn and the majority of his editorial board pressed ahead in their insistence on freedom of expression. They published articles critical of repression during the 1950s when Joe McCarthy was alleging Communist infiltration in the federal government, schools and libraries were censoring books, and state legislatures were requiring public universities to impose loyalty oaths on their faculties. The majority of the board occasionally went too far to suit all the board members. One or two of them opposed the acceptance of an essay on "Loyalty and Freedom" by the liberal Archibald MacLeish. And board member Arthur Schlesinger, Jr., a Harvard professor of history, sent Haydn a letter angrily protesting the publication of an exposé, "Do You Know the Nature of an Oath?" by Howard Mumford Jones, a Harvard professor of the humanities. Schlesinger, as Haydn related, called Jones "an emissary from the Kremlin" and attacked Haydn as "either an accomplice or too naïve to be an editor."

The *American Scholar* was merely insisting on open discussion, not following a particular ideological line. Haydn and his co-editors constantly strove for diversity in both subject matter and points of view. This guiding principle was the same as that of their predecessors. "We have tried," as Haydn put it, "to fill the gap between the magazines of special interests and intellectual cliques and those of middlebrow, semi-mass-market circulation."

They could not neglect circulation, not if the magazine were ever to pay for itself. Like Shimer before him, Haydn could make a single cancellation a matter of personal concern. "I wonder whether you would be willing to write to us and explain what it was which caused you to feel so strongly that you didn't care to receive THE AMERICAN SCHOLAR any longer," he wrote to a Cambridge, Massachusetts, man who had asked that his subscription be canceled immediately. The man's reply bewildered Haydn, and he responded:

> All that I can ascertain for certain from your letter is that THE AMERICAN SCHOLAR is playing pander to some kind of commercial aesthetics and that this "spells plain death to the careless, laughing fluid spirit of aesthetic enjoyment." Furthermore, I discovered that this tendency of THE AMERICAN SCHOLAR is a "product of the American temper, the temper of a vast, crude, wasteful and brutally powerful, unscrupulous people"; however, I am also naïve enough to believe that I am acquainted with American people who are subtle, economical, honest, and even gentle. In short, I do not have much faith in generalizations about the American temper.
>
> . . . At any rate, probably you are wise to cancel your subscription if what you primarily require from a magazine is that it cater exclusively to the "careless, laughing fluid spirit of aesthetic enjoyment." That spirit, which strikes me as perhaps a cross between a spirit of a fairly youthful fawn and that of George Santayana, is doubtless admirable, but I confess that I should not want it for a steady diet unless seasoned frequently with other condiments. However, such an exclusive emphasis is certainly not the province of THE AMERICAN SCHOLAR. May I suggest finally that I believe there is a quite defensible middle ground between the advertising of a popularized aesthetics and the citadel of the ivory tower—that it is quite possible to write for an audience which is not at all a group of specialists, and at the same time not to do the Will Durant sort of thing.[10]

Most subscribers stayed with the magazine, and new ones kept signing up, so that the circulation steadily increased, though not as fast as Haydn thought it should. The magazine's financial condition, he reported in 1946, had become "increasingly healthy during the past few years" but still "would not be in the black" if account were taken of the fact that several employees on the society's payroll gave part of their time to "one aspect or another of the *Scholar* work." The magazine, he thought, was "deserving of much wider support from the Society in general." Only 20 of the 149 chapters and sections and only a "pathetically small" percentage of the individual members subscribed.

After the war the circulation began a rapid growth, going from 3,200 in 1944, to 10,000 in 1953, and to more than 20,000 in 1962. Then the executive committee decided on a campaign to double the number once again. By 1966 the goal of 40,000 had been reached, and in the 1970s the total rose to more than 45,000. With the largest paid circulation, the *American Scholar* had become the country's leading general quarterly.

The magazine had attained its position, however, at considerable cost. Expenditures on "promotion and circulation" had escalated from $2,000 in 1946–1947 to eight times as much in 1961–1962 and forty times as much in 1975–1976. The executive committee noted in 1966 that "the expense of the campaign" had "narrowed the favorable margin between subscription and advertising income and direct publishing expenses." To maintain the increased circulation, promotional expenditures had to keep on going up, since many of the new subscribers were not really potential *American Scholar* readers, and they soon canceled or failed to renew their subscriptions. Eventually the society "found a subscription level of about 25,000 to be manageable." The magazine had failed to become quite self-sustaining, even at its greatest circulation. In some years it had returned enough to pay for its direct costs, or a bit more than that, but never had it covered both those costs and its share of the overhead.

Haydn did not claim for himself or for the editorial board the entire credit for the magazine's success. Hard at work in the book-publishing business, while writing books of his own, he had to look after *Scholar* matters in odd hours. This was comparatively easy to do as long as the United Chapters office remained in New York, where his book business was located, but after the society moved its office to Williamsburg and then to Washington, he was obliged to make fairly frequent trips to those places. Bearing the day-to-day responsibility for the magazine were a succession of managing editors and a manager of circulation and promotion.

One of the managing editors ably carried on for several months in 1967 when Haydn was recuperating from a heart attack he had suffered in Santa Fe, New Mexico. He afterward resumed his killing pace and kept going until December 2, 1973, when he suddenly died.

"His has been a remarkable record: his thirty years at the helm of *The American Scholar* is surely the most impressive one-man-stand in the history of American intellectual journalism." So declared Joseph Epstein on accepting appointment as Haydn's successor in 1974. Epstein, then thirty-seven, was teaching English at Northwestern University while working on his second book. After graduating from the University of Chicago, he had gained experience as an editor at the *New Leader,* at *Encyclopedia Britannica,* and at the publishing firm of Quadrangle Books. "A magazine takes its character from the issues it confronts, from the personality of its editor (the amalgam of his strengths, weaknesses, and idiosyncrasies), and above all from the writers he is able to attract to its pages," he added at the time of his appointment. "My goal is to edit a magazine that will be critically judgmental, lively and unpredictable—and, in the process, will convey the intellectual excitement of being alive in our time."[11]

On December 16, 1941, just nine days after the Japanese attack on Pearl Harbor, the Phi Beta Kappa Senate met at the Yale Club in New York City for what the *Key Reporter* called one of the most important meetings in the society's history. The next day the Senate unanimously adopted the report of a Committee on Policy whose mission was to "consider what ΦBK ought to be, what it really is, and what it pretends to be."

The report reaffirmed the purpose of the society as stated in its constitution—"to recognize and encourage scholarship, friendship and cultural interests"—and the long-standing corollary aim to promote the humanities and the liberal arts. There was nothing new or startling so far, but the report went on to note that the most serious criticism of Phi Beta Kappa was the inactivity of its chapters and to recommend that a standing Committee on Chapter Activities be set up to submit "a patterned program designed to quicken the activities." Then, in reaction against the contemporary Nazi persecution of the Jews and other "inferior races," the report affirmed the society's faith in freedom of conscience and its opposition to discrimination on the basis of religion or race. Finally, in recognition of the fact that the country was at war, the report declared:

> . . . the motto . . . implies that the Society has an obligation to translate the significance of liberal arts education into terms of living in a democracy; i.e., this implies that at the present time ΦBK may properly pledge its support to the President and the government of the United States and, while recognizing the need in times of crisis for training in special skills, urge that we continue educating as many young men and women as possible in the liberal arts as a necessary preparation for more effective service to our country both during the war and in the peace that must follow it.

Thus, according to the Senate's declared policy, the society would devote itself in the long run to increased activity in the cause of humane scholarship.

For the duration of the war, however, the society would have to subordinate its purposes to those of a nation in arms. The United Chapters served as a recruiting agency of sorts. Members could "save lives by getting technicians for the armed forces now," urged the *Key Reporter,* which still went free to all members, though reduced from eight to four pages to conserve paper. The Red Cross needed educators and social welfare workers at home and abroad; "for further information write THE KEY REPORTER." From the Navy Department came a request to "lend aid in supplying capable linguists to undertake the study of Japanese."[12]

So long as the war lasted, the Committee on Chapter Activities did not attempt to do anything. Then in 1946 it went so far as to inquire of the various chapters what activities they were planning or carrying on. "More than one secretary in reply said not only that chapter activities had been interrupted or suspended," the committee reported, "but that the faculty members of the chapters were exceedingly tired, that the burdens had not been relieved, that they had been even worse than during the war period, and it was virtually impossible even to reply, to say nothing about a resumption of a normal program of activities at this time." The committee hoped that more could be done in the future to stimulate the local chapters, for in these "the vitality and effective functioning of the Society must center."

In 1946 the United Chapters undertook to provide some stimulus by resuming the ΦBK Associates National Lectureship, which it had founded in 1942 but had discontinued in 1944 because of wartime conditions. This lectureship made speakers available to chapters and associations in the hope of producing more addresses of the kind that had "made such contributions to American letters as Ralph Waldo Emerson's 'The American Scholar' and Woodrow Wilson's 'The Spirit of Learning.'" Emerson had said the scholar is "Man Thinking," and in 1962 the United Chapters demonstrated that excellent addresses were still possible by sponsoring the publication of *Man Thinking: Representative Phi Beta Kappa Orations, 1915–1959.*

"Chapters and Associations DO Things," the *Key Reporter* kept insisting in varying phraseology during the 1950s and 1960s. The newsletter eagerly collected examples of what they did. They sponsored lectures, colloquia, coffee hours; awarded scholarships and prizes to high-school seniors; gave college students an opportunity to visit noted faculty members and see how scholars work.

Despite the *Key Reporter's* optimism and encouragement, however, comparatively few of the chapters did much on their own initiative except to elect and induct members and hold an annual dinner with an after-dinner speech. The great majority of members did not participate even to that extent. As the president reminded the 1970 Council, "most of the individuals who are elected to Phi Beta Kappa never thereafter really belong as active members of a chapter anywhere unless they happen to be college or university faculty members at an institution where there is a chapter." Some of these chapterless people belonged to one or another of the alumni associations, of which in 1976 there were more than fifty with memberships ranging from a dozen or two to nearly a thousand. All these groups reported regular activities at least to the extent of sponsoring dinners and speakers. Many of the groups also encouraged scholarly achievement by awarding scholarships, prizes, or other honors to outstanding high-school and college students.

One of the most active and by far the largest was the Alpha Association of Phi Beta Kappa Alumni in Southern California. From 1948 on, this association had provided grants to foreigners doing graduate work in American universities, and by 1979 the grants totaled more than $250,000, the recipients more than four hundred. Corporations as well as individuals contributed to the international scholarship fund. The association distributed a well-printed newsletter to its own members and, on occasion, to all the twelve thousand ΦBK members living in southern California.[13]

No doubt the "essence of Phi Beta Kappa" inhered in the local chapters, as one faculty member assured her colleagues in 1956. "There is, I understand, a central organization which performs various useful functions and from time to time appeals for funds," she said. "But few members of Phi Beta Kappa have any dealings with the central office."

Nevertheless, the central organization—the Senate, Council, and secretariat—was responsible for the great expansion of activities that took place during the decades following World War II. What activities to undertake was a question from the start. A New Jersey rhododendron grower had some suggestions, which he vouchsafed to Secretary Haydn in 1944.

> Why do you suppose there are 32nd degree Masons? Because the Masonic system is adapted to human nature. Then why not 64th degree Phi Beta Kappas? Why not a scheme of honors for intellectual attainments—so many points for a scholarly book, so many for a course of reading, a task of memory, the points to be awarded by democratically organized graduate chapters? Why not honorary and honored titles for those who cannot go through the grind for a Ph.D., and even for those who have the Ph.D. and want to learn something?
>
> Why not a democratic Academy of the Arts, like the French Academy enormously enlarged and subdivided?

"I believe, myself, that there are many specific projects which Phi Beta Kappa might back with highly significant results," Haydn replied. "But I question very seriously whether handing out degrees, honors, and marks and points is the most satisfactory road to those results."[14]

During the secretaryship of Haydn's successor Billman, the United Chapters nevertheless proceeded to hand out, if not degrees, at least honors, marks, and points of various kinds. In 1951 it presented the first of the annual book awards. This was the Christian Gauss Prize of $1,000 given for the "best" of the year's books of literary scholarship or criticism and named in memory of a distinguished scholar of modern languages, dean of Princeton University, and president of Phi Beta Kappa. In 1958 the society established the Phi Beta Kappa Award in Science for the work best illustrating the relationship between the sciences and the humanities and in 1959 the Ralph Waldo Emerson Award for "interpretive historical, philosophical, and religious studies in the great tradition of humane learning." These prizes were presented each year at a banquet on December 5, the anniversary of the founding of the society.

At the 1970 Council the president gave a cash prize and the Jaffe Medal for Distinguished Service to the Humanities. This award, made possible by the generosity of Mr. and Mrs. William B. Jaffe, was to be given at each triennial meeting thereafter.[15]

A more direct encouragement to scholarship, one operating on students as well as professors, was the Visiting Scholar Program. The famous geologist and popularizer of science Kirtley F. Mather, while a ΦBK senator, proposed the program, and the Council authorized it in 1955. According to the plan, one or more outstanding scholars every year were to visit colleges under the auspices of the local chapters to appear in classes, meet informally with student groups, and give public lectures. A visiting scholar would spend a few days at each of approximately a dozen institutions assigned to him or her. Priority was to be given to relatively small and isolated institutions where opportunities to hear and meet such scholars were relatively few.

During the first academic year of the plan's operation, 1956–1957, five men visited a total of twenty-nine campuses. By 1976–1977 the number of visitors and visits for the year had tripled. Over the two decades, the participants had included authorities in a variety of fields: literature, history, philosophy, religion, anthropology, economics, psychology, African culture, computer languages, space science, and submarine agriculture. Among the participants were poets, composers, industrial designers, chemists, business leaders, physicians, and a filmmaker. "More directly than almost any other activity of the United Chapters," the secretary of one chapter declared after a scholar's visit, "this program allows and encourages the individual chapter to take an active role in the intellectual life of its institution."[16]

"In the year 1957 what does Phi Bet Kappa do?" William T. Hastings, long a leader of the society, had asked at that time. "Well," he answered, " . . . it is not a pressure group, and it declines to become a formal associate of such a group; it does not even memorialize Congress." But that very same year something happened that caused the society to change its policy and join other organizations in a lobbying campaign. What happened was the Soviet Union's launching of *Sputnik,* the world's first man-made satellite. This seemed to demonstrate that American space technology was lagging behind Russian, and the next year Congress passed the National Defense Education Act, which was to provide large subsidies for scientific and technological training. Already the federal government was financing scientific studies through the National Science Foundation, established in 1950.

In 1962 the Phi Beta Kappa Senate voted to support a resolution of the American Council of Learned Societies (ACLS) for setting up a commission on the humanities. The object was to get federal support for the humanities comparable to that being given to science and technology. In 1963 Phi Beta Kappa combined with the ACLS and the Council of Graduate Schools in the United States to form such a commission under the chairmanship of Barnaby Keeney, president of Brown University (who was to be awarded Phi Beta Kappa's first Jaffe Medal for Distinguished Service to the Humanities).

In 1964 the Keeney commission drew up a report urging the establishment of a National Humanities Foundation, which would parallel the National Science Foundation, to assist in the teaching and study of the humanities and thus countervail the recent "overemphasis" on the sciences. Keeney, on presenting a copy of the report to President Lyndon B. Johnson, found him encouraging; Johnson thought Congress ought to give it careful study. When Keeney addressed the Phi Beta Kappa delegates at their 1964 Council, he acknowledged that some feared the plan would lead to federal thought control, and he gave assurance that the proposed foundation would help to finance activities but would refrain from trying to control them. The Council endorsed the project.

Congress created the National Foundation on the Arts and the Humanities in 1965. This consisted of two "endowments": the National Endowment for the Humanities (NEH) and the National Endowment for the Arts.[17]

In 1967 Phi Beta Kappa submitted to the National Endowment for the Humanities a proposal for a National Humanities Faculty, and the NEH granted funds for a one-year pilot program to begin in 1968. Phi Beta Kappa directed the program in cooperation with the ACLS and the American Council on Education. The purpose was to "bridge the gap between scholars in the various areas of the humanities and teachers in the secondary schools." Scientists from colleges and universities had been working with high-school teachers to improve instruction in science; other professors would now be encouraged to do likewise for instruction in the humanities.

The National Humanities Faculty program got off to a good start, though it ran into a certain amount of resistance. "At times there has been confusion in the schools as to our purpose in being there," the director of the program noted at the end of its first year; "we encounter some questioning as to the importance of the humanities in our technological age." Year after year the NEH increased its grants, some of them on the condition that they be matched by funds from other sources. After three years, some 130 college and university faculty had worked in thirty-two school systems, and the director hoped that some would have worked in each of the fifty states by the end of the 1973–1974 school year. But in 1973 Phi Beta Kappa and its associates ended their relationship with the National Humanities Faculty, which then became NHF, Inc., an independent, not-for-profit corporation still funded mainly by the NEH.[18]

For some time, using the income from a bequest of Isabelle Stone of the Wellesley chapter, Phi Beta Kappa had been providing the Mary Isabel Sibley Fellowship to help women pursue advanced study in French or, in alternate years, Greek literature and culture. The first recipient had to return from France, her work unfinished, upon the outbreak of war in 1939. Most of the later recipients were, like the first, graduate students. A few were holders of the doctorate and were already engaged in full-time teaching. Former Sibley Fellows have held professorships in such institutions as Smith, Stanford, and Swarthmore.

In 1971, looking toward the celebration of the society's bicentennial, the society offered a quite different kind of fellowship—five of them at $20,000 apiece. The winners of the competition for the fellowships were to write books dealing with "the cultural crisis of our time" for a series to be published during 1976 under the general title *Man Thinking in America.* Three of the awards brought forth books, which were published in 1977. These were *Black Culture and Black Consciousness* by Laurence W. Levine, *Many Dimensional Man* by James Ogilvy, and *Crooked Paths: Images of the American Future* by Peter Clecak.[19]

15

Academics, Athletics, Ecclesiastics

World War II having halted the granting of new charters, the United Chapters afterward had to deal with a backlog of applications, and these continued to pile up during the postwar decades. For a while, the supply of charters lagged farther than ever behind the demand for them, as the society made it more and more difficult for most colleges to qualify. To do so, they must now eschew not only the evils of vocationalism but also those of athleticism. And they must avoid infringements of academic freedom, whether coming from state governments, from business interests, or from ecclesiastical authorities. Roman Catholic institutions were late in seeking and in gaining the approval of the Committee on Qualifications, the Senate, and the National Council. Hence arose a rumor that Phi Beta Kappa was anti-Catholic.

In 1942 the Committee on Qualifications called off its scheduled inspections of institutions seeking charters, as the wartime disruption of colleges seemed to make it useless to proceed. The Twenty-First Triennial Council, due to meet the following year, postponed its meeting for another triennium. When the Senate held its first peacetime session in 1945, the senators voted to recommend no charters to the Council when it met in 1946. They would wait until 1949, since the committee would need that long to complete its work.

The task of screening institutions, already an overwhelming one, became even more difficult during the postwar decades. Institutions were multiplying, expanding, and changing as never before. Junior colleges were becoming senior colleges. Teacher colleges, having ceased to be normal schools, were transforming themselves into state colleges and then into branches of state university systems.

Accrediting agencies also proliferated, and the major associations of colleges and universities set up the National Commission on Accrediting to accredit the accrediting agencies. By 1966 the commission had approved six regional and thirty professional ones, and by 1972 these had accredited some 2,700 of 3,000 institutions of higher education. Of the 2,700, about 1,700 were four-year arts-and-sciences colleges (either free standing or incorporated in universities) of a kind that already had a ΦBK chapter or could conceivably aspire to one.[1]

When in 1947 the Senate finally directed the Committee on Qualifications to go ahead, there were already forty-two institutions on the list for detailed investigation—all left over from before the war. The total, with later applications, came to more than a hundred. The six committeemen, two or three of them visiting each campus, would obviously be unable to make that many investigations in less than two years. So the Senate authorized them to inspect only twenty-four of the institutions. When the Council met in 1949—to consider applications for the first time since 1940—it approved a mere ten.

In subsequent triennia the chartering rate fell still lower. During 1952–1955 the committee recommended and the Council approved but two applications, and during 1955–1958 exactly one! Thus, a grand total of three institutions qualified in the course of six years. Then during the 1958–1961 triennium the rate rose somewhat. Having received thirty-three requests for consideration, the committee selected eleven institutions for scrutiny, visited ten of them (one having withdrawn its application temporarily), and recommended six, to all of which the Council granted charters—an average of two a year.

While cautious about giving out new charters, the society was also careful to keep from stretching old ones to cover the added campuses of consolidated universities. The Wisconsin chapter felt some concern in 1971 when the state government began to merge the Wisconsin State University system and its nine campuses with the University of Wisconsin and its various centers and branches. Already the Madison chapter had disallowed the claims of center or branch students to eligibility. Then there was the case of the Oshkosh State University graduate who thought his grades entitled him to membership in the Madison chapter. And now queries were coming from officials of the Wisconsin State University campuses that were being metamorphosed into University of Wisconsin campuses. One administrator "thought they would automatically become a part of Alpha Chapter."

The Wisconsin Alpha secretary wrote to United Chapters Secretary Carl Billman: "Is there some help you can give us in clarifying procedures in a situation of this kind? What, for instance, happened when state schools in New York became part of one system, or was that process different?" Billman replied:

> It is easy to believe how many questions are sure to be raised as a result of the changes now taking place in the organizational structure of the state

university system in Wisconsin. Fortunately some of them can be answered in the light of policy decisions that have been in effect, with Council approval, for the past twenty years or so. The customary solution has been separate chapters on each campus of a multi-campus university. That, for example, is the situation in California, as you probably know. It is also the situation in North Carolina, where there are separate chapters on the state university campuses at Chapel Hill and at Greensboro.

Unfortunately, there was a bit of ambiguity in the United Chapters by-laws, and the policy was "not as clear as it might be," Billman added. On occasion the United Chapters had, of course, authorized a section of an existing chapter for a coordinate women's college or for a school of general studies. But such a college or school had usually been "physically adjacent" to the institution already sheltering the chapter. Billman guessed that "neither the Senate nor the Committee on Qualifications would be enthusiastic about the idea of recommending the establishment of a section on a campus that is not within a very few miles of the campus on which the parent chapter is located."[2]

Meanwhile, the society was becoming considerably more free with charters. By 1961 the number of colleges and universities with chapters had reached a total of 170. That was about 10 percent of all regular four-year institutions. By 1970 there were 205 of them with chapters, but the percentage had risen only a little. Addressing the Council that year, President H. Bentley Glass said:

> . . . I am wondering if Phi Beta Kappa, in spite of its rather large growth during the past triennium, is in any realistic way keeping up with the growth of higher education in the United States. The Committee on Qualifications is a very hard working committee, and during the past triennium has made many more visitations than ever before and come to you with a larger slate of recommended institutions for approval than ever before. Nevertheless, during the past twenty years there have been 362 new four-year colleges or universities established in the United States—362, or an average of 18 per year. And the establishment of new chapters of Phi Beta Kappa has nowhere begun to keep up with such a pace of increase. There are now 1,670 four-year institutions in the United States—the last figures I have seen at least—and 12.3% of these have chapters of Phi Beta Kappa. Is that a proper proportion? Are there only 12% of the institutions of higher learning in the United States that are really worthy of having chapters of Phi Beta Kappa?

At the end of the following triennium, the committee and the Senate set another record when they recommended twenty-four institutions to the 1973 Council. It approved fifteen of them. After the fairly generous action of the Councils of 1976 and 1979, the number of colleges and universities with chapters totaled 228. That was nearly 15 percent of all four-year institutions.[3]

In considering applications, the society had imposed its standards more strictly than it might have done if its leaders had not perceived new and terrifying threats to everything the society stood for. These threats came from advocates of "general education," from college alumni and other boosters of big-time athletics, and from right-wing groups demanding their own brand of anti-Communist orthodoxy. In *General Education in a Free Society* (1945) and in other writings, Harvard President James Bryant Conant argued for a college program consisting of two years of general or cultural and then two years of specialized or vocational courses. With the spread of television in the 1950s and after, professional football became more and more a big business, and so did intercollegiate football, the university teams serving as farm clubs for the pros. Reacting to both World War II and the Cold War, legislators and trustees undertook to impose "loyalty" on administrations and faculties.

One Phi Beta Kappa leader expressed the fears of many others when he discussed "Phi Beta Kappa Today" in 1957. "It is not too much to say that powerful forces are consciously or unconsciously working toward the disappearance of the four-year liberal arts college, in fact if not in name," William T. Hastings then warned his fellow members. The Conant plan, he charged, would lead to the following result: "the two-year package of predigested 'cream' skimmed off the four-year liberal arts program; to be followed by another year or two of 'relevant' studies, i.e., vocational or pre-professional work—job preparation, on a higher or lower plane, not preparation for living as a man [or as a human being]." Hastings continued:

> That it is necessary for us to take the stand against commercialized athletics is clearer every day. Read the newspapers with their unconscious exposures: the making of contracts with schoolboys, the "gift" rackets operated by alumni, the bull market for successful coaches that swings salaries to double and triple a professor's "take" (as high as $18,000, with winter sunbathing at Saratoga thrown in), and now the professional teams reaching down into the colleges and signing up sophomores two years ahead of time.

The society must also take a stand, Hastings said, against "any interference by boards of trustees or legislative bodies" that would endanger freedom of teaching and learning.[4]

To insure that charters went only to good liberal-arts institutions, the society needed only to apply its long-standing principles. But these were somewhat vague, as the society conceded in a leaflet it sent on request to interested groups in the 1950s. "Because of the great difference among institutions and even among the various aspects of an institution," the leaflet read, "such as the number and kind of books in the library, the nature of the teaching and the publications of the faculty, the character of the students, the careers of the graduates, and the general attitude toward scholarship, *no absolute standards can be formulated*" (italics added).

Did the requirements of a liberal education include the teaching of Latin or Greek? Kalamazoo College was offering neither language when the Council considered its application in 1952. For that reason, some delegates opposed granting the application, and it failed to get the necessary two-thirds vote. By 1958 the college was offering both languages, and the Council this time did not hesitate to award it a charter.

Mills College, with a chapter dating back to 1929, was providing no Greek or Latin in 1963, as a Latin professor at nearby Berkeley rather officiously informed the Washington office. "I think that you will agree that no Liberal Arts college is worth the name unless it teaches Greek and Latin in the original," the concerned classicist averred. He went on to note that the Davis and Santa Barbara campuses of his own university had recently been denied charters. "Classics, which had been hanging fire for several years at the Davis campus, were soon added to the curriculum at both places, and I suppose that Phi Beta Kappa has followed." (It did follow in 1967.) Secretary Billman replied:

> Something is lacking, I agree, in a liberal arts curriculum that does not include the opportunity to elect courses in Latin or Greek. But I have to add that course offerings in the Classics are not required for a chapter of Phi Beta Kappa. It is doubtful, however, whether the Phi Beta Kappa Council is likely to approve a charter application from an institution that does not offer elective courses in Latin or Greek. About a decade ago [longer than that], in fact, a liberal arts college in the midwest [Kalamazoo] was turned down by the Council for that very reason. But that action reflected the opinion of the delegates who voted against a chapter, rather than an express policy of the Society.

Thus the standards for an acceptable liberal arts institution were rather indefinite, and the society was not always consistent in applying them.[5]

In dealing with the problem of athletics, however, the society had to develop new standards, and these turned out to be quite specific. Before the 1950s, the United Chapters had never taken an official position on athletics—not even in the early 1900s, when college football degenerated into such brutality as to become a national scandal; or in the 1920s, when coaches first began to excel professors in both plaudits and pay. But in 1951 the Senate instructed the Committee on Qualifications to inquire into their policies regarding intercollegiate sports when appraising applicants. In 1954 the Senate put into the leaflet for interested institutions a statement of what Phi Beta Kappa expected of them in the control of athletics and in the recruitment, eligibility, and support of athletes. "This means," the leaflet emphasized, "that financial assistance for athletes, including scholarships, grants-in-aid, loans and student jobs, will be in approximately the same ratio to the number of athletes in the student body as all financial assistance is to the total number of students."

The statement on athletics came before the 1955 Council for its approval. A delegate from the University of Florida raised an objection: the new requirements would be too strict if they should prevent an otherwise deserving institution from qualifying. A delegate from the Knoxville, Tennessee, association agreed. He said his own institution, the University of Tennessee, had an ambitious athletic program that could not meet the new standards, but this had no adverse effect on the educational program of the college of arts and sciences. So, he contended, the standards would be unrealistic and unfair. Despite these arguments, the Council approved the statement by a voice vote.

Opposition to it continued. The Florida professor later restated the negative case quite eloquently. Phi Beta Kappa was wrong-headed, he argued, in its "illiberal insistence that all parts of a modern complex state university system conform to the liberal arts ideal appropriate to the purely liberal arts colleges of an older day." If the society should persist in this policy, "there would soon be only two types of institutions that could expect to receive charters of Phi Beta Kappa: women's colleges and small liberal arts colleges with no particular ambition in athletics." This prophecy seemed to be coming true when in 1958 the Council chartered only one chapter, the one at Kalamazoo College.

The strongest argument against the new policy, as senators acknowledged, was the charge of unfairness. Henceforth, all the universities with topnotch football teams would be excluded from Phi Beta Kappa. But look at the highly respected institutions that had such teams and also had Phi Beta Kappa charters—for example, Ohio State, UCLA, Michigan, Oklahoma, and Duke! To this charge the senators had an answer:

> The Senate recognizes that inequalities between the standards required of applying institutions and those of institutions already sheltering a chapter are real—and deplorable. But it is also aware that applying institutions have advanced the argument of a double standard on other occasions in the past—in comparing their foreign language offerings, for example, with those at some institutions that shelter a chapter, or their library resources, or the faculty salary scale, or some other point on which the comparison was favorable to the applying institution.

"Institutions not recommended for a chapter may point with natural bitterness to the Society's 'inconsistency,'" the Senate stated at another time. "It has been necessary, however, for the Committee on Qualifications to take the position . . . that the Committee cannot attempt a general investigation of the sheltering institutions, but will act only when incidents occur which suggest that academic standards are impaired."

Criticism of the policy on athletics nevertheless led the senators to relax it a bit. They agreed in 1957 to drop the specific rule that aid to athletes must be no greater per capita than aid to other students. But the senators retained the general principle that, in an applying institution, physical prowess must not have priority over scholarly talent and inter-collegiate athletics must not distort the educational process.[6]

The Committee on Qualifications did not investigate any sheltering institution to see whether its athletic program impaired its academic standards and whether its ΦBK charter ought therefore to be sus-pended. The committee did look into cases, however, where academic standards were thought to be endangered for other reasons. Of particu-lar concern were apparent threats to the freedom of teaching.

During World War II, the committee investigated the University of Georgia, and the Senate resolved that, "in view of political interference of the rankest nature," the university might have "lost some of its qualifi-cations for the maintenance of a chapter of ΦBK," and that the commit-tee should continue to investigate. That was as far as the society went. At the same time, the original home of the society, the College of William and Mary, was temporarily under a cloud of a different kind. "Since the Committee on Qualifications normally considers for charters only in-stitutions unconditionally approved by the Association of American Uni-versities," ran the committee's 1942 statement on the case, "the suspen-sion of the College of William and Mary from the A. A. U. list is a matter of serious embarrassment to the Committee on Qualifications and to the Phi Beta Kappa Senate."

After the University of Texas regents, for political reasons, had re-moved Homer Rainey as president, the Committee on Qualifications in 1944 declined to recommend the suspension of the university's ΦBK charter. The chairman of the committee, himself the president of a Southern college, said that suspension would do the society more harm than good. "For Phi Beta Kappa to punish the regents by suspending the chapter would be a carpetbaggerlike action that would be deeply re-sented by all the other Southern chapters." The first postwar Council in 1946 instructed the committee to make annual checks of the University of Texas for three years and empowered the Senate to suspend the Texas charter "if at any time within the triennium serious deterioration be found."[7]

As the postwar red scare intensified in 1950–1951, University of Cal-ifornia faculty members were facing dismissal if they refused to take a loyalty oath, which the state legislature had required over the opposition of Governor Earl Warren. The Phi Beta Kappa Senate adopted a resolu-tion expressing "grave concern" about academic freedom and tenure at the university. Governor Warren sent his thanks to the United Chapters.

By this time the United Chapters had made it a practice to refuse charters to institutions on the list of "Censured Administrations" of the American Association of University Professors (AAUP). Not until 1968, however, did the Council get around to giving an official endorsement to the AAUPs 1940 *Statement of Principles on Academic Freedom and Tenure.* (In 1980 the society decided that, while continuing to "take due note of AAUP censure," it would "make its own judgment" as to whether, in a particular case, the censure should mean disqualification.) During the 1970–1973 triennium, after having gained the Senate's approval for a charter, Southern Illinois University appeared on the AAUPs censure list. The Senate therefore rescinded its approval of SIU before the 1973 meeting of the Council.

At the 1973 meeting, the chairman of the Committee on Qualifications reminded the delegates of the committee's painstaking procedure. First, every member of the committee went over all the applications from colleges and universities throughout the country. Next, the members met and discussed the applications, decided which of the institutions they should consider further, and asked these for detailed statements of their qualifications. The committee then divided into teams of two or three, one or another of which visited each of the selected institutions, inspected its facilities, and talked with its trustees, administrators, faculty, and students. Finally, the committee reassembled, considered the team reports, and made its recommendations to the Senate for approval (or disapproval).

By the time the 1973 Council met, the delegates had had a chance to look at copies of the committee's evaluations of the approved colleges and universities, so the delegates were ready when the committee chairman gave his report. "There followed extensive questioning to the committee members," the Council's minutes read, "of several of the recommended institutions regarding such matters as library facilities, experimental programs, athletic scholarships, religious restrictions on academic freedom at church affiliated schools and the Ph.D. ratio among faculty members."

"The question of whether Phi Beta Kappa had ever expelled chapters when they had come under AAUP censure was raised," the minutes continued. "It was noted that while there had never been such an expulsion, the Senate had passed a resolution to take up the whole question of academic freedom as it applies not only to prospective chapters but to current chapters as well." Though the society might conceivably suspend a charter on account of AAUP censure, it could not expel a chapter for that or any other reason—unless the constitution of the United Chapters should be so amended as to give it the power of expulsion.[8]

"Phi Beta Kappa and Catholic Colleges"—by 1958 this had become a touchy issue among Roman Catholic educators, an issue they brought out in the open in the pages of the church magazine *America.* There were then 250 Catholic colleges and universities in the country, the Jesuit Neil G. McCluskey pointed out, and only two of them possessed Phi Beta Kappa chapters, though five had chapters of Sigma Xi. Since 1947 twenty-three had applied but none had been accepted. True, Catholic institutions could claim their own honor society, Delta Epsilon Sigma with eighty chapters, and schools belonging to the various orders had other honorary organizations. "Yet the fact remains that in American society—academic or nonacademic— Phi Beta Kappa has a meaning that no Catholic substitute has thus far been able to achieve."

While denying charters to Catholic colleges, McCluskey contended, Phi Beta Kappa had been conferring them on less deserving non-Catholic institutions. For instance, it had given one in 1955 to the University of Connecticut at Storrs, which until 1933 had been Connecticut Agricultural College and as late as 1939 Connecticut State College. "One may be excused for musing over the relative depth of the Storrs liberal-arts tradition and, say, the one that has fed a certain Catholic institution on the Potomac whose foundation as a liberal-arts college dates from 1789, and another Catholic institution which is the oldest college in the Louisiana Purchase territory and the first university west of the Mississippi." The plain implication was that Phi Beta Kappa was discriminating against Georgetown University and Saint Louis University, along with other Catholic institutions.[9]

More explicit than the Jesuit critic was a non-Catholic Washington woman, a Smith College ΦBK graduate of 1939, who in 1961 happened to hear that few Catholic institutions possessed chapters and that Georgetown had repeatedly been turned down while inferior non-Catholic institutions had been accepted. "In the light of this," the woman wrote indignantly to the United Chapters office, "it is impossible to refrain from raising the question of an anti-Catholic bias on the part of Phi Beta Kappa."[10]

In fact, the paucity of Catholic colleges with chapters was no more due to anti-Catholic bias on the part of Phi Beta Kappa than it was due to anti-Protestant prejudice on the part of the Catholic colleges, as a history of the matter will show. The society had long had an odor of Protestantism about it, the secretary having been for many years a Protestant minister and the central office his parsonage. Throughout the first century of the society's existence, almost every sheltering institution was affiliated with some denomination—William and Mary with the Episcopalian, Yale with the Congregationalist, Harvard with the Congregationalist and then the Unitarian, Brown with the Baptist, Wesleyan with the Methodist, and so on. Many of the denominational colleges required the president and some or all of the trustees to be members of the affiliated church.

"But, except as regards the Roman Catholic institutions," James Bryce observed in the 1880s, "there is seldom any exclusion of teachers, and never of students belonging to other churches, nor any attempt to give the instruction (except, of course, in the theological department, if there be one) a sectarian cast." Thus the secularization of the Protestant colleges had gone a long way by the time the United Chapters was formed, and the trend continued and accelerated after that. In the early 1900s the Carnegie Foundation gave added impetus to secularization by offering funds for faculty pensions only to institutions free from ecclesiastical control. To be eligible for the Foundation's grants, a college must have *education* as its aim, not proselytization or "denominational empire."

This requirement put some of the church-related schools in a dilemma, as they tried to obtain Foundation money and at the same time to retain church support. A question arose as to whether the University of Chicago, for example, was under the control of the Baptist church. "The Carnegie Foundation says it is, basing its decision on the stipulation that two thirds of the trustees and the president must be of that denomination," the author of *Great American Universities* reported in 1910. "But . . . the hard-shell brethren say it is not a Baptist institution but decidedly heretical and dangerous." It imposed no "creedal restrictions" on professors, and the faculty contained Catholics, Jews, atheists, and Protestants of various other denominations as well as Baptists.[11]

The University of Chicago and other denominational institutions of Phi Beta Kappa quality steadily lost their religious identity and, by or before mid-century, retained only a nominal connection with a church or no connection at all, being classified simply as "private" colleges or universities. Exceptions among Phi Beta Kappa institutions were few. One of them was Davidson College, which received its ΦBK charter in 1922. As late as 1965, Davidson remained completely under the control of the Presbyterian Church in the United States (the Southern Presbyterian Church). According to the Davidson bylaws, the trustees, the president, and three fourths of the full professors had to be Presbyterians. All other faculty members must belong to "some evangelical church." The college was "a teaching arm of the Church," and all teaching was to be done "in the light of the Christian faith."

By the 1950s, no Protestant college under strict church control and with a missionary objective would have been considered eligible for Phi Beta Kappa. A professor at one institution of that kind, Sterling College, a Presbyterian school in Sterling, Kansas, assumed that no such truly "Christian" college could claim a charter as of 1979. The professor did not blame this on religious prejudice. Writing in the magazine *Christianity Today*, he said: "Until a Christian liberal arts college achieves the status of Phi Beta Kappa excellence, or something equivalent, the non-Christian world is entitled to bypass all rhetoric and ask: 'Is there really an excellent Christian college?'"

When an application from Furman University came before the Council in 1973, many of the delegates were concerned about the following paragraph in the visiting team's report to the Committee on Qualifications:

Though Furman has strong ties with the South Carolina Baptist Convention, there would appear to be little denominational interference with the process of a liberal education. The Trustees have directed that staff members will be selected "without regard to race, color, sex or religious affiliation," to quote the official report to Phi Beta Kappa. Each faculty member's contract contains a statement that those employed by the University to teach are expected to recognize their obligations by exhibiting an exemplary Christian life before the students and the public both inside and outside the college. Furthermore, they are to be especially careful to avoid making or approving any statements that run counter to the historic faith or the present work of Baptists, "and that so far as is consistent with the professor's conscientious views and professional duties, he shall advocate and advance the causes fostered by the said denomination."

This passage seemed self-contradictory when it indicated "little denominational interference" and yet conceded that faculty members were expected to refrain from expressing disagreement with the Baptist faith. Did this imply a religious restriction on academic freedom? So many of the delegates thought it did that the application was voted down. Then the next day, after receiving assurances from Furman that academic freedom indeed prevailed there, the Council reconsidered and approved the application.[12]

While the originally Protestant institutions with "the status of Phi Beta Kappa excellence" were losing their "Christian" character—and while the secular state universities were rising to excellence—the Catholic colleges were insisting on their Catholic mission and resisting the secularization trend. In 1904 a Catholic archbishop declared: " . . . let us have no deal, no alliance with the promoters of godless education, either in the primary schools, or in the intermediate schools, or in the universities. At every stage, we must have God, and Christ, and the Pope, and our ancient faith, no matter what the consequences."

The consequences eventually proved to be rather disturbing to many Catholics. A study in the 1920s showed that Catholics were grossly underrepresented in *Who's Who in America*. They were similarly underrepresented among scientists, according to a survey in the 1927 edition of *American Men of Science* (and according to later surveys in the 1950s). This poor showing provoked some Catholic educators to urge that Catholic institutions adopt some of the professionalism and scholarly spirit of the secular universities. But clerics, trained in seminaries, long continued to dominate the administrations and predominate on the faculties of Catholic institutions.[13]

The values of Phi Beta Kappa were not the values that had prevailed among the Catholic colleges, and not until the 150th year of the society did people at any Catholic college even ask to be considered for a charter. Then—in the reform spirit that called for an approximation of the secular model—a few administrators and professors at Catholic institutions began for the first time to make overtures to Phi Beta Kappa

In 1926 the newly founded School of Foreign Service at Georgetown University contained as many as three Phi Beta Kappas on its faculty. One of them, the famous international lawyer James Brown Scott, indicated his wish, on behalf of the "selected group of graduate students" at the school, for a charter of the fraternity. In response, Secretary Voorhees explained that charters were customarily granted not to a graduate school but to the institution itself and, more specifically, to its undergraduate college of arts and sciences.

In 1929 the Georgetown registrar, Walter J. O'Connor, writing on the stationery of the dean of the College of Arts and Sciences, sent an inquiry to Voorhees, who pointed out that an application must originate with the Phi Beta Kappas on the faculty. There were none on the faculty of the College, O'Connor responded. "This is due to the fact that our College faculty is composed of members of the Society of Jesus and men who have not attended colleges where they would have an opportunity to become members of Phi Beta Kappa." The registrar was looking, he added, for Phi Beta Kappas on the faculties of the law and medical schools "to initiate the necessary correspondence." But nothing further developed at that time.

In 1933, when some of the Georgetown men were showing interest again, one of them appealed to Joseph Q. Adams, director of the Folger Shakespeare Library in Washington, and Adams wrote in their behalf to ΦBK President Clark S. Northup. "I am told that some years ago Georgetown professors started a movement to make an application for a charter," Adams related, "but that the President vetoed the idea; he was a strong Jesuit, with, I presume, anti-Protestant feelings. But the new President is extremely enthusiastic." It was too late, however, for consideration in the 1931–1934 triennium. As if to console the Georgetown people, Secretary Shimer confided to Registrar O'Connor that the Catholic University of America was on the list for investigation. This must have been cold comfort to O'Connor, for he was firmly convinced that Georgetown, as the oldest Catholic university in the United States, ought to be the first to have a Phi Beta Kappa charter.[14]

On behalf of the Catholic University of America, Professor Roy J. Deferrari of the Greek and Latin department had inquired about a chapter just a few weeks after O'Connor did so on behalf of Georgetown in 1929. "We consider it rank negligence that we never took steps toward this before," Deferrari wrote to Voorhees. Eager to expedite the Catholic University application, Voorhees sought the assistance of E. G. Swem of the William and Mary chapter, since the district system was still in effect and an applicant would need the endorsement of existing chapters in the same district. "It never occurred to me before that there is no Catholic institution in Phi Beta Kappa," Swem replied. "I am ready to help you if I can." But his help was not enough; the Senate declined to recommend Deferrari's university. "I quite agree with you," Deferrari, now dean of the graduate school, wrote to Shimer, " . . . that religious prejudice, rather than a consideration of academic qualifications, played a major part in the whole procedure."

In the next triennium, though gaining a place on the list for investigation, Catholic University again failed to get a charter. Deferrari applied once more in the very next triennium, the one ending in 1937. Early that year the university inaugurated a new rector, the Right Reverend Joseph M. Corrigan. In his inaugural address, Corrigan intimated that he meant to turn the university back from all secularizing tendencies. He spoke of the "disaster brought about by the failure of theology to dominate university education" and said "theology must send down its confidence of hope into every branch, major and auxiliary, that is to be taught in these halls." His remarks gave pause to the Committee on Qualifications, whose chairman told Deferrari "there seemed to be uncertainty about the educational aims of the new Rector." The committee declined to recommend Catholic University.[15]

At the same time, the committee did recommend another institution in the District of Columbia—George Washington University. Seven ΦBK faculty members of GWU had presented an application as early as 1907, then had deferred it because of the university's "economic vicissitude" following the Panic of '07. Finally, in 1935 GWU people applied again, though quite aware that, according to one "body of opinion" in the society, "chapters should be restricted to campus universities and colleges" and "none should be established in urban institutions with large bodies of evening students."

Dean Deferrari was exasperated to learn that the committee was approving George Washington University while disapproving Catholic University. He protested to Shimer:

> . . . when these high standards are held before us and at the same time an institution is selected as meeting these which we know by direct contact is definitely inferior to our own in these very standards we are somewhat bewildered. You speak "of the scholarly and of the liberal or cultural aspects of the institution." The institution I have in mind is notoriously a night school. It has not given a course in Greek for years, and Latin is in such little demand that for a period of over a year it offered no Latin courses at all. Whereas our own institution has been a supporter for years of the American Schools for Classical Studies both at Rome and at Athens, this university in question has never contributed financially or in a scholarly way to the development of these two schools. As for the atmosphere of the institution, it is distinctly a "bread and butter" one. Everyone is looking for credits with a view to bettering their economic status, either in the school system or in the various government offices. Classes are so large that in some of them loud speakers are used to carry the voice of the instructor to all parts of the hall.

While bypassing Deferrari's university, the Committee on Qualifications was—for the first time ever—giving its endorsement to a Catholic institution. This was the College of Saint Catherine, a women's college in St. Paul, Minnesota. The committee's decision troubled members of the Dartmouth chapter so much that their secretary complained to the United Chapters office. They were convinced that Catholic University would have been a much better choice. In reply, Shimer said that if the Dartmouth members did "not care to accept the judgment" of the committee, they should themselves make a "careful study" of Catholic University, and they "could get very significant data from the North Central Association," the accrediting agency. "The question isn't whether St. Catherine is the best of the Catholic institutions, but rather whether it belongs in the list of the 8 best colleges now interested in obtaining a charter."[16]

The 1937 Council voted to give charters to all the eight on the committee's list, including both Saint Catherine and George Washington. In the very next triennium, the committee and the Council made amends to Catholic University, granting it a charter in 1940. After that, there were two Catholic institutions with Phi Beta Kappa chapters, but more than twenty years were to pass before there was a third. This would be not Georgetown but Fordham University.

By 1938 the Fordham president, Robert I. Gannon, had become interested in obtaining a charter, and the Committee on Qualifications readily agreed to visit the Fordham campus during the triennium. The committee concluded, however, not only that the university had a problem with its library but also that it spent too much on athletes and too little on other students. In those days Fordham was a football power, famous for its "Seven Blocks of Granite" (one of whom, Vince Lombardi, maintained later when coach of the Green Bay Packers: "Winning isn't everything. It's the only thing").

In 1948, when President Gannon tried a second time, he could not understand why the Committee on Qualifications was not even going to examine his university. Its football and scholarship programs had changed, he told Shimer, and its library had improved. "The only point you mention is emphasis on curricula definitely liberal in character. Our classical tradition of 400 years is unchanged." In response, Shimer explained that the committee had to be highly selective with the more than a hundred applications that had piled up since the war. The committee was studying only those institutions "with definitely liberal curricula *which on the whole seem most likely to be recommended for a charter.*"

Once more, in the 1955–1958 triennium, the committee passed over Fordham. The university's very recent "de-emphasis on football," Secretary Billman admitted, was promising for the university's future prospects, but the present situation was uncertain. Currently the university was giving athletes more than twice as much as other undergraduates in scholarship aid—an average of nearly $800 to $362. This was grossly at variance with the 50/50 ratio that the committee had been insisting upon.

Finally, in 1960 the committee again sent a visiting group to Fordham. As "an organ of the Jesuit order," the visitors noted, the university had an administration of Jesuits and a faculty consisting almost entirely of Catholics, nearly half of them with a degree from Fordham, many others with a degree from Woodstock (the Maryland seminary for priests), and nearly all with degrees from Catholic institutions. Only about a dozen were members of Phi Beta Kappa. All but a few of the students were Catholics, and all classes began with prayer. "There may still be a massive thought control," one of the visitors reported. "Are the 'liberal' subjects in fact liberally studied?"

A partial answer to this question came later that same year, when Fordham students were rehearsing for two French plays, *Death Watch* by Jean Genêt and *No Exit* by Jean Paul Sartre. After almost a month of rehearsals, the vice president for student affairs stepped in to ban the plays. "They were considered as unrepresentative of Catholic philosophy," the *New York Times* elucidated. Nevertheless, the Committee on Qualifications, the Senate, and the Council approved Fordham for Phi Beta Kappa, and it received its charter in 1961.[17]

Georgetown was next in 1964. From the viewpoint of the Phi Beta Kappa authorities, Georgetown had long suffered from "serious weaknesses." One of these was an overemphasis on athletics. In the three seasons 1938–1940, the Hoyas lost only one football game, though playing teams of the caliber of Boston College and Syracuse. Then in 1951 Georgetown dropped out of intercollegiate football but continued intercollegiate competition in basketball and other sports. "To attract academically qualified athletic talent," a Georgetown spokesman conceded in 1958, "the College awards a small number of scholarships which—in necessary response to a highly competitive situation—are relatively generous."

Another alleged weakness of Georgetown was the "inbreeding" of the faculty. "As our Faculty must necessarily contain many Jesuits," Registrar O'Connor had protested in 1935, "it would be a matter of some years before we could overcome inbreeding." The lay membership gradually increased, and in 1964 the visitors from the Committee on Qualifications found that 112 of the 146 faculty members were laymen, 21 percent of whom were non-Catholic. Nearly 60 percent of the faculty held doctorates, and these came from an "admirable" distribution of institutions— almost as many from Harvard as from Georgetown, the same number from Yale as from Catholic University, and three times as many from Wisconsin as from Woodstock.

According to the visitors' report, the Georgetown president anticipated that Jesuit institutions would "inexorably become less and less Jesuit," as they were already doing. The Georgetown catalog still contained reassertions of Christian doctrine—statements of a kind that could make the reader wonder whether Georgetown was "a school for indoctrination or one of questioning scholars devoted to research and

teaching in the liberal arts and sciences." The Phi Beta Kappa representatives concluded, however, that in practice Georgetown now met "the
tests of a true university."[18]

Other Catholic institutions followed in obtaining ΦBK chapters: Notre
Dame and Saint Louis in 1968, Boston College, Manhattan, Marquette,
and Trinity (D.C.) in 1971, Holy Cross in 1974, and Santa Clara in 1977.
By 1977 there were still only a dozen of them all told.[19]

16

Race, Ethnicity, and Scholarship

"In considering the application of colleges and universities for charters," the Phi Beta Kappa Senate declared in 1942, "the religious affiliation (if any) shall be disregarded and all chapters should be informed that membership in ΦBK must be open on the same terms to students of all races and of all religious beliefs." Nevertheless, the United Chapters continued to face accusations of anti-Catholic prejudice, as has been seen. At the same time, a few of the individual chapters felt compelled to deny charges of anti-Semitism, and the central office took pains to refute allegations that the society was anti-black. Whatever the validity of the complaints, the society eventually became quite comprehensive with respect to blacks, Jews, and other minorities.

In 1947 President Harry S Truman's Commission on Higher Education produced a six-volume report, *Higher Education for American Democracy.* "For the great majority of our boys and girls," the report stated, "the kind and amount of education they may hope to attain depends, not on their own abilities, but on the family or community into which they happen to be born or, worse still, on the color of their skin or the religion of their parents." Even after victory in the war fought presumably for the Four Freedoms—freedom of speech and religion, freedom from fear and want—American youths could still find their educational opportunities limited by poverty or ethnicity.

Ethnic bias was reflected in the admissions quotas that the more prestigious Eastern private institutions had begun to impose on Jewish applicants shortly after World War I. Usually the quota system was hidden. Ostensibly, the discriminatory institutions merely added "character" to scholarship as an entrance requirement, or began to look for "regional

balance" in admissions, or set up an alumni committee to screen out
unpromising applicants. Columbia, for example, had a student body 40
percent Jewish when the university imposed its quota in 1920. In two
years the Jewish enrollment at Columbia dropped to 22 percent. At both
CCNY and Hunter College, it remained in the range of 80 to 90 percent.[1]

A special Mayor's Committee on Unity in New York concluded that
the situation had become worse instead of better during World War II,
the war to overthrow Naziism in Germany. In 1946 the *American Scholar*
published an article, "College Quotas and American Democracy," in
which the chairman of the Mayor's Committee highlighted the contra-
diction between democracy and the quota system. In 1948 the *Key Reporter* called attention to the rise in anti-Semitism as shown by the wartime
decline in the percentage of Jewish students admitted to professional
and graduate schools.[2]

While Phi Beta Kappa publications deprecated anti-Semitism, one or
two of the individual chapters may not have been above practicing it. At
any rate, there were occasional charges of discrimination against Jews.
Such discrimination could hardly occur in those chapters that elected
students strictly on the basis of their academic record. But to keep out
"grade-grabbers" and "grinds," some chapters in the 1920s and 1930s
had begun to take into account not only classwork but also "character"
and other personal qualities. If these qualities were discussed by the
chapter as a whole, a candidate might risk rejection simply because of
some members' likes and dislikes.

The University of Pennsylvania chapter seemed to have found a way
out of the dilemma when it set up an "electoral board" to interview
candidates. This pleased Secretary William A. Shimer, who thought
"such a method may avoid both the evils of depending solely upon grade
averages and [the evils of] favoritism and prejudice."

Just a few years after that, in 1940, Shimer received a letter from a
college president who, at a conference in Philadelphia, had heard confidentially from a University of Pennsylvania professor that there were
"certain abuses in the elections to Phi Beta Kappa" at the university.
"The specific charge was that Jewish students were clearly discriminated
against in the elections, one professor reminding his fellow members
before an election that while a student who was being considered had a
name which seemed to indicate he was a Gentile, he was in fact a Jewish
student." Shimer assured his informant that he, too, would be greatly
disturbed if the allegation were true. He would also be greatly surprised.
"Occasionally," he explained, "such charges have arisen through the disappointment of one or two students who were not elected for perfectly
good reasons." He made inquiries, which apparently satisfied him.[3]

Rumors of anti-Semitism at Pennsylvania continued to reach the
United Chapters office from time to time. In 1947 one of the senators,
on a visit to Philadelphia, talked with professors there and reported back

that he had been unable to "turn up any evidence of discrimination," as Secretary Carl Billman later recalled. "We decided to let the matter drop, but to make further inquiries if anything new turned up."

Nothing new turned up for a couple of years, and then in 1949 there came a letter from a Jewish woman who complained that her son had been unfairly treated at Pennsylvania. After being honored as a high-school valedictorian, the pre-med student had completed four college years in three with grades high enough (this mother said) for Phi Beta Kappa. But he was asked to appear before a committee for a personal interview. The committee asked him only one question: "Who is Major George Fielding Eliot?" The young man did not happen to know that Eliot was a contemporary writer on military affairs. To his "surprise and chagrin," as his mother put it, "he did not receive the most coveted of prizes, the 'Key.'"

Fully two years later, Billman sent a copy of this letter to the William and Mary president, who was going to Philadelphia to give the Phi Beta Kappa address at the university. The president reported back to Billman:

> I discussed the . . . letter with the officers of Delta of Pennsylvania and learned that there is nothing to it, as I thought. Indeed, the initiates whom I saw last night were about seventy percent Jewish. The practice of interviewing borderline candidates . . . was (fortunately) abandoned two years ago. It was the opinion of the officers of Delta that it did not prove anything and was generally misinterpreted by the candidates.

Billman replied: "It's good to know there is nothing amiss there."[4]

The impression that the initiates at Pennsylvania in 1951 were "about seventy percent Jewish" was in line with a general perception that Jews were disproportionately represented in Phi Beta Kappa. Their over-eagerness for scholarly distinction was said to be a reason for hostility to them. In *Christians Only: A Study in Prejudice* (1931) Heywood Broun and George Britt attributed campus anti-Semitism to the intellectuality of the Jewish student, to his striving for a Phi Beta Kappa key in preference to a football letter. If the answers on senior questionnaires can be taken as evidence, however, students in general rated the key a higher honor than the letter.

It is hard to say to what extent—if any—Jewish students surpassed non-Jewish students in elections to Phi Beta Kappa. Studies of the question seem to have led to confusing and contradictory results. A professor at CCNY compared 145 ΦBK members with 149 other students (a random sample) from the classes of 1929–1931. He found that Jews constituted 86.3 percent of ΦBK membership and 90.6 percent of the control group.[5]

In 1964 a student of names, Nathaniel Weyl, presented the results of his attempt to find the ethnic origins of the society's membership by analyzing the names of those elected. Weyl derived two ratios: first, the ratio between particular ethnic or national groups in the society and the entire membership; second, the ratio between the number of each group in the United States and the entire population of the country. If the first ratio should equal the second, the "coefficient" would be 100, and it would mean that the proportion in the honor society was the same as the proportion in American society as a whole.

Obviously, this kind of calculation involves a lot of guesswork. Who knows, for instance, whether *Johnson* is a British, an Irish, or a Scandinavian name? Who knows whether *Jacobson* is Jewish or Norwegian? How can anyone tell the number of people of a particular national origin there are in a population as mixed as that of the United States?

Anyhow, for the period 1776–1922, Weyl's coefficient for the British (English, Scottish, Welsh) in Phi Beta Kappa proved to be 98, and for the Jewish, 84. For the period 1922–1961, however, the coefficients were 142 and 695. If correct, these figures would mean that during those forty years Phi Beta Kappa elected almost four times as many Jews, proportionally, as gentiles of British descent and almost seven times as many as Americans in general. Next to the Jewish Americans, incidentally, the people with the highest coefficient for 1922–1961 were the Chinese Americans with a rating of 229.

As a partial explanation of the Jewish showing in Phi Beta Kappa, Weyl referred to the disproportionate number of Jewish students, citing 1955 statistics to the effect that 62 percent of Jews of college age were in college, but only 26 percent of all Americans of college age were in college. Even so, if Weyl's calculations were accurate, the Jews among the undergraduates were being elected at more than three times the rate of non-Jews.[6]

Subsequent investigations came to somewhat different conclusions. Basing their study on 1970 statistics, the sociologists Everett Carll Ladd, Jr., and Seymour Martin Lipset found that 80 percent of college-age Jews and 40 percent of the college-age population in general were enrolled in colleges or universities. "Jews perform much better as students," Ladd and Lipset inferred; ". . . they have been represented in the membership of Phi Beta Kappa at about twice their proportion in the undergraduate population"—only about *twice* the proportion; not three times, as Weyl had indicated.

Still another researcher in the 1970s, Stephen Steinberg, uncovered data that seemed to disagree with those of Ladd and Lipset. On the basis of his sampling, Steinberg concluded: "The proportion reporting an undergraduate grade-point average of A is about the same for Jews as for non-Jews (17 versus 19 percent)." Indeed, these figures would suggest that slightly fewer Jews, proportionately, were making straight A's. Steinberg found, however, that a much higher proportion of Jewish students (40 as compared with 13 percent) were attending what he classified as "high-ranking" institutions. He did not go into the question of

Jewish membership in Phi Beta Kappa, but if it was in fact disproportionate, perhaps this could have been due partly to the disproportion of
Jews in the high-ranking colleges and universities—the institutions that
would be most likely to have Phi Beta Kappa chapters.[7]

These investigations leave many questions unanswered. Nevertheless,
they would seem to indicate that whatever might have been the anti-
Semitic policies of some institutions at one time, Jewish students by the
1970s—if not long before that time—were quite well represented in the
"best" colleges and universities and also in Phi Beta Kappa.

When sociologists looked into the eth-
nic composition of Phi Beta Kappa, they omitted consideration of one
important group: Americans of African descent. Sociologists could not
very well include blacks in a statistical study of that kind, as Harry W.
Roberts discovered when he thought of making such a study in 1968.
Roberts, head of the sociology department at Virginia State College, a
predominantly black institution in Petersburg, wrote to the United
Chapters office to request the following information: the first person
elected, the first white female, the first black male, and the first black
female—all with dates; also the total black and total white membership.
In reply, Secretary Billman explained that the society did not keep records of a kind that would enable him to answer all the questions. He did
provide information as to the first five members, the first two women,
and the first *known* black person.

The Vermont chapter, which had elected the two women, also elected
the black. He was George Washington Henderson, born a slave and still
an illiterate youth when a Union army officer, a Vermont alumnus,
brought him north after the Civil War. Henderson graduated from the
University of Vermont at the head of his class in 1877. After studying at
the Yale Divinity School, he served as a Congregational minister and, at
Straight, Fisk, and Wilberforce universities, as a professor of theology
and the classics. At the time of his death in 1936, his wife was quoted as
saying he was not the first but the second Negro elected to Phi Beta
Kappa. His predecessor, if any, remained unidentified. Edward Bou-
chet, a black Yale graduate of an earlier class, became a member of the
society, but he was not elected until 1884, seven years after Henderson.[8]

A number of early black members may have been lost to history
because chapters were hesitant to identify them and give them publicity.
The historian of the Denison chapter, for example, noted that the chapter elected ninety-five alumni upon its organization in 1912. "To indicate
how entirely the standard of scholarship was accepted for eligibility," the
historian commented, "it is interesting to note that one Negro graduate
was elected at this time."

The *Phi Beta Kappa Key* first gave conspicuous publicity to a black
member in 1929, when it ran a biographical sketch and a photograph of

Ernest E. Just. Born in Charleston, South Carolina, in 1883, Just had studied at the State College in Orangeburg and then had worked his way through Kimball Academy in Meriden, New Hampshire. He graduated from Dartmouth with Phi Beta Kappa in 1907 and gained a doctorate from the University of Chicago in 1916. A Howard University professor and a world-renowned marine biologist, he was doing research in Italy at the time the *Key* took notice of him.[9]

In 1938 the editors of the *Key Reporter* proposed an article about Charles T. Davis, a black Phi Beta Kappa graduate of that year's Dartmouth class. "Davis might be embarrassed by any such publicity," the secretary of the Dartmouth chapter objected; ". . . as there is no distinction here between white and colored members, [the executive committee] felt that the Society should not make any such distinction. Furthermore, it is felt that there might be a little antagonism on the part of some of the southern chapters if this matter was over-emphasized."

The *Key Reporter* gave attention to other black members now and then, particularly when something about them besides their race was newsworthy. The periodical acknowledged Paul Robeson in 1940, when he was scheduled to sing at a Phi Beta Kappa "national dinner" in New York. It ran a photograph of Darwin T. Turner in 1947, when, at fifteen, he graduated from the University of Cincinnati as one of the youngest persons ever elected to the society. (Turner was to be a ΦBK senator in the 1980s.)[10]

While there was no way of ascertaining how many blacks had made Phi Beta Kappa at most of the institutions with chapters, it was easy enough to know how many had done so at quite a few of the institutions. At those in the South, the total was zero, since none of them admitted blacks.

Segregation, making no exceptions, was a bit of an embarrassment to the United Chapters. Once, in 1936, Shimer received a complaint that impelled him to send a polite inquiry to the William and Mary chapter. "I know of no policy or incident that may have occasioned a conclusion 'that Negroes were not welcomed at the Phi Beta Kappa Building at Williamsburg,'" the chapter secretary replied. "It will be understood, of course, that the College of William and Mary is a state institution and that it follows the laws and customs of the State of Virginia in not admitting negroes to matriculation and instruction or to its social affairs." Thanking the chapter secretary for his "good letter," Shimer wrote back: "This should justly appease our critic."

Some members of Phi Beta Kappa, though living in segregated states, looked upon segregation with less insouciance than Shimer did. "The members at Atlanta, Louisville, and Wilmington, Delaware, have invited the Negro members to attend their association meetings and take part in the dinners," it was reported in 1938. "And they are able to find hotels in the South that will permit these joint meetings." In 1942 the association of Miami and vicinity surveyed the schools of Dade County, Florida, to reveal how far the educational opportunities for blacks fell below those for whites. The association was "making these facts known in order to arouse public opinion sufficiently to demand betterment."[11]

More and more Phi Beta Kappa members (like Americans in general) became concerned about racism at home as they kept on supporting a war against racism abroad. In 1944 the *American Scholar* published an article, "Democracy and the Negro," in which Otelia Cromwell challenged the nation's intellectual leaders to point the way toward eliminating segregation and inequality. One reader responded: "My father . . . member of Alpha of Ohio, Phi Beta Kappa, and I, member of Epsilon of Ohio, are both gratified to find Phi Beta Kappa leading again as it faces the stirring issue of the American brand of race prejudice as it confronts us today."[12]

But Phi Beta Kappa did not speak with a single voice, as members showed only too plainly at the end of the war, when they debated the issue of segregated education in the pages of the *Key Reporter*. The debate began with an antisegregation letter from Broadus Mitchell, a Kentucky-born economist who had taught at various universities, including Howard, and was currently director of research for the International Ladies Garment Workers Union. "In order to present simultaneously both sides of the question," the editor of the *Key Reporter* noted, "we invited Virginius Dabney, liberal editor of the *Richmond Times-Despatch*, to write a letter defending segregated education." Mitchell led off with a specific proposal:

> Often there is strong criticism of Phi Beta Kappa because it does not take a more constructive part in guiding educational policy. Before granting a charter (which is highly coveted by non-Phi Beta Kappa institutions), why should it not demand that a condition for receiving one be that no discrimination on account of race, creed, color or national origin be practiced? In the case of chapters already established, there could be an earnest exhortation from the national and local officers for reform, and these pleas, taken in the total, would be impressive. Old chapters could not be disestablished because their institutions discriminate, but the Society could refuse to set up new ones where such offense to education is given. By this action, the Society would raise a standard of democracy to which many would rally.

In refutation, Dabney argued that such action by Phi Beta Kappa "would probably rouse still more resistance . . . to better educational and other facilities for the Negro" in the South and would have only a negative effect on the segregating colleges and universities; it would "not cause those institutions to admit colored students."

Later a Ho-Ho-Kus, New Jersey, woman joined the negative argument by reminding her fellow members that many Northern institutions, with their quota systems, also discriminated against certain groups. "If all the colleges now holding ΦBK charters were thoroughly investigated, how many would meet the standards which Dr. Mitchell sets up for colleges seeking new charters?" the Ho-Ho-Kus woman asked. "If equality is good for the South, it is also good for the North." A man in Houston, Texas, had another question. "I may be a bit mixed on what constitutes the bases for awarding ΦBK, but I thought it was good character and high scholarship," he wrote. "What has the Negro question to do with ΦBK?"[13]

With members thus at loggerheads, the National Council took no action on the issue until the Supreme Court had declared segregation laws unconstitutional in 1954. Then at their meeting the next year, the delegates adopted the following resolution: ". . . the 1955 triennial Council of the United Chapters of Phi Beta Kappa urges all its chapters and members throughout the land to lend their active support to the effectuation as soon as practicable, in good faith, of desegregation of public education in the United States." That wording was innocuous enough, presumably, to satisfy all but the most die-hard segregationists among the delegates. Certainly it was a far cry from the definite anti-segregation policy that Broadus Mitchell had proposed ten years earlier.

In 1957 the CCNY chapter attempted to revive the Mitchell principle. A CCNY petition called on the Senate to deny a charter to any publicly supported institution that excluded students on account of race. The petition also asked the Senate to consider denying charters to private institutions that did the same. These proposals the Senate rejected after due deliberation as unwise and undesirable. It merely reaffirmed the Council's 1955 resolution urging all chapters and members to support desegregation.

When Southern colleges and universities finally began to desegregate, they did so in response to pressures other than those coming from Phi Beta Kappa as an organization. Once the chapters at Southern institutions had admitted blacks, they treated them as eligible for membership. In 1963 the University of Georgia chapter elected two of them.[14]

Most of the blacks attending college in the South continued to go to all-black or predominantly black institutions and could not be elected in course until their institutions obtained a ΦBK charter.

To readers of the *Saturday Evening Post,* after the appearance of its February 19, 1949, number, Phi Beta Kappa looked like a thoroughly racist organization. The *Post* was still the most prestigious of American weeklies, full of information and entertainment, when magazines of that kind had not yet been superseded by television; and here in it was an article bearing the byline of a highly reputable and authoritative journalist, a Pulitzer-prize-winning biographer, a former Columbia University professor—Henry F. Pringle. In collaboration with his wife, Pringle was now telling the world about "America's Leading Negro University," Howard University, in Washington, D.C. The Pringles had the following to say about Phi Beta Kappa:

> Having gained some appreciation of the quality of Howard, we were astonished to learn that there is no chapter of Phi Beta Kappa, the honorary scholastic fraternity, there or at any other Negro college. A score or more of the faculty, who did their undergraduate work in Northern institutions, are members. Howard—along with such other leading Negro institutions as Fisk, in Nashville, Tennessee; Talladega, in Alabama; and North Carolina College at Durham, North Carolina—is certainly just as eligible as any of the 141 colleges where chapters of Phi Beta Kappa have been established.
>   A Phi Beta Kappa key is a definite financial asset to the scholar who plans to teach. The versions of why Howard graduates are denied this asset seem obscure. Applications for a charter may be made by the faculty members who belong to the society. In 1942 the Phi Beta Kappas on the Howard faculty filed a tentative application, but no action was taken on it because of the war. The fraternity's national organization denies any discrimination and points to the number of Negroes who have been elected at Northern colleges. The fact remains that Howard still does not have its chapter. If it gets it, Phi Beta Kappa's lack of prejudice will be demonstrated.[15]

These paragraphs, rousing Phi Beta Kappa members from the Atlantic to the Pacific, brought a flurry of protests to the United Chapters office. The Mills College chapter demanded to be promptly informed "as to the present status of the Howard University application." A Barnard College alumna wrote: "I was reading with interest a discussion of discrimination in education in the Key Reporter and it occurred to me that I had read of an alleged discriminatory practice which seems to have been adopted by P B K itself." A Union College alumnus, teaching at Fresno State College, was "deeply shocked." A quite peremptory Cornell University alumna wanted Howard to be granted a charter "at once and without further 'red tape.'"[16]

Patiently, Billman tried to placate the protesters and set the record straight. Howard, he repeatedly explained, had made preliminary inquiries but no formal application since the temporary suspension of the chartering process because of the war. "Last year Dr. Alain Locke, who is Professor of Philosophy at Howard, was elected to the Editorial Board of *The American Scholar*," Billman told one questioner. "In conversation here at the office, he has stated that Howard did not feel ready to apply in 1947, though he has indicated that the University will in all probability ask to be examined in the next triennium." To another correspondent, Billman wrote:

> The officers and Senators are concerned about the implication in newspaper stories like the one recently published in the *Saturday Evening Post*. They do not believe, however, that Phi Beta Kappa has been discriminatory against Negro institutions. Nearly all of them are located in states where

segregated education is enforced by law. A glance at the highly vocational and technical nature of the curricula at most Negro institutions, and a comparison of their financial resources with those of colleges for whites, will show that America has shamefully failed to support equality of educational opportunity for all of its citizens.[17]

This was certainly an accurate summary of the past and prevailing status of segregated higher education for blacks. Booker T. Washington had advocated vocational training; William E. B. Du Bois, broad liberal and humane learning; and Washington's views had won out. Through the first decades of the twentieth century, gifts to black liberal-arts colleges declined, while the endowments of "industrial schools" increased. Appropriations for state institutions for blacks were scanty. Howard, founded in 1867, was supported by the federal government, but for decades Congress treated the university like a stepchild.[18]

Howard was still a struggling, straggling institution when people first thought of its seeking a ΦBK charter. As early as 1911, the young biology instructor Ernest E. Just, then only four years out of Dartmouth with his key, made overtures to the United Chapters office. During the 1920s Professor Just encouraged students to make direct approaches. "Beta Beta Chapter of Delta Sigma Theta Sorority is desirous of organizing at Howard University at Washington, D.C., a chapter of the Phi Beta Kappa Society," one student wrote. And another: "We have an honor society here at Howard, Kappa Mu, built on the same requirements of scholarship as the Phi Beta Kappa. This organization . . . is desirous of becoming a chapter of Phi Beta Kappa." To all such correspondents, Secretary Oscar M. Voorhees politely explained the process by which charters were granted.[19]

In 1924 the Howard dean of men, Edward L. Parks, asked Voorhees for application blanks, saying the ΦBK members on the faculty wished to start the necessary proceedings. Voorhees replied that it was too late: "The National Council will meet in September, 1925, and the chapters are now considering the institutions that they will recommend for charters." Parks wondered why the Howard men could not take the initiative and present their case to the chapters in the district. Did Voorhees mean they could never apply? In response, Voorhees described the workings of the district system—first, the chapters in each district make nominations; next, each of the five or six institutions receiving the most votes is invited to submit a statement of its "strength and standing"; then the Council makes the final choices, "not to exceed three from each district." From all this, Parks could infer that Howard would simply have to wait for some of the all-white chapters in the South Atlantic district to nominate it![20]

Voorhees made the point explicit when in 1930 the dean of Howard's college of liberal arts, E. P. Davis, took up the cause that Parks apparently had abandoned in despair. Voorhees told Davis: ". . . before an institution can be an active candidate it will need to be nominated by the chapters in the District." But Shimer, after succeeding Voorhees, informed Davis in 1931 that the society was discontinuing the district system and was setting up the Committee on Qualifications. Thereafter Howard, like any other institution, could make direct application to the committee.

The new procedure, it seemed to Shimer, would allow prejudice less scope than the old system did. Nevertheless, the United Chapters soon faced charges much like those the Pringles were later to make. In 1934 Shimer wrote to Davis:

> Recently from two sources reports have come to this office to the effect that racial prejudice has prevented Howard University from receiving a chapter of ΦBK. I know of absolutely no basis for such reports. In fact, I had always thought that ΦBK deserved commendation for having avoided such prejudices. A number of negroes have been elected to membership. I wish you would send me a list of all such members of whom you happen to know. The proposed revision to the general Constitution includes the following statement: "There shall be no discrimination because of such characteristics as sex, race, color, residence, or religious, political, economic, or other belief."
>
> Will you kindly let me know whether you feel that there is any justification for this criticism? Do you know how the reports got started?

"I have neither made nor encouraged unfavorable reports concerning Phi Beta Kappa's attitude toward Howard University," Davis replied. "But our standing and the character of the work we are doing are such that the criticism of this kind by others is inevitable." True, the number of blacks elected was growing from year to year; the list could be found in the *Negro Year Book* published at Tuskegee Institute. "But the election of individual Negroes to existing chapters is a very different thing from chartering a chapter in a college administered and attended wholly or almost wholly by Negroes."

"I suppose it is true that the standing and character of the work you are doing makes criticism of this kind inevitable," Shimer conceded. "Criticism of other kinds occurs in connection with our failure to establish chapters in other institutions such as Catholic University, Georgetown University, University of Pittsburgh, Pennsylvania State, Elmira, Clark University, and the University of Florida." Shimer continued with another question for Davis: "Do you think that the standing and character of your work would rank Howard University ahead of all of the above institutions?" Davis would only say: "I have great respect for them and their work."[21]

Then fourteen years later the Pringles spent some time on the Howard campus, talking with administrators, faculty, and students to gather material for the *Saturday Evening Post* article that was to give wide circulation to the local grievance. Before writing the article, Pringle inquired of the Phi Beta Kappa secretary how chapters were established, whether Howard had applied, and if so, why it had been turned down. In a prompt reply, Billman detailed the procedure. He pointed out that colleges were not accepted for examination unless they were on the approved list of the Association of American Universities, that Howard had been on that list only since 1939, and that in 1942 the ΦBK members there had submitted a request for examination too late for the Committee on Qualifications to consider it before the committee suspended its investigations because of the war. (The AAU list was discontinued in 1948.) The Howard people, he assured Pringle, would have another chance in the 1949–1952 triennium.

In putting together their article, the Pringles managed to avoid mentioning any of the points that Billman had so carefully made in his letter to them. Instead, they set themselves up as experts who knew better than Billman or any other Phi Beta Kappa representative just what should qualify an institution for one of the society's charters. "The only true standard for evaluating a university," they pontificated, "is the quality of the men and women who teach there." And they listed some of the shining lights on the Howard faculty, among them John Hope Franklin, who at thirty-three was the author of the "standard work" on the history of American blacks.

But the Pringles also discovered—and revealed in their article—a number of "defects" at Howard. Distinguished though the professors were, they suffered under a "work load" so heavy that they could not find time for conferences with undergraduates, whose most frequent complaint was that they did "not get enough personal attention from their instructors." There was a "woeful lack" of scholarship funds. Because of the "inadequacy of Southern segregated grammar and high schools," there was an "abnormally high rate of probations and failures." Too many students, the president himself believed, came to Howard "to prepare for specific vocations rather than to obtain a broad general culture." The president, Mordecai Wyatt Johnson, was still laboring mightily to overcome the handicaps of the university, which had been "in a chaotic state when he took over" in 1926.[22]

These serious defects, which the Pringles themselves noted, did not keep them from asserting that Howard was "certainly just as eligible as any of the 141 colleges" possessing Phi Beta Kappa charters in 1949. But comparable defects, if found in institutions other than Howard, could have been expected to give the Committee on Qualifications pause.

In 1951 the committee chairman, William T. Hastings, visited the Howard campus. Hastings was an advocate of equal opportunity for blacks in higher education and in Phi Beta Kappa. At Howard he was struck by the "high mortality figures—40% drop at the end of the first year and another 20% at the end of the sophomore year." With a positive attitude, he took these figures to mean that "proper standards in instruction were maintained." He noted "a few weaknesses," however. The li-

brary holdings were inadequate, and the students did not "make sufficient use" of what was there, partly because of the "type of course
assignments."

More serious, the major-minor requirement was such that it could lead
to "an illiberal degree of specialization," and the college of liberal arts
contained "some technical or vocational subjects, with highly developed
programs" that were "undesirable from the point of view of Phi Beta
Kappa." One of the professors "said he was not sure that Howard met
the standards" of the society. He was referring to "the presence in the
College of Departments which did not belong to Liberal Arts"—practical
art, business administration, drama, education, home economics, and
physical education.

In reporting to his committee, Hastings proposed recommending
Howard on the condition that the chapter incorporate the following
stipulations in its bylaws: Majors in the nonliberal departments of the
college would be ineligible; no more than fifteen semester hours of
technical or applied subjects could be accepted, and the grades in these
courses would be disregarded; each candidate's program would have to
be examined to see that the candidate had not "carried specialization so
far as to result in an unbalanced, illiberal course of study."

The Council approved the Howard application in 1952, and the chapter was organized in 1953. It became the Gamma of the District of
Columbia, the third to be established in D.C. Georgetown University was
still waiting for a chapter, and some of Georgetown's advocates were still
complaining that inferior institutions were being favored above their
own. In his 1958 article in *America,* the Jesuit critic of Phi Beta Kappa
gave several examples. One of them he did not mention by name; he
merely referred to it as a university that had received a charter in 1952
even though its entrance requirements were so low that 40 percent of
the students dropped out at the end of the freshman year and another
20 percent at the end of the sophomore year.[23]

Fisk University received a charter at
the same time as Howard, in 1952, and Morehouse College in 1967. The
chapters in these three predominantly black institutions stood on terms
of perfect equality with all the rest. Like the others, each of the three
could exercise its voice and its vote in respect to the granting of new
charters.

The Fisk chapter, Delta of Tennessee, expressed some doubts about
authorizing a University of Tennessee chapter—which was not to become the Epsilon of Tennessee until twelve years after the Delta had
been formed. When the university at Knoxville was under consideration
in 1962, the Fisk chapter received an invitation to comment, since Fisk
was in the same district. Delta would "welcome the creation of another
chapter in Tennessee," its secretary informed Billman, but had "serious

misgivings about the ratio and amounts of scholarships granted by the University of Tennessee in behalf of athletes." Also, Delta had reservations as to "the record of the rural controlled legislature" in its appropriations. "That is, Delta chapter would hope that, if a charter is granted, one of the considerations should be contingent upon whether said legislature tends to 'starve or restrict' the dependable income of the liberal arts college as compared, say, to the college of agriculture." Fisk itself did not have a very dependable income, but no matter. A predominantly black institution now had a perfect right to judge the qualifications of a predominantly white one.[24]

And blacks could now hold high positions, even the highest one, in the Phi Beta Kappa organization. Not only Paul Robeson and Alain Locke but also Saunders Redding and Ralph Ellison served on the editorial board of the *American Scholar.* John Hope Franklin, an alumnus member of the Fisk chapter, became secretary of the Howard chapter in 1955. In 1966 he was elected to the Phi Beta Kappa Senate. In 1970 he won election to the vice presidency, and in 1973 he succeeded to the presidency of the United Chapters.

Franklin was then John Manly Distinguished Service Professor of American History at the University of Chicago. He held a Ph.D. from Harvard and was on the way to collecting honorary degrees from seventy-odd colleges and universities. A past president of the Southern Historical Association, he served as president of the Organization of American Historians during his Phi Beta Kappa term and was afterward to be president of the American Historical Association.

At the 1973 Council meeting in Nashville, Tennessee, where years earlier Franklin had gone with other segregated students to Fisk, the delegates resolved: "We welcome Professor Franklin to his presidentiad cordially and with every expectation that he will bring us triumphantly into the third century of our existence." He was to preside at the bicentennial council in 1976.

No doubt the William and Mary founders of Phi Beta Kappa had had in mind only white males of British stock when they declared that their society should not be "confined to any particular place, Men or description of Men" but should be extended to "the Wise and Virtuous of every degree." In his centennial address to the Massachusetts Alpha in 1881, Wendell Phillips referred quite unselfconsciously to the "Saxon foundations" of the American republic and to the "gift which the Anglo-Saxon race" had made to humanity in the republic's founding. During its first two hundred years, the republic underwent a metamorphosis in some respects, and so did Phi Beta Kappa.[25]

17

The Flattery of Imitation

Phi Beta Kappa was the ancestor or prototype of all Greek-letter fraternities and sororities and of all scholastic honor societies. In choosing their names and symbols they followed its example more or less closely—sometimes too closely for comfort. Profit-making organizations also adopted or adapted its insignia to promote merchandise of unimaginable variety, ranging from crib notes to brassieres. The United Chapters was rather slow to standardize the insignia of the various chapters and still slower to protect the key and the name from the growing host of imitators, many of them quite unscrupulous.

At their very first meeting on December 5, 1776, the founders of Phi Beta Kappa agreed on a square silver medal as an emblem. The Yale branch, after its organization, adopted a medal identical with William and Mary's. The Harvard branch made a slight change, the substitution of its own founding date, September 5, 1781, for the original December 5, 1776. Harvard kept its silver medal for more than a century, but Yale soon began to experiment with variations. By 1798 the Yale medal had acquired a ring in a pivot on the top edge. By 1806 the medal was being made of gold instead of silver, and a steel watch key was attached to the bottom of it. Later, as watches came to be wound by their own stem, the key ceased to be functional and the whole thing was made of gold. As new chapters were formed, a few of them used a silver medal like Harvard's for a while, but sooner or later every chapter followed Yale's example by authorizing a golden key.

The medals and keys were remarkably consistent over time in that all of them (so far as is known) retained the original inscriptions of *S. P.* and

ΦBK, together with a hand and some stars. But great variations developed in size and shape, ornament, styles of lettering, the depiction of the hand, and most significantly in the number of stars included. The number of stars remained three on the Yale, Harvard, and Dartmouth emblems until after 1817, when the Union College branch was organized. The Union brothers assumed that the three stars stood for the Alphas of Connecticut, Massachusetts, and New Hampshire, so they added a fourth star for the Alpha of New York. When Maine got its Alpha at Bowdoin, the Union chapter added a fifth star—and a sixth when Rhode Island got its Alpha at Brown. The other Alphas followed Union's lead.

By the 1850s the majority of chapters were using seven stars, and despite the proliferation of chapters, a majority of them were still doing so in 1900. At that time, according to United Chapters Secretary Eben B. Parsons, they had "fixed upon the number seven as the symbol of completeness"—the magic number. But not all the chapters conformed. The count of stars on the various emblems ranged all the way from one to a full two dozen.[1]

When chartering new branches, before the creation of the United Chapters, the existing Alphas had tried to maintain uniformity with regard to insignia as well as other matters. For example, the Connecticut Alpha stipulated in the 1847 charter it prepared for Western Reserve: "Each branch of the Society is to have the same devices on its Medal, the same Badge of Membership, the same Tokens of recognition, etc." But the society made no serious or successful effort to impose uniformity until long after the formation of the United Chapters.

In 1903 Secretary Oscar M. Voorhees sent each chapter this communication:

> It would seem advisable that some regulations should be adopted respecting the medal or key of Phi Beta Kappa. Many shapes and sizes are in use, almost as many in fact as there are jewelers engaged in their manufacture. Some have rounded corners, though the early ones were square. The number of stars varies from three, as at first, to ten [actually, from one to twenty-four] as required by the Alpha of New Jersey. If any definite regulations are to be adopted, they should grow out of the history of the medal, and not be arbitrarily enacted.

In a delayed response, the National Council of 1907 appointed a Committee on the Key. Making a preliminary report in 1910, the committee recommended the following: Each chapter should be allowed to choose between a medal and a key. The inscriptions should closely follow the original, but there could be three, seven, or nine stars and such border or other ornamentation as might be desired. No manufacturer was to be "designated badge-maker to the Fraternity"; instead, each chapter could go on making its own arrangements with a jeweler.

Without accepting or rejecting this preliminary report, the 1910 Council instructed the committee to prepare a design and to base it on the early keys. The committee promptly did so, and Voorhees presented illustrations of the chaste result, in four sizes, in the *Phi Beta Kappa Key* for January 1911. "The purpose in view when the committee was first appointed was the adoption of 'a design of key that should be accepted as standard,'" Voorhees now declared. "Hence the design herewith submitted is by authority of the Council, and is binding upon new chapters. It is also commended to the other chapters for favorable action."

The Council lacked the power to force the standard key upon existing chapters, and many of them held on to their traditional insignia. They found it difficult to do so, however, after the 1916 Council voted "that the Secretary be authorized to secure from the leading fraternity jewelry houses propositions for furnishing keys as exclusive jewelers to Phi Beta Kappa; and that on his announcement of a satisfactory arrangement every effort will be made by the members of this Council to secure the cooperation of the chapters." After competitive bidding, Voorhees gave a monopoly on the United Chapters' business to a particular firm. From 1917 on, orders from the cooperating chapters passed through the central office, which retained a modest royalty, for several years the office's only regular source of income except for the chapters' annual registration fee.

Some of the established chapters continued for a time to use their own designs and their own jewelers. Middlebury did so at least until 1921; Nebraska and Princeton, until 1925; Bowdoin, until 1941; and Brown, as late as the 1960s.[2]

While the letters ΦBK remained throughout on every medal or key, the pronunciation of the letters took even longer to be standardized than the key. The original pronunciation, "Fie Beeta Cappa," persisted into the twentieth century, but by this time it had to compete with "Fee Bayta Cappa," which Voorhees himself apparently used. Either of the two, the English or the classical, seemed to have the virtue of linguistic consistency, though the latter supposedly had more of the virtue when amended to "Fee Bayta Cahppa." The pronunciation that came to have the widest usage, however, was "Fie Bayta Cappa."

By the 1930s, some members had begun to worry about the diversity of pronunciations and the illogicality of the prevailing one. The Senate recommended to the 1934 Council a declaration in favor of "Fie Beeta Cappa," but a majority of the delegates thought common usage rather than logical consistency ought to govern. The next Council, the one of 1937, was expected to make "Fie Bayta Cappa" official. One at least of the society's classical scholars favored it, pointing out that Greek pronunciations changed over time and that there were no sounds in English exactly equivalent to those of the Greek Φ, B, and K.

Of the 276 delegates at the 1937 Council, 123 voted for "Fie Bayta Cappa," 31 for "Fie Beeta Cappa," 11 for "Fee Bayta Cappa," and none for "Fee Bayta Cahppa." In response to the question whether the pronunciation should be uniform, 111 said yes and 48 no. But only 22 thought the Council should take any formal action on the matter; 137 preferred simply an announcement of the results of the informal poll. Fewer than 60 percent of the delegates bothered to vote, and one of them merely marked his ballot: "All d____ foolishness."[3]

Collegiate organizations with Greek-letter names multiplied rapidly after 1900. These included both social and professional fraternities and also honor societies. Some of the honor societies recognized excellence mainly or exclusively in scholarship; others, in "leadership" or "citizenship." Some, the largest number, were departmental, limiting their membership to students who distinguished themselves in a particular subject. Others were general, embracing a fairly broad field of learning. Some were local, confined to a single campus; others, though beginning as locals, became national or at least intercollegiate.

At the turn of the century, there were only about a half-dozen honor societies of more than local significance. Much the oldest and most prestigious was, of course, Phi Beta Kappa, which originally had covered the whole of collegiate learning but, with the expansion and specialization of higher education, had chosen the liberal arts and sciences as its field. Next in age was Tau Beta Pi, established at Lehigh in 1885 for engineering, and then Sigma Xi, started at Cornell in 1886 for scientific research. Broader and more directly competitive with Phi Beta Kappa was Phi Kappa Phi, which the presidents of three state universities launched in 1898 to give their institutions an honor society in the absence of Phi Beta Kappa and which later enlarged its scope to include all of a university's colleges and schools.

Within the next few decades, dozens of new honor societies appeared. In 1923 the tenth edition of *Baird's Manual of American College Fraternities* listed forty-two of them, in addition to sixty-seven professional fraternities. In 1928 the fraternity magazine *Banta's Greek Exchange* offered an alphabetical list of fifty that it called "national honor societies." Two claimed to be "general" and one, "inter-fraternity." The rest represented the following subjects: agriculture (2), architecture, art, athletics, biology, chemistry (3), civil engineering, classics, commercial, dentistry, dramatics (2), economics, education, electrical engineering, electrical science, engineering (2), forensics (4), forensics-literature, forestry, French, industrial education, journalism (3), history, law, literature, medicine, military training, music (2), optometry, pharmacy, Romance languages, Romanic languages, science (2), and social science (1).

After surveying this list, Secretary Voorhees was moved to remark: "A sort of delirium seems to have overtaken the college world, with this plethora of organizations seeking National recognition as one result."

Something, he had long been convinced, ought to be done about it. Voorhees had readily accepted when in 1925 the executive committee of Sigma Xi invited him to lunch to discuss the problem of setting standards for honor societies and qualifying them for recognition. The 1925 Council authorized Phi Beta Kappa to cooperate with Sigma Xi and other societies in a series of conferences. One of these was held in Williamsburg at the time of the Phi Beta Kappa sesquicentennial celebration in 1926, with representatives from the following societies present: Phi Beta Kappa, Tau Beta Pi, Sigma Xi, Phi Kappa Phi, Alpha Omega Alpha (medicine), and the Order of the Coif (law). These representatives then and there completed the organization of the Association of College Honor Societies. The ACHS was to serve as a kind of accrediting agency for honor societies.

"Perhaps Phi Beta Kappa can help to greater discrimination by announcing that it is not interested in placing chapters in colleges that are already over supplied with these societies," it seemed to Voorhees. In particular, the 1928 Council "will need to decide if there is room for Phi Beta Kappa Chapters in institutions where Phi Kappa Phi already flourishes." But the United Chapters did not adopt a policy of excluding institutions on the basis of the number or kind of organizations they had, nor did it remain permanently a member of the ACHS. Honor societies continued to spring up, one of them, Alpha Sigma Lambda, presuming to be the "evening college equivalent" of Phi Beta Kappa. The United Chapters came to follow the practice of neither approving nor disapproving such societies officially.[4]

Individual chapters were themselves responsible for encouraging one of the new organizations. This was the Phi Society, which the Phi Beta Kappa chapter at Denison University organized in 1926 for freshmen. The chapter elected the Phi members and gave them a banquet at the start of their sophomore year. By 1935 there were Phi Societies also at Colgate, Ohio Wesleyan, Western Reserve, Cornell College, the Florida State College for Women, and Rollins College and Stetson University (though neither Rollins nor Stetson yet had a ΦBK chapter). "The Council never opposed the establishment of these groups, but it did turn down a suggestion that the United Chapters encourage other chapters to follow suit and to supply insignia for members of the Phi groups then in existence," Secretary Billman related in 1958. "The Council took the position that probationary memberships, or 'pledges,' are undesirable, and that there is no real need for formally organized 'junior' groups."[5]

Members of some chapters had no desire to encourage the spread of honor societies. In 1928 Lloyd W. Taylor, head of the physics department at Oberlin, declined an invitation to introduce there a chapter of Sigma Pi Sigma, a recently formed physics fraternity. "Our chief objection is that of becoming a party to the multiplication of 'honor' fraternities which are trailing in the wake of the Phi Beta Kappa and the Sigma Xi," Taylor told the national secretary of Sigma Pi Sigma, a member of the physics department at Penn State. "There can be no denying that the new organizations are affecting the prestige of these old ones much as the application of the title 'Professor' to magicians and corn doctors has affected the popular attitude toward the calling in which you and I are engaged."[6]

From time to time, educators looked to the United Chapters for advice about honor societies. "You are familiar, I am sure, with the junior college honor society, Phi Theta Kappa. We have been told that this organization has the approval of Phi Beta Kappa," a Ward-Belmont School official wrote to Voorhees in 1931. "Will you please tell us if Phi Beta Kappa has made any recommendations relative to junior college honor societies or if there has been any definite discussion regarding the organization of a junior college branch of Phi Beta Kappa?" Assistant Secretary William A. Shimer had to reply that, no, Phi Beta Kappa had never approved Phi Theta Kappa, had never made any recommendations regarding such societies, and had never considered setting up a junior college branch.[7]

In 1952 an assistant to the president of Texas Christian University asked whether Sigma Delta Pi was on the "approved list" of Phi Beta Kappa and whether a Sigma Delta Pi chapter would help pave the way for a Phi Beta Kappa chapter. In reply, Secretary Billman summarized the current United Chapters policy, which was not what Secretary Voorhees had once hoped it would be. Billman wrote:

> Phi Beta Kappa neither endorses nor objects to any of the Greek letter societies that have taken a single field as their province. During the 20's and 30's, when these societies began to multiply very rapidly, Phi Beta Kappa was concerned by a potential threat to standards, because each society functioned without reference to the others and there was a considerable amount of overlapping. This danger was largely overcome after the organization of the Association of College Honor Societies. The Association has explicit standards and does not admit applicant organizations unwilling or unable to meet them. Although Phi Beta Kappa prefers not to affiliate with the Association, it does look with favor on the Association's objectives.
>
> The presence of another honorary society on the campus of an institution seeking a chapter of Phi Beta Kappa would not help or hinder an application to Phi Beta Kappa. I cannot recall a single case in which the fitness of an institution for a chapter of Phi Beta Kappa has been questioned on the ground that there were too many other honorary groups on campus, or one that was objectionable from our point of view.[8]

One society that Phi Beta Kappa certainly did not object to was Sigma Xi. At least as early as 1918, the chapters of the two societies at the University of Pennsylvania were holding an annual joint meeting. From 1958 on, the United Chapters and Sigma Xi co-sponsored an address at every midwinter meeting of the American Association for the Advancement of Science.

The number of honor societies continued to increase, as did that of social and professional fraternities. Between 1967 and 1976, the number of honor society chapters grew from 5,100 to 7,211, of which 5,691 belonged to the ACHS. Social fraternities represented in the National Interfraternity Conference lost ground, the number of chapters decreasing from 4,022 to 2,962; but fraternities as a whole gained, the chapter total going from 4,493 to 4,899. By 1976 the grand totals reported for fraternities, sororities, and honor societies of all kinds were 23,363 active chapters with 11,652,097 living members.

These organizations recognized Phi Beta Kappa as their progenitor. The four national associations of fraternities and sororities observed the bicentennial of the American fraternity at Williamsburg in 1976. While there, the representatives toured Raleigh Tavern, which they looked upon as the fraternity's birthplace, their own ancestral home. Phi Beta Kappa delegates were meeting in Williamsburg at the same time but kept their bicentennial observances separate.[9]

Fraternities, sororities, and other college societies had imitated Phi Beta Kappa not only in taking Greek-letter names but also in adopting insignia similar to the familiar key. The oldest of the social fraternities had done so—Kappa Alpha, founded in 1825, almost half a century after Phi Beta Kappa. Generally, the others followed suit, as did the honor societies that multiplied so quickly after 1900. Many of the keys could readily be told apart from Phi Beta Kappa's, but some were such close copies as to be complete look-alikes at a little distance. Some of the names were downright travesties, such as Beta Kappa Phi, Phi Theta Kappa, and the Phi Beta Kappa chapter of Beta Sigma Pi.

Besides the college societies, a great many nonacademic organizations and individuals adapted the name or the key for their own purposes, all of them unauthorized. During the 1950s, 1960s, and 1970s, the ΦBK symbols were used to promote enterprises and commodities of a truly remarkable variety. The exploiters included an evangelist, a gospel center, a service club, a restaurant, a race track, and an ecological lobby (the Independent Phi Beta Kappa Environmental Study Group), a bank, a real estate broker, a travel agency, a novelty company, a department store, and a men's clothing store. Among the goods promoted were hats, shoes, hosiery, dresses, suits, ties, jewelry, men's and women's underwear, cigars, cigarettes, cosmetics, vitamins, books, magazines, phonograph records, student aids, draperies, lighting fixtures, tools and other hardware, and even industrial chemicals. Phi Beta Kappa unwillingly cooperated in selling all those things and more.[10]

Some people made downright fraudulent use of the Phi Beta Kappa connection. "Such use is by no means infrequent, particularly by unscrupulous persons seeking employment," Secretary Shimer noted in

1936. "Keys are sometimes found among the equipment of criminals." A 1954 example of an unscrupulous person seeking employment was the upstate New York high-school principal who was indicted for perjury after having told the board of education he belonged to Phi Beta Kappa. In his defense, he said he had misunderstood the board's question at the time of his job interview.[11]

As early as 1901, the United Chapters had begun to take steps—rather halting ones—to protect the society's insignia from misuse. The Council that year passed a resolution calling upon manufacturers and jewelers to refuse a key to anyone "except upon an order countersigned by an officer of a Chapter."

But the United Chapters did not try to get legal protection until honor societies had started to proliferate. In 1925 a committee advised the Council to do something to prevent other societies from adopting keys "closely resembling" Phi Beta Kappa's. "The Phi Beta Kappa emblem, either as a medal or as a key, has been in use for a century and a half," the committee pointed out; "and, while no copyright of its shape and design has been sought, it has certainly earned the right to protection against imitation by other societies." The committee recommended that the officers of the United Chapters be empowered to "enter a protest"—as the secretary had already done in the case of Theta Kappa Nu—and to "take such steps as may be found necessary to make such protest effectual."

The Council of 1925 directed the Senate to proceed in that direction. After some deliberation, the senators decided to apply for a trademark. They could not get the Patent Office to register the key's distinctive shape, since that had been in use by so many societies for so long. Finally on July 24, 1928, the Patent Office did register the three-letter combination ΦBK.

Complaints kept coming to the United Chapters office from members who objected to key copying. One complaint in 1936 particularly aroused the concern of Secretary Shimer. It came from a member of the Brown chapter who was an officer of a Rotary club and who sent Shimer a jeweler's advertisement illustrating honorary keys for Rotary officers. One of the keys was practically identical with Phi Beta Kappa's in size and shape. Shimer protested to the jeweler, who responded: "A great many other manufacturing jewelers are making very similar keys for any number of various uses." Shimer had no recourse except to appeal to the Phi Beta Kappa membership:

> Obviously a key having approximately the shape and proportions of the ΦBK key is likely to be considered as such, no matter what its other distinguishing marks may be. This is detrimental and embarrassing both to ΦBK and [to] the imitating organization. On the other hand many societies have shown admirable originality in adapting the key motif to general designs easily distinguishable from ΦBK keys. This may be desirable, for

the prestige of ΦBK has made the key design the accepted symbol of scholarship and high honor. The members of ΦBK, by registering annoyance and reporting individual cases of too close imitation, can do most to protect the ΦBK key. The United Chapters voices protests and appeals. It will welcome suggestions concerning the stand the Society should take and how it should take it.[12]

The United Chapters decided to seek further legal protection after commercial firms had begun to infringe upon the society's name. In the early 1950s the Bates Shoe Company, advertising its shoes as "Phi Bates," attempted to get that term registered as a trademark. The United Chapters filed a protest, and the Patent Office rejected the application. The society successfully applied for new trademarks of its own, to cover the term "Phi Beta Kappa" and also the inscriptions on the key. But the society's attorneys advised that its success and the Bates Company's failure would not necessarily prevent the company from continuing to advertise "Phi Bates." To stop the company, the society would have to bring suit. "We decided against that," Billman recalled, "partly because of the expense, but chiefly because we didn't want to give the Bates Company a lot of free publicity."[13]

Even more annoying than "Phi Bates" shoes were "Fybate" lecture notes. These were published in Berkeley, right across the street from the University of California campus. Intended to enable students to pass courses without going to class, the notes were hardly in keeping with the spirit of a true Phi Bete. By the time they were brought to the attention of the United Chapters, however, it was too late to do anything about them. By then, about 1959, Fybate Lecture Notes had been available for a quarter of a century, and the words had long since been trademarked.[14]

"Phi Beta Jantzen," though perhaps beneath Phi Beta Kappa's dignity, was in a way as flattering to the society as it was, supposedly, to the female figure. Here was a 1963 advertisement for a "back to school wardrobe" of panties and brassieres on models sketched inside the outlines of enlarged Phi Beta Kappa keys. Surely such an association of learning and loveliness ought to go far toward dispelling the false stereotype of the dowdy, undesirable woman Phi Bete (though, to be sure, that was not the advertiser's purpose). But Billman's assistant, Diane Threlkeld, wrote to the company and objected to the ad as offensive.

An attorney for Jantzen, Inc., thanked Mrs. Threlkeld for her letter and expressed the company's regrets. "I am sure you will agree, however, that there was no trade mark infringement involved because of the dissimilarities of the goods and services involved." Anyhow, the Phi Beta Jantzen campaign would last only through the fall season. "You have our word that it will not be repeated in future years."[15]

Max Factor & Co., though somewhat less conciliatory than Jantzen, Inc., was also responsive to Phi Beta Kappa's protest. This had to do with the designation of an eye makeup in 1964. "We are sorry you don't approve of our use of the name 'Eye Beta Kappa' and assure you that should a similar promotion ever be discussed, your expressions will be given careful consideration," the Max Factor general counsel informed Billman. "At present, however, our Company has no intention of repeating its Eye Beta Kappa Kit, though our decision in this regard was arrived at without the benefit of your Society's views." Whether sincerely or sarcastically, the counsel closed with the line: "It was nice of you to write to us."[16]

The magazine *Business Week* used a picture of a Phi Beta Kappa key in a 1978 billboard advertisement and, despite objections from the United Chapters, repeated the ad in the *New Yorker* and the *New York Times*. The Senate Executive Committee considered suing McGraw-Hill, Inc., the parent company of *Business Week,* for $100,000 damages if the magazine should persist in its unauthorized use of the insignia. The magazine soon desisted.

But Bloomingdale's then set up small shops in its department stores and advertised them under the name of "Phi Beta Caper." Phi Beta Kappa had retained an attorney for such trademark infringement cases, and he promptly put a stop to this caper with a letter to the company. Another of his letters to the Riggs National Bank of Washington, D.C., brought a promise to quit using a picture of the key, with the word "Riggs" superimposed, on a pamphlet promoting an educational loan program.[17]

These business corporations were more cooperative than one of the infringing academic institutions proved to be. Capuchino High School in San Bruno, California, a San Francisco suburb, set up an honor society and named it "Phi Beta Cap." In response to Billman's protest in 1962, the Capuchino principal said he had talked with his associates at the school. "They are in complete agreement with me that we should in no way whatsoever alter the name of our honor society." The principal went on to boast that their society had "received wide publicity in the great neighboring institutions of Stanford University, San Jose State College, and San Francisco State College, as well as such private institutions as St. Mary's, University of San Francisco and the University of Santa Clara."

Billman wrote again to reiterate his objection and to threaten legal action. "The name you are using is a clear violation of our rights," he said. "As such, it is open to punitive steps on our part." He wished the Capuchino honor society every success in achieving its aim. "But at the same time, we can hardly believe that the most auspicious start for an organization dedicated to the encouragement of scholarship is to plagiarize the name of Phi Beta Kappa. Nor are we ready to acquiesce in the plagiarism."

Billman next heard from the San Mateo County district attorney's office, the legal representative of Capuchino High School. "In reviewing the law on this matter we have concluded that the letters Phi Beta, being of common usage, are not the sole property of any organization or fraternity," the deputy district attorney stated. "The question then resolves as to whether the terms 'Cap' and 'Kappa' are the same or similar enough as to be misleading. In our humble opinion they are not." But, he said, Phi Beta Cap's legal representatives would be glad to discuss the matter with Phi Beta Kappa's legal representatives.[18]

The Capuchino people had called Billman's bluff. In the circumstances, there was little he could do—except perhaps to ponder the question whether there was any more honor among honor societies than among thieves. A lawsuit, he thought, would probably cost more than the United Chapters could afford, and the outcome would be uncertain. He was hoping to get an act of Congress that would make possible better protection of the identifying marks of group membership, but he made no headway with that project.[19]

18

"Relevance"—or Rationality

When Phi Beta Kappa reached its 175th birthday in 1951, its leaders were more concerned than ever about its mission of promoting liberal education, since the advocates of vocational training appeared to be rapidly gaining ground. There was also a widespread anti-intellectualism to contend with but, as yet, no hint of the way it was soon to invade academe itself. Then in the 1960s campus revolutionaries began to denounce "elitism" and to demand "relevance." To some of them, Phi Beta Kappa seemed the most irrelevant of academic organizations because it was intellectually the most elite. By the time it attained its 200th anniversary, however, the society had survived a period of campus revolt with undiminished prestige. Phi Beta Kappa had met and overcome many other challenges in its long life, and it entered its third century with excellent prospects for dealing successfully with whatever challenges it was yet to meet.

Before the Pearl Harbor attack, a Committee on the Status of the Liberal Arts and Sciences had started to gather information from the chapters but had suspended its work on account of the war and never made a final report. The society's leaders nevertheless continued to give thought to the matter and particularly to the bearing of the war upon it.

"All times of strain, such as war or depression, tend to shake men's faith in liberal education," ΦBK Senator Frank Aydelotte told a university audience in 1942. Yet, he said, a liberal education, cultivating the "freedom of the mind" as it did, constituted the "very foundation of our democracy." The first aim of such an education was "not to enable a man to make a living but rather to teach him how to live."[1]

Another senator, Christian Gauss, discussed the wartime policy of Phi

Beta Kappa in a 1943 issue of the *American Scholar.* "We all must recognize," he said in emphasizing the importance of liberal studies, "that the perversion of systems of education into instruments of propaganda has been one of the causes which facilitated the rise and strengthened the hold of the totalitarian states." Now that the war was interrupting most Phi Beta Kappa activities, there was a need for the older members, the women, and others not preoccupied with the war effort to concentrate on a "reexamination of the criteria of a liberal education." The Senate had long been conscious of this need to redefine the liberal arts. If the humanities were declining, the reason was perhaps not only the "crass materialism" of their foes but also the "blind traditionalism" of their friends. Too many professors fought against new courses and defended traditional ones—sticking to the letter rather than the spirit of the humanities—because of a "vested interest."[2]

Six distinguished educators focused on the liberal-vocational conflict when in a 1944 issue of the *American Scholar* they discussed "The Function of the Liberal Arts College in a Democratic Society." What was needed, John Dewey said, was to give a "humane direction" to the teaching of practical and technical subjects. Agreeing with that, Arnold S. Nash put it this way: "The study of the humanities should be carried on not as an addendum to but within the context of and giving meaning to a student's professional studies." Similarly, Ernest Earnest declared: "The liberal arts college cannot educate some sort of mythical men of vision; it must educate chemists and sociologists and journalists with vision." Kenneth C. M. Sills added that the college must "combine the ideal and the practical, the general and the vocational, the useful and the liberal." Only two of the six discussants saw no place for vocational subjects in the liberal-arts curriculum. Alexander Meiklejohn argued that vocational and liberal education should be completely separate, and the liberal should be identical for all students, without the "intellectual anarchy" of the elective system. And Scott Buchanan quoted Robert M. Hutchins: "Liberal education is the same for everybody everywhere always."[3]

A member of the *American Scholar* board of editors, the Harvard astronomer Harlow Shapley, expressed confidentially a view that indicated some dissension among the advocates of liberal as opposed to vocational education. Writing to the magazine's editor in 1945, Shapley said he had thought of

asking the hounds for humanism why they become so narrow and intolerant when they talk about "the future of the humane tradition." To be sure, those who have been driven into a defensive position because of the rather impenetrable technicalities of science and scientific thought are entitled to advocate the maintenance of the discipline to which they have become hardened. . . . I am not going to ask them if life and living are humane; and if they think that life is constructed of iambic pentameters and the analysis of the obscurities of Carlyle with his stomach ache. . . . No,

I am weary of these verbally alert "humanists" who in their anti-scientific attitude reveal that science to them is technology, gadgets and nostrums, and not the clean cut intellectual adventure that we try to make it.[4]

For all their wartime thinking and planning, the Phi Beta Kappa advocates of a purely liberal education could not halt or even slow the advance of what they considered "illiberal tendencies" during the postwar years. William T. Hastings, chairman of the Committee on Qualifications, complained in 1951: "Educational theorists are never at rest, and curricula are in a state of permanent instability." Both the Truman commission's report and the Harvard report encourage curricular experimentation. The presidential commission proposed "to redefine liberal education in terms of men's problems . . . to invest it with content that is directly relevant to the demands of contemporary society."

Apart from exhortation, Phi Beta Kappa could resist these trends only through the United Chapters' power to grant or withhold charters and through the individual chapters' policies in electing members. In 1952 the National Council required, as a "condition of eligibility," that a student must have taken at least two years of mathematics and two years of a foreign language either in high school or in college. The Council refused to eliminate this requirement when in 1964 the Committee on Qualifications and the Senate recommended that it do so.[5]

In the course of the campus revolt, which began at about that time, its participants repudiated the liberal arts as irrelevant and obsolete. But they also repudiated, on the same grounds, the practical or vocational programs that were supposed to be "directly relevant to the demands of contemporary society."

On the occasion of Phi Beta Kappa's 175th anniversary, which occurred amid the rising tensions of the Cold War, the *New York Times* paid the society this compliment:

There have been periods in our history when it was thought good fun to lampoon the "long-haired professors," "impractical intellectuals" and similar stereotypes. But the lessons of history point clearly to the key roles of intellectuals in our fantastically complex society. The President of the United States would be lost without large numbers of them to advise him in every key area of policy determination. A tycoon of industry may boast that he never went to high school, but he depends upon his laboratory scientists, his production engineers and his statistically minded sales executives for the successful operation of his business. In short, without the aristocracy of intelligence which Phi Beta Kappa has always sought to foster our democracy would be far poorer materially and spiritually and the free world would be far less competent to meet its tough contemporary responsibilities.

That editorial was written at one of those periods of prevailing anti-intellectualism to which the writer referred, anti-intellectualism being a large ingredient of Joe McCarthyism. Friends of Phi Beta Kappa were often on the defensive even before the outbreak of campus anti-intellectualism in the 1960s.

One friend of the society accused it of being too intellectual. Recounting the remarkable careers of the first fifty members, the friendly critic called upon the colleges of the 1950s, "mindful of their shortcomings in preparing their graduates for citizenship and public service," to emulate the William and Mary of the 1780s. "Will Phi Beta Kappa itself, as a contributor to avoiding chaos, arouse itself from its complacent worship of mere intellect as reflected in college grades and seek to recapture its long-neglected glory of distinguished public service?"[6]

The role of Phi Beta Kappa members in public life—both the actual and the ideal role—remained rather moot. During the war a British observer had been impressed by the large number of them in the American government (most of them holdovers from the New Deal bureaucracy) but thought they were declining in relative importance as more and more businessmen joined the wartime administration. President Dwight D. Eisenhower had a high regard both for the society and for one of its members, Congressman Charles Halleck, a graduate of Indiana University. Regarding Halleck, Eisenhower noted in his diary in 1953: "He is a Phi Beta Kappa, which means at least that he is highly intelligent and mentally adept." In that respect, however, Halleck seemed quite unlike other politicians Eisenhower knew. "This man is a different type."[7]

In 1958 (when there were still only forty-eight states and ninety-six U.S. senators), Senator George Aiken of Vermont raised the question: "Who would want to see 96 Phi Beta Kappas in the Senate?" James Reston, the Washington correspondent of the *New York Times*, was moved to reply. First, he obtained from the United Chapters office a list of the society's members who had figured prominently in national politics. Then—under the headline "Is Intelligence Always a Handicap?"—he discussed the significance of the list, which included five of the last nine presidents of the United States. Finally, he put a question of his own: "But if Senators scoff at Phi Beta Kappas and Presidents prefer the company of retired distillers, steel tycoons and mud-pack artists, who is to put an end to this silly and derisive national joke about intellectuals?" The *Key Reporter* reprinted Reston's article.

Like earlier listings of famous Phi Betes, however, this one failed to distinguish between those elected on the basis of their undergraduate performance and those honored for later achievements. When one of the senators acknowledged that he had received his key "by grace and not by works," the *Key Reporter* consoled him with an assurance that many of the men on Reston's list were honorary members. Indeed, only one of the last nine presidents (Theodore Roosevelt)—and only two others of all the presidents (John Quincy Adams and Chester A. Arthur)—had been elected in course. Two more were alumni members (Rutherford B. Hayes and Calvin Coolidge), and all the rest of the presidential Phi Betes were honorary.[8]

Society members appeared to assume their largest role in government when John F. Kennedy succeeded Eisenhower in 1961. "In the new Administration the accent is on scholarship," a news magazine reported. "Sixteen members of the top echelon of the new Administration belong to Phi Beta Kappa, the national scholastic honorary society." Three of these men were department heads, among them Dean Rusk of State and Robert S. McNamara of Defense. Rusk had earned his key at Davidson College; McNamara, at Berkeley. (Remaining in the Lyndon B. Johnson administration, both Rusk and McNamara were to have a large part in making and defending Vietnam policy—the policy that provoked the campus rebels who included Phi Beta Kappa among their enemies.) Twenty-five members of the society were also members of Congress, the Oklahoma delegation boasting the largest number with four.[9]

There remained some difference of opinion as to how well Phi Beta Kappas were doing in occupations other than politics. Certain answers for the early 1950s were provided in a study that an expert at the Columbia University Bureau of Applied Social Research made on the basis of data from a *Time* survey. According to this study, the Phi Beta Kappa student and the "Big Man on Campus," generally thought to be "two antithetical types," were actually "not so far apart." The "all-A students" were "more likely than anyone else to be the campus leaders." These A students in later life had the "best earnings record" in every occupational field, even in government jobs. They stood out most of all in the "learned (and low-paid) professions" into which the largest numbers of them went, but they also held a "clear advantage" in the high-paying professions of medicine, dentistry, and the law. Even in business their advantage was "fairly clear-cut," though "much more tenuous."

Nevertheless, some educators continued to view the economic prospects of the A students as comparatively poor. A *Key Reporter* review of the recent career study concluded: ". . . the man who plans to enter the business world can well argue that grades mean nothing at all." And the president of Birmingham Southern College reportedly wished his institution to be "dedicated to quality" but not to be merely a "breeding ground" for Phi Beta Kappas. "We want a good sprinkling of C students," he was quoted as saying in 1961. "They're the ones who go on to make the money with which to endow our college in the future."[10]

In the academic world at least, Phi Beta Kappas unquestionably were successful, as was indicated by a 1968 volume analyzing the careers of the 6,082 men elected to the Yale chapter between 1780 and 1959. The study showed a considerable shift during that period from education, law, medicine, and the ministry to business, industry, and science. Within each of these vocations, as well as in government service, the ΦBK graduates stood high. From this group, constituting less than one fifth of all the graduates, had come four fifths of Yale's top officers—presidents, provosts, secretaries, and treasurers. The data demonstrated that "academic leadership in undergraduate college is definitely correlated with that in later life."

As of 1970, the Yale president, Kingman Brewster, Jr., was not a Phi Beta Kappa by virtue of undergraduate election. Neither was the Harvard president, Nathan M. Pusey, though he was an honorary member. But Robert F. Goheen of Princeton, Edward H. Levi of Chicago, and many other college and university executives had earned their membership as students.[11]

Colleges and universities continued to covet a Phi Beta Kappa chapter, and those possessing one often referred to it as a sign of institutional excellence. Even to apply for a charter was taken as evidence of high standing, at least in the case of Parsons College in Iowa. One visitor in 1961 hailed Parsons, which was being "run on business principles" to make a profit, as a "great success story" (in fact, the enterprise was to go out of business before long). "And as a symbol of this proud new era in its history, the college is applying this year for a Phi Beta Kappa chapter."

Ostensibly, Parsons was taking marginal students and making real scholars out of them. The college librarian offered himself as an example of that kind of scholarly rehabilitation. "He had flunked out at the University of California, then went on to earn a Phi Beta Kappa key at Kenyon College, an equally demanding institution," the 1961 visitor wrote. "'I'm very proud of this key,' he said, fingering the gold symbol of redemption hanging from his key chain." In thus flaunting the key, the Parsons librarian was quite exceptional.[12]

Most academics seldom if ever displayed it. Even the national president of Phi Beta Kappa, as of 1970, kept his key in a dresser drawer along with his socks and shirts and took it out only for special occasions. The secretary of the Harvard chapter left his in a jewelry box. The secretary of the Bowdoin chapter had received a chain for his key from his mother-in-law when he graduated in 1959, but he had never worn either the key or the chain and could not remember where they were in 1970.

These three scholars confessed their negligence to a *New York Times* reporter, and he interpreted it as a response to the campus anti-intellectualism of the moment. "The advantage of membership is hard to tell, except that one gains the privilege of wearing the small square key, a status symbol of no small magnitude that has adorned vests for generations," the reporter wrote. "Even that privilege is seldom exercised today. The display of such honors is widely considered arrogant and ostentatious."

In fact, the privilege had been seldom exercised for many years. In 1959 a member in Cincinnati protested to the *Key Reporter:* ". . . how quickly these keys disappear from sight, for our social order still seems to require young people to hide signs of intellectual ability." It was hard to understand "why our culture encourages the wearing of foot-high letters on the bosoms of athletes . . . and at the same time frowns on the 'vulgar display' of symbols of scholastic excellence." Phi Beta Kappas seemed to fear the anti-intellectualism around them—and perhaps even to share it.[13]

In the 1960s and 1970s, the campus revolt presented a twofold challenge to Phi Beta Kappa. For one thing, it led to a lowering of the educational standards that the society existed to maintain. For another thing, it caused some of the brightest students, seemingly the best candidates for membership, to repudiate the society itself.

The revolt began with the "free speech" movement at Berkeley in 1964, when students demonstrated against a rule prohibiting political activity on the campus. Broadening to include additional goals, the movement soon spread to other institutions. The New Left, organized as the Students for a Democratic Society, called for a variety of reforms both inside and outside academe. With SDS support, blacks demanded not only civil rights but also recognition and respect. Women on and off the campus joined the cause of "women's liberation." Many young men and women took up the "counterculture," showing their contempt for middle-class values by their unconventional dress, speech, and general life-style. Students belonging to none of these groups combined with them in opposing, often violently, the draft and the Vietnam war.

Campus rebels turned their hostility against the system of higher education because it was a bulwark of the "establishment," because it provided research and other support for the military, and because it failed to meet what they considered their most pressing educational needs. They wanted courses that would be "relevant" to their interests of the moment and would give them an immediate sense of well-being. When the Carnegie Commission on Higher Education sampled student opinion with a questionnaire, 91 percent of the seventy thousand responding indicated a desire for greater "relevance."

While condemning the traditional curriculum, the rebels also denounced the traditional grading system. Grades themselves were irrelevant. Even worse, they made artificial distinctions of a kind that had no place in a truly democratic society. They smacked of "elitism," and that was bad.

In demanding reforms, the students had the backing and often the leadership of faculty members, especially the younger ones. The reformers succeeded in bringing about a number of significant changes. New curricula appeared, such as black studies, other ethnic studies, and women's studies. Distribution requirements were relaxed, so that students could avoid taking courses they considered particularly irrelevant—or particularly difficult. The grading system also was relaxed, so that students could avoid being graded in some, or even all, of the courses they did take. They were marked only "pass" or "fail," or only "satisfactory" or "no credit," with no risk of a failing grade.

Colleges and universities yielded not only to demands from their own faculty and students but also to pressure from the federal government, which by 1972 had become the largest source of funds for higher education. The government could withhold funds from colleges and universities that provided insufficient "affirmative action" under the Civil Rights Act of 1964. To avoid the appearance of racial discrimination, some institutions lessened their admission requirements, and in 1970 CCNY virtually eliminated them, adopting a policy of "open admissions." According to the 1974 report of the Carnegie Commission on Higher Education, many colleges and universities were being compelled to lower their own standards of academic excellence in order to meet the government's standards of equal opportunity.

One consequence of the changing educational policies was an inflation of grades. According to a statistical study, the national average of undergraduate marks (on a scale of 4.0 for an A) went from 2.36 in 1960 to 2.75 in 1975. Approximately two thirds of the increase occurred between 1968 and 1973. According to another study the average rose, between 1963 and 1974, from C plus/B minus to B, and the percentage of A grades more than doubled. This did not signify an improvement in student performance, as other statistics clearly indicated. During the period from the middle 1960s to the early 1980s, scores on the verbal part of the Scholastic Aptitude Test, taken by high-school seniors, declined by 14 percent. During the same period, scores on the verbal part of the Graduate Record Examination, taken by college and university seniors, declined even more—by 16 percent. Meanwhile, scores on the English and history achievement portions of the Graduate Record Exam also were falling.[14]

Reflecting the campus spirit of the time, some of the abler and more vocal students began to express their disillusionment with honor societies in general and with Phi Beta Kappa in particular. In 1967 a Cornell senior related in the *Cornell Daily Sun* how he had been elected to Phi Eta Sigma, the freshman honor society, then to Phi Kappa Phi, and finally to Phi Beta Kappa—each time paying a fee, attending a banquet, and listening to a speech. "There seems something wrong with a university and an educational system," he wrote, "which honors its students by charging them outrageously to belong to worthless honoraries which meet but once for a chicken dinner and a lousy lecture." He preferred to believe that "scholarship, at least at Cornell, is its own reward."[15]

In 1968 the graduating senior with the highest grade-point average in the College of Letters and Sciences at Berkeley made the Phi Beta Kappa banquet an occasion for renouncing the "grade-point game" at which he had succeeded so well. In a speech that drew much applause from his fellow Phi Betes, he declared:

> It was foolish of me not to realize that by playing the grade-point game I was compromising my ideals about education and copping out to a dehumanizing system. But the university helped me to fool myself. Its homage to the grade-point standard made it easy for me to assume that because of my grades, I was really learning about the world around me. . . .
>
> I think American politicians and educators should stop their platitudes about the greatness of college education, the broadening of the mind and the development of the individual. The truth, unfortunately, is that most middle-class students attend college today because they are under parental or internal pressure to prepare themselves for a well-defined career. Their parents are obsessed with making their way up the socio-economic ladder to the promised land of the "good life," the affluent society with steak on every grill and three cars in every garage. These parents, and sometimes their children, have equated happiness with the earning power that a college education promises to bring.

In saying all this, the young man was no doubt commenting perceptively on popular attitudes in his time. He was also confessing to having been a grade-grabber and a grind, the kind of student for whom Phi Beta Kappa really had no place.[16]

Two years later, another top senior at Berkeley confessed her own unworthiness. In her column in the student newspaper, she said her "kind of knowledge" was only "how to get around the system." She felt "shame and guilt" for having accepted election to Phi Beta Kappa, and she had nothing good to say about the society or about the "rotten degenerate system" of which it was a part. Several of her classmates, among the 460 who were elected, did not have to share her guilt: they refused to join.

"Phi Beta Kappa: Who Needs It?" the *Harvard Crimson* had asked as early as 1964. "The University already offers enough carrots to the donkey of academic achievement," the editorial chairman of the paper wrote. "Realizing this, several undergraduate Phi Beta Kappa members have suggested abolishing the chapter." Two of those elected in 1970 accepted membership in the hope of converting the chapter into an agency for campus reform. Disappointed, one of the two resigned. "I think the organization may have gone past the point of usefulness," he said.

At no place in 1970 were Phi Beta Kappa faculty members more shocked and embarrassed than at William and Mary. There, in Raleigh Tavern, in the very room where (supposedly) Phi Beta Kappa had first met, one of the initiates took occasion to deliver a tirade against the society. "Usually it's a short thank you," the chapter secretary drily remarked when reporting the incident.

Incidents of this kind probably did not reflect the attitude of the great majority of students being elected—a total of more than twelve thousand in 1970. Even so, more students than ever before were declining membership or accepting it without enthusiasm. The society's leaders grew more and more concerned about this revulsionary trend, as they also did about the general deterioration of academic standards.[17]

The United Chapters responded to the crisis by appointing study committees, discussing committee reports, considering society opinion, making various recommendations, and taking no definite action.

In 1967 the Triennial Council set up the Pass-Fail Study Committee to report on ungraded courses and their implications for Phi Beta Kappa elections. The committee sent a questionnaire to the chapters and received replies from 121, of which 97 reported some kind of pass-fail option at their institutions. The committee also looked into detailed studies of the system at particular colleges and universities. And it exchanged information with an Association of College Honor Societies committee, which had sent a questionnaire to college registrars.

Advocates of pass-fail, the investigators discovered, justified the option on the following grounds: It would allow students to learn for the sake of learning, not for the sake of gaining the reward or avoiding the punishment of a grade. And it would encourage them to broaden their interests by enabling them to take courses in unfamiliar fields without fear of a lowered grade-point average. While the ΦBK committee found some validity in these arguments, it also encountered much evidence to the contrary. Apparently, many if not most students taking a course on a pass-fail basis slighted the work in that course in order to concentrate on graded courses they considered more important. Thus the effect was to narrow rather than to broaden their intellectual focus.

The pass-fail option, the committee concluded in 1969, was so little used that it was not a serious problem for most chapters in electing members—though it might become a problem if it should spread. But according to the committee, most chapters put too much emphasis on the grade-point average anyhow. They ought to take into account other evidences of scholarship, such as honors work, original research, opinions of instructors, and the difficulty of the courses taken. There should be a "study in depth" of each candidate. Accordingly, the committee recommended that chapters include in their bylaws a statement to the effect that a particular grade-point average "confers no right to consideration for membership or favorable action thereon." The committee also called upon the chapters and their delegates at the 1970 Council to continue the discussion.[18]

Meanwhile the situation worsened, and the society's leaders became increasingly alarmed. At the 1968 Senate meeting, Irving Dilliard brought up the question: "What can Phi Beta Kappa do to help the college and university world now under assault from so many quarters?" He suggested that the society might help by "issuing a statement expressing its concern and encouraging responsible and constructive responses." At the 1969 meeting, Dilliard, on behalf of the Committee on Policy, presented a statement deploring the troubles but proposing no action in regard to them.

By 1970, the troubles were approaching a climax, with violence on several campuses and actual bloodshed at Wisconsin, Jackson State, and Kent State. On June 7, the *New York Times* published a long front-page article under the headline "School Ferment Roils Phi Beta Kappa." The reporter had interviewed (among others) H. Bentley Glass, president of the United Chapters and professor of biology and academic vice president at the State University of New York at Stony Brook. "We need to reappraise the nature of the society and redefine its objectives during a period of rapid change and crisis," Glass told the reporter. "We must make sure our scholarly ideals are tied in with social needs."[19]

When the Council met in September 1970, Glass did what no previous president had done: he appeared before the assembled delegates to deliver what he called a "state of the Union" address. "There is a great deal of questioning going on upon the campuses of the land today about the necessity of having honor organizations such as Phi Beta Kappa," he said, reminding the delegates of what they already knew. "There are questions, in fact, whether such societies as ours are simply relics of the past and whether they can any longer play an effective role in higher education." He added: "I do think . . . that in these times when so many student leaders are insisting that our higher education is not relevant . . . they have a point."

Glass informed the delegates that, on his recommendation, the Senate was setting up a committee to consider the future of Phi Beta Kappa as an organization. Many of the delegates, especially the younger ones, were not completely satisfied with that, and they secured the adoption of the following resolution:

> Be it resolved that since there is a widespread feeling on the part of many thoughtful persons, particularly undergraduates, that Phi Beta Kappa has not been adequately concerned with critical issues of our time, many of which influence the course of intellectual life and higher education in America, and since we share this feeling, we urge the formation of a special study group to explore means by which Phi Beta Kappa can provide leadership in certain areas of traditional humanistic concern. The intent of this resolution is to narrow the gap between existing Phi Beta Kappa membership and eligible students. To satisfy that intent we hope that the committee will be equally composed of established and developing scholars.

Other resolutions called for committees to redefine a liberal education, to study the standards for election, to draft new guidelines for the chapters, and to propose programs for enhancing the society's influence.[20]

At their meeting later the same year, the senators decided to appoint a single Committee on the Role of Phi Beta Kappa and to merge with it their own recently formed ad hoc committee. Chairing the combined group was Glass, who had just completed his term as United Chapters president. To start, the Glass committee sent a questionnaire at random to five thousand sustaining (that is, older, contributing) members and to one thousand newly elected ones. The questionnaire asked, among other things, what the members found valuable in belonging to the society and what they thought could be done to improve it.

More than 30 percent of the older members and fewer than 19 percent of the younger ones responded. The former opposed by three to one and the latter by two to one the deemphasizing of grades. Statements on the questionnaire and in other connections revealed the extremes of disagreement. "I see the Phi Beta Kappa tradition as a repository of excellence," one professor wrote. "Let's not become hung up on 'validity,' 'relevance,' 'concern,' 'redeeming social significance,' and the rest of the catchwords of the moment, even though these are all perfectly legitimate humanistic aspirations, not simply recent discoveries of the very young." But a 1969 graduate of Barnard College, who had been elected to ΦBK in her junior year, thought membership ought to be open to everyone who possessed "a love of learning."[21]

Reformers seemed to be holding the upper hand when in 1971 the Glass committee made a preliminary report to the Senate. The report included proposals to elect students before the end of the senior year, to give their fellow students a say in the elections, and to extend eligibility to those majoring in engineering, business administration, and other vocational fields. "Phi Beta Kappa, the scholastic society that one member has termed 'a haven for those seeking to indulge in self-congratulation,' may soon admit more, younger, and less purely intellectual members," the *Washington Post* predicted. The paper pointed out, however, that the society would not act upon the proposals until the next meeting of the Council in 1973. "Much of the intervening time may be spent trying to convince die-hard members who like their 'haven' of intellectual elitism to accept proposed reforms."[22]

Instead, the "die-hard members" convinced the committee that its proposals went too far. Copies of the preliminary draft, when circulated among the chapters, drew a great deal of criticism. In response, the committee deleted several "controversial elements." The main feature of the final draft was a revised model chapter constitution. This redefined a liberal education so as to involve, in Glass's words, "a concern for the human condition, for the moral responsibility of the individual in a complex society." The model constitution also specified that other criteria should supplement the grade-point average in the evaluation of prospective members.

By the time the 1973 Council met, the sense of urgency had decreased considerably. The United States, after signing a peace accord, had re-

moved its combat forces from Vietnam, and the campuses of the nation
had returned to comparative calm. When the United Chapters president, Rosemary Park, introduced the Glass committee to the delegates,
she recalled the circumstances that had prompted its formation: "The
last triennial Council met at a time of widespread campus unrest when
many students questioned whether learning could be justified apart
from human need—whether, as one student put the question, 'I have
the right to be an astronomer when there are urgent problems of urban
disorganization, racial injustice, health care delivery, etc.'"

Glass proceeded to give the report of the committee that had set out to
"examine the ways in which Phi Beta Kappa can become more relevant
in higher education." The report evoked a long discussion, and final
consideration was put off to the last day of the Council. When the time
came to vote on the model constitution, there was no longer a quorum
present.[23]

When the Thirty-First Triennial
Council met in Williamsburg in 1976 to conduct its ordinary business
and to celebrate the society's two-hundredth anniversary, there was
cause for apprehension as well as celebration. True, the campus turmoil
was receding into the past, and the society had endured, its reputation
essentially intact. But its long familiar problems were still present and
were intensifying—its problems of maintaining academic standards,
evaluating scholarly achievement, defining a liberal education, and defending it against the inroads of careerism.

At the 1976 meeting, there was a wide-ranging discussion of "Grade
Inflation and Academic Standards." Most of the chapters, Secretary
Kenneth Greene pointed out, were reacting to grade inflation by raising
the minimum grade-point average for the consideration of candidates
for membership. As a result, the total of those elected was decreasing,
even though the number of chapters and the overall enrollments in the
liberal arts and sciences were increasing. From 1952 to 1973, the numbers elected had grown each successive year, but they then began to
shrink (so that, for 1973–1974, the figure was 17,184; for 1974–1975, it
was 16,173; and for 1975–1976, it was 15,754).

This trend indicated that, in some sense, the chapters were maintaining or even raising their election standards. But it provoked one of the
delegates to question whether, in consequence of the chapters' pushing
the minimum grade-point average higher and higher, it was not becoming harder and harder "to differentiate between the student who is a
grind, a good memorizer, and the student who is really a creative, productive scholar." Greene explained that a "great number" of the chapters
were not relying on grades alone but were taking other kinds of evidence
into account.

Some were finding ways to distinguish between the more inflated and the less inflated grades—and thus to distinguish between the grade-grabber and the real scholar. The Indiana University chapter devised an "academic performance index," to be used for the first time in electing from the class of 1977. This index corrected the grade-point average by "overweighting" high grades in difficult courses and "underweighting" them in easy ones. "The necessary information is obtained from a computer program that traces the grades earned by each senior student in the College of Arts and Sciences who is otherwise qualified for possible election and whose GPA is 3.65 or above," a delegate from Indiana explained. The program "compares each individual student's grades in each course he has completed with the average grade given in that course."

More troubling than the consequences of grade inflation were the causes of it. Among these were, on the one hand, the indifference to grades on the part of students and faculty who considered them more or less meaningless and, on the other hand, the excessive attention to grades on the part of students who hoped to get into law or medical school and who put pressure on often compliant professors. And there was, in some places, the elimination of failures. There was also the temptation to keep grades up so as to attract and retain students, particularly in institutions where funds for faculty salaries depended largely on the student head count.

The most important cause of grade inflation, however, appeared to be the widespread abandonment of distribution or general education requirements. This made it possible for students to concentrate on subjects they found easy or interesting and to avoid taking a broad range of courses that might seem distasteful or difficult, such as languages, sciences, and mathematics. The result was a massive retreat from the kind of broad, liberating education that Phi Beta Kappa stood for. The chapters faced the task of campaigning for the retention or restoration of a liberal arts curriculum on their respective campuses.[24]

Besides discussing the liberation of the curriculum, the delegates at the bicentennial Council took a positive if modest step toward the liberation of women. Already in 1972 the separate male and female Phi Beta Kappa associations in New York had merged, and the two in Philadelphia were soon to do so. But grievances remained. One was the paucity of women selected for the Visiting Scholars program. Another was the society's use of "sexist" terms. From time immemorial, though often referring to the fraternity itself as "she," its male members had referred to the student in the abstract as "he," to a liberal education as one that made a "man" free, and to a committee head, male or female, as the "chairman." In 1976 the Council raised the society's consciousness at least to the point of "desexing" the language of the constitution.[25]

The bicentennial of the Harvard chapter (1981) provided another occasion for congratulations. "Two Centuries of Elitism: Phi Beta Kappa Still Flourishes Amid Changing Standards," ran the headlines in *Time* magazine. "There is much to celebrate. Less than ten years ago, on many college campuses, ΦBK was regarded as odiously 'elite,'" *Time* noted. "Today everybody is eager to join, partly because undergraduates again think the distinctive gold ΦBK key may help unlock the door to worldly success."[26]

Actually, there remained the old doubt as to whether the key would really unlock the door to success in business. But there could be little doubt that the key opened the way—if no more than that—to ambition of another kind. As one Phi Beta Kappa senator put the matter,

> There are many powerful currents in our [American] society . . . that tend toward the lowering of the demands that professors make upon themselves and upon those they teach, currents that tend toward easing requirements and toward mediocrity. But Phi Beta Kappa stands, as the pointing hand on our key symbolizes, for aspiration, for reaching higher. When William James, that quintessential American philosopher, was asked in the 1890s to talk on the topic of the role of the college-educated person in a democratic society, he said, in sum, that the role of college education was to give its students a sense of excellence, a sense of what is superior in the works of the mind and in human life. It is that spirit that guides the deliberations of Phi Beta Kappa—in the chapters, in the Senate, and in the Council.[27]

Afterword

At a time when elitism has been on the wane, Phi Beta Kappa, assumed by many to be its most vigorous advocate, became neither overly defensive nor needlessly confrontational. It has nevertheless been unwavering in its support of the highest standards of scholarship and of intellectual integrity. In 1976 at its bicentennial celebration, when the Society's president reaffirmed its commitment to those principles as the best possible kind of elitism, even he was surprised at the enthusiastic approval of the audience. The problem, then as now, was not active anti-intellectualism translated into opposition to high standards or to so-called irrelevance, so much as it was plain ignorance of what Phi Beta Kappa is or what it stands for.

When the Postmaster General's Committee on Commemorative Stamps rejected Phi Beta Kappa's request for a stamp to mark its bicentennial, it did so on the grounds that its policy precluded its honoring social fraternities and sororities. That was a shattering experience for an organization that had an exalted view of itself. No amount of pleading by such distinguished members of Phi Beta Kappa as Senators Hubert Humphrey, Frank Church, and John C. Stennis, and Representatives Brock Adams, Claude Pepper, and Ella Grasso could persuade the Committee that Phi Beta Kappa was not just another Greek letter social organization. Likewise, while it is true that during the 1960s and 1970s many college students declined election to Phi Beta Kappa because they rejected any and all forms of elitism, some, albeit a much smaller number, declined election in the 1980s because they simply did not know anything about the organization.

Thus, the task of Phi Beta Kappa in the 1970s and 1980s was clearly one of educating its various constituencies—students, the university community, and the general public—regarding the importance and, indeed, the relevance of the basic principles of Phi Beta Kappa to the intellectual and academic life of the late twentieth century. Obviously this required Phi Beta Kappa to become involved in what came to be known in certain circles as "outreach programs."

To its credit, Phi Beta Kappa saw an opportunity in its program of visiting scholars to enhance both its image and its influence; and it seized upon it. It significantly increased the number of visiting scholars, not confining the program to members of Phi Beta Kappa, and encouraged and commended distinguished scholarly work wherever it found it. If the program could help institutions sheltering chapters somehow to increase the awareness of students of Phi Beta Kappa's existence, to say nothing of its importance to their academic life, perhaps its impact could become considerably greater before the end of the twentieth century. Already, both students and institutions have manifested an increased interest in Phi Beta Kappa as a worthy institution on campus and one that would be attractive to able and serious students. When a chapter of Phi Beta Kappa was established at the University of Tulsa in May 1989, the occasion turned out to be a gala event in which many parents of members of the new chapter made it clear that this was something of which they were extremely proud.

The enhanced image of Phi Beta Kappa could be seen not only in the rather humorous, even frivolous, attempts by various commercial firms to use the key and other symbols of the organization in their promotion campaigns but also in the quite serious efforts of colleges and universities to secure chapters of Phi Beta Kappa for their own campuses. All members of the campus community agreed that a chapter of Phi Beta Kappa was a real asset. Many members of Phi Beta Kappa, while making no compromises with high standards, were eager to reach out to institutions without chapters and encourage them to meet the requisite qualifications. The president, the Council, the Senate, and the Committee on Qualifications all offered advice and assistance to institutions seeking to qualify for chapters. Often this assistance and advice paid off as such institutions strengthened library holdings, revised curriculums, and fulfilled all other requirements.

Meanwhile, neither the Committee on Qualifications nor the triennial Council was so ardent in its outreach programs as to relax standards or dispense with the most careful scrutiny of applying institutions. All groups having to do with the creation of new chapters were especially anxious to impress upon would-be chapters the critical importance of making certain that excessive emphasis on intercollegiate athletics must not take precedence over other, more important considerations. The preponderance of athletic scholarships over other forms of student awards and assistance, the maintenance of special facilities and amenities for athletes, and the attraction of huge revenues from nationally televised sporting events served, in too many instances, to undermine the basic purposes of the institution or dilute its academic program. One distinguished educator recently lamented that many colleges and universities were being turned into entertainment centers. Recent Councils have taken notice of these trends by calling them to the attention of applying institutions faced with such problems. While Phi Beta Kappa's resolve in the matter was unmistakable and unequivocal, its impact on the larger academic community was not quite so clear.

Much more difficult than holding institutions seeking Phi Beta Kappa chapters to high standards was the task of making certain that institutions already with chapters continued to adhere to those standards. Questions of academic freedom persisted and, if anything, increased during the 1970s and 1980s when students and members of faculties challenged the prerogatives and the policies of governance under which some institutions operated. Whenever an institution sheltering a chapter of Phi Beta Kappa was placed by the American Association of University Professors (AAUP) on its list of censured institutions, it was of immediate concern to Phi Beta Kappa. Serious questions arose. Should it rely on the findings of the Committee on Academic Freedom and Tenure of the AAUP, or should it conduct an independent investigation of such alleged grave offenses? As the number of censured institutions increased, it became quite clear that Phi Beta Kappa had neither the machinery nor the resources to conduct its own investigations. Meanwhile, it seemed unfair to censured institutions to place them under a cloud by relying solely on the findings of the AAUP. The Committee on Qualifications made several attempts to look into specific cases, but it soon discovered that it could not perform an effective task in that area while fulfilling its primary mandate to make certain that institutions applying for new chapters met the standards required by Phi Beta Kappa. The Committee on Qualifications, the Senate, and the triennial Council debated the matter, but reached no definitive conclusions beyond recognizing the gravity of the problem, and determining to do whatever could be done, even on a limited basis. The organization faced a similar situation whenever an institution sheltering a chapter of Phi Beta Kappa was placed on probation or suspended by the National Collegiate Athletic Association for alleged infractions of some of its regulations. Even if it was unable to invoke sanctions against offending institutions, the Council was unequivocal in decrying the creeping—some would say leaping—professionalization of intercollegiate athletics.

Phi Beta Kappa has held firmly to the view that the liberal arts can best be strengthened and preserved through time-honored programs that have served well the liberal-arts tradition. It is fair to say that most members are not pleased with the reduction or elimination of requirements in such subjects as literature, philosophy, history, mathematics and foreign languages. At the same time, they have not opposed curricular innovations that seek to preserve the liberal arts in new, exciting, and relevant ways. At meetings of the Committee on Qualifications as well as at meetings of the triennial Council, debates over curricular matters have been intense and, indeed, have caused some to draw the inference that opposition to any change is strong. The debates, however, more accurately reflect a disinclination to change merely for the sake of change and a determination to make certain that any change that takes place will be not only constructive but a genuine improvement as well.

Thus, Phi Beta Kappa has welcomed new ways of teaching the classics, for example, including the introduction of new insights and new interpretations of materials inspired by the current interest in such subjects as women and minorities. It has recognized the general importance of the humanities in American life as well as in the curriculum by establishing the Phi Beta Kappa Award for Distinguished Service in the Humanities that has been made to such outstanding humanists as Barnaby Keeney, Louis B. Wright, Robert Lumiansky, and Daniel J. Boorstin. It has singled out philosophy for special celebration by endowing a professorship, through the generosity of the philosopher Patrick Romanell, to which a distinguished philosophy professor is appointed annually. For a number of years, moreover, Phi Beta Kappa has sponsored a public lecture at the annual meeting of the American Association for the Advancement of Science, thus indicating its strong commitment to the study and diffusion of scientific knowledge.

As far as human rights and human relations are concerned, Phi Beta Kappa can well be regarded as no better or no worse than some other organizations, although it can be argued that it should have been ahead of most organizations in such matters. Conceivably, its commitment to the humanistic tradition could have influenced the direction of its policies; but here its policies reflected rather accurately the social and political views of its members who, after all, represented America's many different, conflicting attitudes regarding race, religion, and ethnicity. As Phi Beta Kappa saw the social ferment all around it, there were indications that it wished the ferment would somehow disappear without involving Phi Beta Kappa. There were some things, however, that it could not avoid. As far as chapters at historically black institutions were concerned, for example, there were no serious applications since three such chapters had been authorized at Fisk, Howard, and Morehouse in the 1950s. This could well be an area which would benefit from active involvement by Phi Beta Kappa through an outreach program. With remarkable strides in faculty strength, student performance, physical facilities, and financial stability in many historically black institutions, Phi Beta Kappa can well anticipate applications from some of them in the near future.

In its internal organization, Phi Beta Kappa reflected its own commitment to new policies as far as women and minorities were concerned. Beginning in the 1950s, one could observe both a heightened sensitivity as well as a quickened concern that Phi Beta Kappa would not be left at the starting gate, as it were, as the nation moved toward greater racial and sexual equality. If Phi Beta Kappa's record was not spectacular in such matters, it was nevertheless respectable. By 1991, three women and one male African American will have served as presidents of Phi Beta Kappa within the previous two decades. Meanwhile, in 1990, the principal agency of governance, the Senate, consisting of twenty-four members, had nine women and one male African American. While the remaining fourteen Senators were white males, not all of them were by any means Anglo-Saxon Protestants.

The management of Phi Beta Kappa's affairs in the past fifteen years has been admirably efficient. In 1974, Joseph Epstein succeeded Hiram Haydn, the long-time editor of the *American Scholar,* whose death left the Society's senators quite pessimistic about the future of the journal. The publication of the prestigious *Scholar* was one of the most important and most expensive undertakings of the organization. Could it continue, the senators asked themselves, without becoming an intolerable burden on the Society's resources? As Epstein put together an editorial board consisting of such outstanding writers as Ada Louise Huxtable, William Styron, and Saul Bellow, the intellectual and literary quality of the magazine improved significantly. Some of the nation's most seasoned and distinguished writers contributed to its pages. At the same time, many of the best of the younger writers were encouraged by achieving publication in the *Scholar.* Epstein, ever the stimulating and provocative writer, enlivened each issue with a contribution from "Aristides," his *nom de plume.*

Meanwhile, also in 1974, the position of secretary, the chief executive officer of Phi Beta Kappa, became vacant through the untimely death of Carl Billman, who had served in that capacity since 1947. After a nationwide search, the organization appointed Kenneth M. Greene, president of Lasell Junior College, to the position. Greene immediately took steps to increase the efficiency of the Society's operations, to invigorate the work of the committees, to work closely with the Phi Beta Kappa Associates (a group of members dedicated to promoting the academic and fiscal strength of the organization), to soothe the tempers of those who argued that Phi Beta Kappa was too elitist, and to establish closer contact with its growing membership. The Sustaining Membership Appeal, largely the work of Greene, has brought in more than a million dollars annually, while the endowment fund has grown to over eight million dollars. As a result, Phi Beta Kappa enjoys the greatest financial stability that it has ever had and can contribute more easily to such important enterprises as the *American Scholar,* the *Key Reporter* (a newsletter sent to every member), and to its several prizes awarded annually for outstanding scientific and literary works.

Today there is no single image of Phi Beta Kappa. Its images in the minds of Americans in general seem to range from the view that its membership consists of so-called eggheads with little or no contact with or interest in the real world, to the view that members of Phi Beta Kappa are the very bulwark of the nation's intellectual and creative life. Among the members, there is a self-image that ranges from a studied if only half-serious indifference to the very existence of the organization, to the kind of moral and intellectual support that reflects pride in what it means both for themselves and to the nation to belong to the Society. In its preoccupation with its own wide-ranging agenda and its large and steadily increasing membership, the Society plays no specific or direct role in creating images. It is, nevertheless, bemused by the way in which various commercial enterprises seek to exploit, through huckstering their own goods and services, the very name of Phi Beta Kappa, whether they call it Eye Beta Kappa, Phi Beta Caper, Phi Beta Jantzen, or whatever.

Image aside, Phi Beta Kappa does enjoy wide respect and considerable influence in important sectors of American society. A large assortment of public figures—scholars, educational administrators, politicians, and persons from the world of business—express pride in belonging to Phi Beta Kappa. And if some members do not wear their key or cannot even find it, there are some nonmembers who have been known to acquire it in some devious way and wear it with as much pride as an undergraduate who was elected in his junior year. In 1988, more than one political pundit observed that the outcome of the presidential election could hardly be catastrophic since the nominees of both major parties were members of Phi Beta Kappa. Even those who have limited knowledge of the organization are usually deferential to its members in intellectual matters when they learn that those members can presumably contribute to the solution of problems by using their minds in a certain way!

How is it that an organization, with relatively obscure and inauspicious beginnings and with nothing much in the way of a power base, is able to enjoy considerable respect and a reasonable amount of influence? Surely, one reason is that for more than two centuries Phi Beta Kappa has stood for the highest academic and intellectual standards. In a country that historically has placed such great stock in material and practical things, its people have always been able to muster some respect, however grudging at times, for things that exalt the mind and spirit. The very age of Phi Beta Kappa invites veneration, to be sure, and in a nation given to symbols, nothing epitomizes excellence more than an honors society born just five months after Thomas Jefferson wrote the Declaration of Independence.

Another possible explanation for the respect and influence enjoyed by Phi Beta Kappa is that it is regarded by many not merely as a symbol of excellence but also as the very ideal of excellence to which so many aspire. Whether in the classroom or in the workplace, Americans have the highest regard for excellence, however much they might express public disdain for the outward manifestations of intellectual elitism. Any toleration of mediocrity is more a confused gesture of lip service to the leveling process than any serious adherence to low standards. The American desire for proficiency, even perfection, is consonant with the high standards of Phi Beta Kappa; and most Americans see in high standards, proficiency, and perfection a common ground on which all or most of them wish to stand. In any case, Phi Beta Kappans are not regarded as "grade freaks" so much as they are looked upon as persons setting standards of excellence in the world at large as well as in the classroom.

Finally, the weight of the sheer size of the membership of Phi Beta Kappa has doubtless something to do with its standing. No other group whose primary interest is in the promotion and celebration of the constructive use of one's mental capacities can claim a membership of 425,000. There is a pervasiveness of the membership, moreover, that is a direct consequence of its size. While there are many colleges and universities that do not yet have chapters of Phi Beta Kappa, which most of them would be pleased to have, very few of them have a faculty with no members of the organization. When the numbers are few in a given community—collegiate or otherwise—their visibility seems all the greater, a beacon light, as it were, adding luster as well as stimulation to those communities that want to try a bit harder to excel. And when one adds the numbers, steadily approaching a half million, to the weight of two centuries of unremitting commitment to the principle of excellence and the accumulated tradition growing out of such commitment, it is possible to see how the organization enjoys some respect and even a reasonable amount of influence. It is not too much to expect that this influence will continue to grow in this the third century of Phi Beta Kappa's existence.

Durham, North Carolina
October 1989

John Hope Franklin
Past President
Phi Beta Kappa Society

Appendix A

The Roll of Chapters in the Order of Founding

Chapter Name	School Name	Date of Founding
Alpha of Virginia Inactive 1781–1851 and 1861–1893	College of William and Mary	12/5/1776
Alpha of Connecticut Inactive 1871–1884	Yale College	11/13/1780
Alpha of Massachusetts	Harvard College	9/6/1781
Alpha of New Hampshire	Dartmouth College	8/20/1787
Alpha of New York	Union College	7/22/1817
Alpha of Maine	Bowdoin College	2/22/1825
Alpha of Rhode Island	Brown University	7/21/1830
Beta of Connecticut	Trinity College	7/2/1845
Gamma of Connecticut	Wesleyan University	7/7/1845
Alpha of Ohio Inactive intermittently before 1884	Western Reserve College[1]	10/28/1847
Alpha of Vermont	University of Vermont	3/7/1849
Alpha of Alabama Inactive 1861–1912	University of Alabama	7/14/1851
Beta of Massachusetts	Amherst College	8/9/1853
Beta of Ohio	Kenyon College	6/29/1858
Beta of New York	New York University	12/23/1858
Gamma of Ohio	Marietta College	6/9/1860
Gamma of Massachusetts	Williams College	7/30/1864
Gamma of New York	College of the City of New York	7/24/1867
Beta of Vermont	Middlebury College	8/17/1868
Alpha of New Jersey	Rutgers College[2]	2/22/1869

1. Now Case Western Reserve University.

2. Sections of the Alpha of New Jersey chapter at Rutgers University were established at New Jersey College for Women (now Douglass College) on June 10, 1922, and at Newark College on December 3, 1958.

Chapter Name	School Name	Date of Founding
Delta of New York	Columbia College[3]	4/22/1869
Epsilon of New York	Hamilton College	5/ /1870
Zeta of New York	Hobart College[4]	7/6/1871
Eta of New York	Madison University[5]	6/19/1878
Theta of New York	Cornell University	5/28/1882
Alpha of Pennsylvania	Dickinson College	4/12/1887
Beta of Pennsylvania	Lehigh University	4/15/1887
Iota of New York	University of Rochester	4/20/1887
Alpha of Indiana	DePauw University	12/27/1889
Alpha of Illinois	Northwestern University	2/18/1890
Alpha of Kansas	University of Kansas	4/2/1890
Gamma of Pennsylvania	Lafayette College	4/5/1890
Delta of Massachusetts	Tufts College	11/18/1892
Delta of Pennsylvania	University of Pennsylvania	12/9/1892
Alpha of Minnesota	University of Minnesota	12/13/1892
Alpha of Iowa	University of Iowa	9/30/1895
Alpha of Maryland	Johns Hopkins University	10/10/1895
Alpha of Nebraska	University of Nebraska	12/23/1895
Beta of Maine	Colby College	1/3/1896
Kappa of New York	Syracuse University	2/10/1896
Epsilon of Pennsylvania	Swarthmore College	6/9/1896
Beta of Indiana	Wabash College	11/7/1898
Alpha of California	University of California	12/23/1898
Zeta of Pennsylvania	Haverford College	1/20/1899
Alpha of Wisconsin	University of Wisconsin	2/2/1899
Epsilon of Massachusetts	Boston University	2/8/1899
Mu of New York	Vassar College	4/7/1899
Delta of Ohio	University of Cincinnati	4/7/4199
Beta of New Jersey	Princeton University	6/7/1899
Lambda of New York	St. Lawrence University	6/24/1899
Beta of Illinois	University of Chicago	7/1/1899
Alpha of Tennessee	Vanderbilt University	11/5/1901
Alpha of Missouri	University of Missouri	12/5/1901
Eta of Pennsylvania	Allegheny College	2/18/1902
Alpha of Colorado	University of Colorado	10/18/1904
Zeta of Massachusetts	Smith College	10/19/1904
Beta of California	Leland Stanford Jr. University[6]	11/1/1904
Alpha of North Carolina	College of Arts and Sciences, University of North Carolina[7]	11/7/1904
Beta of Colorado	Colorado College	11/11/1904
Eta of Massachusetts	Wellesley College	11/14/1904
Epsilon of Ohio	Ohio State University	12/8/1904
Theta of Massachusetts	Mount Holyoke College	1/30/1905
Alpha of Texas	University of Texas	2/2/1905
Beta of Maryland	Goucher College	2/24/1905

3. Sections of the Delta of New York chapter at Columbia University were established at Barnard College on June 7, 1901, and at the School of General Studies on June 5, 1952.

4. Now Hobart and William Smith colleges.

5. Now Colgate University.

6. Now Stanford University.

7. Now University of North Carolina at Chapel Hill.

Chapter Name	School Name	Date of Founding
Zeta of Ohio	Oberlin College	11/8/1907
Eta of Ohio	Ohio Wesleyan University	11/9/1907
Gamma of Illinois	University of Illinois	11/11/1907
Alpha of Michigan	University of Michigan	11/13/1907
Theta of Pennsylvania	Franklin and Marshall College	1/20/1908
Beta of Iowa	Grinnell College	4/11/1908
Beta of Virginia	University of Virginia	6/16/1908
Alpha of Louisiana	Tulane University	2/26/1909
Alpha of West Virginia	West Virginia University	12/5/1910
Theta of Ohio	Denison University	11/18/1911
Gamma of Indiana	Indiana University	1/20/1911
Gamma of Virginia	Washington and Lee University	5/5/1911
Iota of Ohio	Miami University	6/14/1911
Beta of Wisconsin	Beloit College	6/19/1911
Gamma of Wisconsin	Lawrence College[8]	2/20/1914
Gamma of California	Pomona College	3/7/1914
Alpha of Georgia	University of Georgia	3/14/1914
Beta of Minnesota	Carleton College	3/31/1914
Alpha of Washington	University of Washington	4/20/1914
Iota of Massachusetts	Radcliffe College	5/11/1914
Beta of Missouri	Washington University	5/13/1914
Alpha of North Dakota	University of North Dakota	6/3/1914
Delta of Illinois	Knox College	2/15/1917
Delta of Virginia	Randolph-Macon Woman's College	5/5/1917
Gamma of Maine	Bates College	5/29/1917
Beta of North Carolina	Trinity College[9]	3/29/1920
Beta of Washington	Whitman College	1/20/1920
Nu of New York	Hunter College	2/11/1920
Alpha of Oklahoma	University of Oklahoma	5/24/1920
Iota of Pennsylvania	Gettsyburg College	1/11/1923
Delta of Maine	University of Maine	1/26/1923
Gamma of North Carolina	Davidson College	3/1/1923
Alpha of Oregon	University of Oregon	4/14/1923
Gamma of Iowa	Drake University	4/19/1923
Delta of Iowa	Cornell College	5/3/1923
Alpha of Kentucky	University of Kentucky	3/12/1926
Beta of Tennessee	University of the South	3/16/1926
Beta of Georgia	Agnes Scott College	3/23/1926
Alpha of South Carolina	University of South Carolina	4/8/1926
Kappa of Ohio	College of Wooster	4/20/1926
Delta of California	Occidential College	5/12/1926
Alpha of South Dakota	University of South Dakota	6/4/1926
Alpha of Idaho	University of Idaho	6/5/1926
Beta of Texas	Rice Institute	3/1/1929
Epsilon of California	University of Southern California	3/14/1929
Zeta of California	Mills College	3/16/1929

8. Now Gamma-Delta of Wisconsin chapter at Lawrence University, formed when Lawrence College and Milwaukee-Downer College merged in 1964.

9. Now Duke University.

Chapter Name	School Name	Date of Founding
Gamma of Georgia	Emory University	4/5/1929
Gamma of Washington	Washington State College	4/6/1929
Epsilon of Virginia	University of Richmond	4/12/1929
Lambda of Ohio	Ohio University	4/26/1929
Kappa of Massachusetts	Wheaton College	3/18/1932
Alpha of Arkansas	University of Arkansas	4/4/1932
Alpha of Arizona	University of Arizona	4/20/1932
Epsilon of Illinois	Illinois College	4/6/1932
Xi of New York	Wells College	5/7/1932
Alpha of Utah	University of Utah	1/3/1935
Delta of Connecticut	Connecticut College for Women[10]	2/13/1935
Alpha of Florida	Florida State College for Women[11]	3/5/1935
Lambda of Pennsylvania	Pennsylvania State College	12/7/1937
Kappa of Pennsylvania	Washington and Jefferson College	10/30/1937
Beta of Alabama	Birmingham-Southern College	11/26/1937
Eta of California	University of California, Los Angeles[12]	1/14/1938
Omicron of New York	University of Buffalo[13]	1/29/1938
Zeta of Virginia	Randolph-Macon College[14]	2/3/1938
Beta of Florida	University of Florida	2/18/1938
Alpha of the District of Columbia	George Washington University	2/22/1938
Beta of Oregon	Reed College	5/6/1938
Gamma of Minnesota	College of St. Catherine	5/17/1938
Mu of Pennsylvania	Bucknell University	11/7/1940
Beta of Michigan	Albion College	11/8/1940
Gamma of Colorado	University of Denver	11/25/1940
Alpha of Wyoming	University of Wyoming	11/26/1940
Pi of New York	Elmira College	11/29/1940
Delta of North Carolina	Wake Forest College	1/13/1941
Beta of South Carolina	Wofford College	1/14/1941
Beta of the District of Columbia	Catholic University of America	1/15/1941
Delta of Wisconsin	Milwaukee-Downer College[15]	1/21/1941
Delta of Minnesota	St. Olaf College	11/4/1949
Gamma of Tennessee	Southwestern at Memphis[16]	12/5/1949
Gamma of Texas	Southern Methodist University	12/12/1949
Eta of Virginia	Hampden-Sydney College	12/13/1949
Epsilon of Iowa	Coe College	12/13/1949

10. Now Connecticut College.

11. Now Florida State University.

12. Originally chartered as a section of the Alpha ofCalifornia chapter at the University of California, Berkeley, on Jun 14, 1930.

13. Now State University of New York at Buffalo.

14. Originally chartered as a section of the Delta of Virginia chapter at Randolph-Macon Woman's College on May 3, 1923.

15. Now Gamma-Delta of Wisconsin chapter at Lawrence Univeristy, formed when Milwaukee-Downer and Lawrence College merged in 1964.

16. Now Rhodes College.

Chapter Name	School Name	Date of Founding
Sigma of New York	Queens College	1/9/1950
Rho of New York	Brooklyn College	1/13/1950
Nu of Pennsylvania	Wilson College	1/20/1950
Theta of Virginia	Sweet Briar College	3/3/1950
Zeta of Illinois	Augustana College	3/17/1950
Epsilon of Wisconsin	Ripon College	12/12/1952
Beta of New Hampshire	University of New Hampshire	12/16/1952
Alpha of Hawaii	University of Hawaii	12/19/1952
Gamma of Michigan	Wayne State University	1/16/1953
Xi of Pennsylvania	University of Pittsburgh	1/19/1953
Lambda of Massachusetts	Clark University	2/2/1953
Eta of Illinois	Rockford College	2/21/1953
Delta of Tennessee	Fisk University	4/4/1953
Gamma of the District of Columbia	Howard University	4/8/1953
Epsilon of North Carolina	University of North Carolina at Greensboro[17]	2/17/1956
Epsilon of Connecticut	University of Connecticut	4/4/1956
Alpha of Delaware	University of Delaware	4/25/1956
Delta of Michigan	Kalamazoo College	12/9/1958
Theta of Illinois	Lake Forest College	2/9/1962
Iota of Virginia	Hollins College	2/20/1962
Tau of New York	Fordham University	3/1/1962
Theta of California	Scripps College	3/1/1962
Mu of Massachusetts	Brandeis University	3/5/1962
Omicron of Pennsylvania	Chatham College	3/12/1962
Gamma-Delta of Wisconsin	Lawrence University[18]	8//1964
Gamma of Maryland	University of Maryland	12/16/1964
Epsilon of Tennessee	University of Tennesse at Knoxville	1/29/1965
Delta of Indiana	Earlham College	2/18/1965
Nu of Massachusetts	University of Massachusetts	3/30/1965
Alpha of New Mexico	University of New Mexico	4/6/1965
Iota of California	University of California, Riverside	4/7/1965
Delta of the District of Columbia	Georgetown University	4/29/1965
Delta of Georgia	Morehouse College	1/6/1968
Gamma of Missouri	St. Louis University	1/7/1968
Epsilon of Michigan	Michigan State University	2/4/1968
Epsilon of Indiana	University of Notre Dame	2/11/1968
Pi of Pennsylvania	Muhlenburg College	2/16/1968
Epsilon of Minnesota	Macalester College	2/22/1968
Kappa of California	University of California, Davis	5/14/1968
Lambda of California	University of California, Santa Barbara	5/18/1968
Upsilon of New York	Manhattan College	2/12/1971

17. Originally chartered as a section of the Alpha of North Carolina chapter at the University of North Carolina at Chapel Hill on December 12, 1934.

18. Previously Gamma at Wisconsin chapter at Lawrence College and Delta of Wisconsin chapter at Milwaukee-Downer College, which became a single chapter when the schools merged.

Chapter Name	School Name	Date of Founding
Zeta of Wisconsin	Marquette University	2/14/1971
Epsilon of the District of Columbia	Trinity College	2/18/1971
Mu of Ohio	Hiram College	2/19/1971
Zeta of Michigan	Hope College	2/19/1971
Phi of New York	Skidmore College	2/20/1971
Kappa of Virginia	Mary Washington College	2/22/1971
Zeta of Indiana	Purdue University	2/23/1971
Delta of Texas	Texas Christian University	2/24/1971
Xi of Massachusetts	Massachusetts Institute of Technology	2/24/1971
Chi of New York	Herbert H. Lehman College	3/11/1971
Beta of Kentucky	Centre College	3/12/1971
Omicron of Massachusetts	Boston College	4/6/1971
Lambda of Virginia	Mary Baldwin College	4/26/1971
Psi of New York	Harpur College of the State University of New York at Binghamton	5/7/1971
Delta of Colorado	Colorado State University	11/15/1973
Gamma of South Carolina	Furman University	12/5/1973
Omega of New York	Hofstra University	12/5/1973
Beta of Arizona	Arizona State University	12/13/1973
Zeta of Iowa	Iowa State University	12/19/1973
Mu of California	University of California, Irvine	1/25/1974
Beta of Kansas	Kansas State University	2/11/1974
Eta of Wisconsin	University of Wisconsin at Milwaukee	2/24/1974
Alpha Alpha of New York	State University of New York at Albany	3/7/1974
Epsilon of Texas	Trinity University	4/5/1974
Zeta of Minnesota	Hamline University	4/24/1974
Rho of Pennsylvania	Temple University	4/26/1974
Nu of California	San Diego State University	4/27/1974
Alpha Beta of New York	State University of New York at Stony Brook	5/2/1974
Pi of Massachusetts	College of the Holy Cross	5/8/1974
Zeta of Texas	Baylor University	4/12/1977
Beta of Rhode Island	University of Rhode Island	4/22/1977
Xi of California	University of Redlands	4/24/1977
Nu of Ohio	Kent State University	4/27/1977
Beta of Louisiana	Louisiana State University	4/28/1977
Pi of California	University of Santa Clara[19]	4/29/1977
Rho of California	California State University, Long Beach	5/3/1977
Mu of Virginia	Virginia Polytechnic Institute and State University	5/8/1977
Omicron of California	San Francisco State University	5/26/1977
Sigma of California	University of California, San Diego	5/27/1977
Iota of Illinois	University of Illinois at Chicago	6/3/1977

19. Now Santa Clara University.

Chapter Name	School Name	Date of Founding
Eta of Michigan	Alma College	4/3/1980
Gamma of New Jersey	Drew University	4/26/1980
Delta of Maryland	Western Maryland College	5/1/1980
Gamma of Florida	Stetson University	11/19/1982
Delta of Florida	University of Miami, Coral Gables	2/5/1983
Eta of Minnesota	Gustavus Adolphus College	4/7/1983
Pi of Ohio	Bowling Green State University	4/17/1983
Eta of Iowa	Luther College	4/19/1983
Tau of California	Claremont McKenna College	4/21/1983
Delta of Washington	University of Puget Sound	4/10/1986
Sigma of Pennsylvania	Villanova University	4/13/1986
Upsilon of California	University of California, Santa Cruz	6/3/1986
Alpha of Mississippi	Millsaps College	3/4/1989
Eta of Texas	University of Dallas	4/29/1989
Beta of Oklahoma	University of Tulsa	5/5/1989

Appendix B

Colleges and Universities with Chapters of Phi Beta Kappa

School Name, Chapter Name, Location	Year of Founding
Agnes Scott College, Beta of Georgia, Decatur	1926
Alabama, University of, Alpha of Alabama, University	1851
Albion College, Beta of Michigan, Albion	1940
Allegheny College, Eta of Pennsylvania, Meadville	1902
Alma College, Eta of Michigan, Alma	1980
Amherst College, Beta of Massachusetts, Amherst	1853
Arizona, University of, Alpha of Arizona, Tucson	1932
Arizona State University, Beta of Arizona, Tempe	1973
Arkansas, University of, Alpha of Arkansas, Fayetteville	1932
Augustana College, Zeta of Illinois, Rock Island	1950
Bates College, Gamma of Maine, Lewiston	1917
Baylor University, Zeta of Texas, Waco	1977
Beloit College, Beta of Wisconsin, Beloit	1911
Birmingham-Southern College, Beta of Alabama, Birmingham	1937
Boston College, Omicron of Massachusetts, Boston	1971
Boston University, Epsilon of Massachusetts, Boston	1899
Bowdoin College, Alpha of Maine, Brunswick	1825
Bowling Green State University, Xi of Ohio, Bowling Green	1983
Brandeis University, Mu of Massachusetts, Waltham	1962
Brown University, Alpha of Rhode Island, Providence	1830
Bucknell University, Mu of Pennsylvania, Lewisburg	1940
California, University of,	
Berkeley, Alpha of California	1898
Davis, Kappa of California	1968
Irvine, Mu of California	1974
Los Angeles, Eta of California	1939(1930)[1]
Riverside, Iota of California	1965

---

1. Originally chartered as a section of the Alpha of California.

School Name, Chapter Name, Location	Year of Founding
San Diego, Sigma of California	1977
Santa Barbara, Lambda of California	1968
Santa Cruz, Upsilon of California	1986
California State University, Long Beach, Rho of California	1977
Carleton College, Beta of Minnesota, Northfield	1914
Case Western Reserve University, Alpha of Ohio, Cleveland	1847
Catholic University of America, Beta of D.C., Washington	1941
Centre College, Beta of Kentucky, Danville	1971
Chatham College, Omicron of Pennsylvania, Pittsburgh	1962
Chicago, University of, Beta of Illinois, Chicago	1899
Cincinnati, University of, Delta of Ohio, Cincinnati	1899
Claremont McKenna College, Tau of California, Claremont	1983
Clark University, Lambda of Massachusetts, Worcester	1953
Coe College, Epsilon of Iowa, Cedar Rapids	1949
Colby College, Beta of Maine, Waterville	1896
Colgate University, Eta of New York, Hamilton	1878
Colorado, University of, Alpha of Colorado, Boulder	1904
Colorado College, Beta of Colorado, Colorado Springs	1904
Colorado State University, Delta of Colorado, Fort Collins	1973
Columbia University, The College, Delta of New York, New York City	1869
Sections: Barnard College	1901
School of General Studies	1952
Connecticut, University of, Epsilon of Connecticut, Storrs	1956
Connecticut College, Delta of Connecticut, New London	1935
Cornell College, Delta of Iowa, Mt. Vernon	1923
Cornell University, Theta of New York, Ithaca	1882
Dallas, University of, Eta of Texas, Irving	1989
Dartmouth College, Alpha of New Hampshire, Hanover	1787
Davidson College, Gamma of North Carolina, Davidson	1923
Delaware, University of, Alpha of Delaware, Newark	1956
Denison University, Theta of Ohio, Granville	1911
Denver, University of, Gamma of Colorado, Denver	1940
DePauw University, Alpha of Indiana, Greencastle	1889
Dickinson College, Alpha of Pennsylvania, Carlisle	1887
Drake University, Gamma of Iowa, Des Moines	1923
Drew University, Gamma of New Jersey, Madison	1980
Duke University, Beta of North Carolina, Durham	1920
Earlham College, Delta of Indiana, Richmond	1965
Elmira College, Pi of New York, Elmira	1940
Emory University, Gamma of Georgia, Atlanta	1929
Fisk University, Delta of Tennessee, Nashville	1953
Florida, University of, Beta of Florida, Gainesville	1938
Florida State University, Alpha of Florida, Tallahassee	1935
Fordham University, Tau of New York, New York City	1962
Franklin and Marshall College, Theta of Pennsylvania, Lancaster	1908
Furman University, Gamma of South Carolina, Greenville	1973
Georgetown University, Delta of D.C., Washington	1965
George Washington University, Alpha of D.C., Washington	1938
Georgia, University of, Alpha of Georgia, Athens	1914
Gettysburg College, Iota of Pennsylvania, Gettysburg	1923

School Name, Chapter Name, Location	Year of Founding
Goucher College, Beta of Maryland, Towson	1905
Grinnell College, Beta of Iowa, Grinnell	1908
Gustavus Adolphus College, Eta of Minnesota, St. Peter	1983
Hamilton College, Epsilon of New York, Clinton	1870
Hamline University, Zeta of Minnesota, St. Paul	1974
Hampden-Sydney College, Eta of Virginia, Hampden-Sydney	1949
Harvard University, Alpha of Massachusetts, Cambridge	1781
Haverford College, Zeta of Pennsylvania, Haverford	1899
Hawaii, University of, at Manoa, Alpha of Hawaii, Honolulu	1952
Hiram College, Mu of Ohio, Hiram	1971
Hobart and William Smith Colleges, Zeta of New York, Geneva	1871
Hofstra University, Omega of New York, Hempstead	1973
Hollins College, Iota of Virginia, Hollins College	1962
Holy Cross, College of the, Pi of Massachusetts, Worcester	1974
Hope College, Zeta of Michigan, Holland	1971
Howard University, Gamma of D.C., Washington	1953
Idaho, University of, Alpha of Idaho, Moscow	1926
Illinois, University of,	
Chicago, Iota of Illinois	1977
Urbana-Champaign, Gamma of Illinois	1907
Illinois College, Epsilon of Illinois, Jacksonville	1932
Indiana University, Gamma of Indiana, Bloomington	1911
Iowa, University of, Alpha of Iowa, Iowa City	1895
Iowa State University, Zeta of Iowa, Ames	1973
Johns Hopkins University, Alpha of Maryland, Baltimore	1895
Kalamazoo College, Delta of Michigan, Kalamazoo	1958
Kansas, University of, Alpha of Kansas, Lawrence	1890
Kansas State University, Beta of Kansas, Manhattan	1974
Kent State University, Nu of Ohio, Kent	1977
Kentucky, University of, Alpha of Kentucky, Lexington	1926
Kenyon College, Beta of Ohio, Gambier	1858
Knox College, Delta of Illinois, Galesburg	1917
Lafayette College, Gamma of Pennsylvania, Easton	1890
Lake Forest College, Theta of Illinois, Lake Forest	1962
Lawrence University, Gamma-Delta of Wisconsin, Appleton	1914[2]
Lehigh University, Beta of Pennsylvania, Bethlehem	1887
Louisiana State University, Beta of Louisiana, Baton Rouge	1977
Luther College, Eta of Iowa, Decorah	1983
Macalester College, Epsilon of Minnesota, St. Paul	1968
Maine, University of, at Orono, Delta of Maine	1923
Manhattan College, Upsilon of New York, New York City	1971
Marietta College, Gamma of Ohio, Marietta	1860
Marquette University, Zeta of Wisconsin, Milwaukee	1971
Mary Baldwin College, Lambda of Virginia, Staunton	1971
Mary Washington College, Kappa of Virginia, Fredericksburg	1971
Maryland, University of, Gamma of Maryland, College Park	1964

2. In 1964, after Lawrence College and Milwaukee-Downer merged as Lawrence University, the Gamma chapter at Lawrence and the Delta chapter at Downer (1941) combined as the Gamma-Delta of Wisconsin.

School Name, Chapter Name, Location	Year of Founding
Massachusetts, University of, Nu of Massachusetts, Amherst	1965
Massachusetts Institute of Technology, Xi of Massachusetts, Cambridge	1971
Miami, University of, Delta of Florida, Coral Gables	1983
Miami University, Iota of Ohio, Oxford	1911
Michigan, University of, Alpha of Michigan, Ann Arbor	1907
Michigan State University, Epsilon of Michigan, East Lansing	1968
Middlebury College, Beta of Vermont, Middlebury	1868
Mills College, Zeta of California, Oakland	1929
Millsaps College, Alpha of Mississippi, Jackson	1989
Minnesota, University of, Alpha of Minnesota, Minneapolis	1892
Missouri, University of, Alpha of Missouri, Columbia	1901
Morehouse College, Delta of Georgia, Atlanta	1968
Mount Holyoke College, Theta of Massachusetts, South Hadley	1905
Muhlenberg College, Pi of Pennsylvania, Allentown	1968
Nebraska, University of, Alpha of Nebraska, Lincoln	1895
New Hampshire, University of, Beta of New Hampshire, Durham	1952
New Mexico, University of, Alpha of New Mexico, Albuquerque	1965
New York, City University of,	
Brooklyn College, Rho of New York	1950
City College, Gamma of New York	1867
Herbert H. Lehman College, Chi of New York	1971
Hunter College, Nu of New York	1920
Queens College, Sigma of New York	1950
New York State University, of, at	
Albany, Alpha Alpha of New York	1974
Binghamton, Psi of New York	1971
Buffalo, Omicron of New York	1938
Stony Brook, Alpha Beta of New York	1974
New York University, Beta of New York, New York City	1858
North Carolina, University of, at	
Chapel Hill, Alpha of North Carolina	1904
Greensboro, Epsilon of North Carolina	1956(1934)[3]
North Dakota, University of, Alpha of North Dakota, Grand Forks	1914
Northwestern University, Alpha of Illinois, Evanston	1890
Notre Dame, University of, Epsilon of Indiana, Notre Dame	1968
Oberlin College, Zeta of Ohio, Oberlin	1907
Occidental College, Delta of California, Los Angeles	1926
Ohio State University, Epsilon of Ohio, Columbus	1904
Ohio University, Lambda of Ohio, Athens	1929
Ohio Wesleyan University, Eta of Ohio, Delaware	1907
Oklahoma, University of, Alpha of Oklahoma, Norman	1920
Oregon, University of, Alpha of Oregon, Eugene	1923
Pennsylvania, University of, Delta of Pennsylvania, Philadelphia	1892
Pennsylvania State University, Lambda of Pennsylvania, University Park	1937
Pittsburgh, University of, Xi of Pennsylvania, Pittsburgh	1953
Pomona College, Gamma of California, Claremont	1914

3. Originally chartered as a section of the Alpha of North Carolina.

School Name, Chapter Name, Location	Year of Founding
Princeton University, Beta of New Jersey, Princeton	1899
Puget Sound, University of, Delta of Washington, Tacoma	1986
Purdue University, Zeta of Indiana, Lafayette	1971
Radcliffe College, Iota of Massachusetts, Cambridge	1914
Randolph-Macon College, Zeta of Virginia, Ashland	1938(1923)[4]
Randolph-Macon Woman's College, Delta of Virginia, Lynchburg	1917
Redlands, University of, Xi of California, Redlands	1977
Reed College, Beta of Oregon, Portland	1938
Rhode Island, University of, Beta of Rhode Island, Kingston	1977
Rhodes College, Gamma of Tennessee, Memphis	1949
Rice University, Beta of Texas, Houston	1929
Richmond, University of, Epsilon of Virginia, Richmond	1929
Ripon College, Epsilon of Wisconsin, Ripon	1952
Rochester, University, Iota of New York, Rochester	1887
Rockford College, Eta of Illinois, Rockford	1953
Rutgers-The State University, Alpha of New Jersey, New Brunswick	1869
Sections: Douglass College	1921
Newark College	1958
Saint Catherine, College of, Gamma of Minnesota, St. Paul	1938
Saint Lawrence University, Lambda of New York, Canton	1899
Saint Louis University, Gamma of Missouri, St. Louis	1968
Saint Olaf College, Delta of Minnesota, Northfield	1949
San Diego State University, Nu of California, San Diego	1974
San Francisco State University, Omicron of California, San Francisco	1977
Santa Clara University, Pi of California, Santa Clara	1977
Scripps College, Theta of California, Claremont	1962
Skidmore College, Phi of New York, Saratoga Springs	1971
Smith College, Zeta of Massachusetts, Northamptom	1904
South, University of the, Beta of Tennessee, Sewanee	1926
South Carolina, University of, Alpha of South Carolina, Columbia	1926
South Dakota, University of, Alpha of South Dakota, Vermillion	1926
Southern California, University of, Epsilon of California, Los Angeles	1929
Southern Methodist University, Gamma of Texas, Dallas	1949
Stanford University, Beta of California, Stanford	1904
Stetson University, Gamma of Florida, DeLand	1982
Swarthmore College, Epsilon of Pennsylvania, Swarthmore	1896
Sweet Briar College, Theta of Virginia, Sweet Briar	1950
Syracuse University, Kappa of New York, Syracuse	1896
Temple University, Rho of Pennsylvania, Philadelphia	1974
Tennessee, University of, Knoxville, Epsilon of Tennessee	1965
Texas, University of, at Austin, Alpha of Texas	1905
Texas Christian University, Delta of Texas, Fort Worth	1971
Trinity College, Beta of Connecticut, Hartford	1845
Trinity College, Epsilon of D.C., Washington	1971
Trinity University, Epsilon of Texas, San Antonio	1974

4. Originally chartered as a section of the Delta of Virginia.

School Name, Chapter Name, Location	Year of Founding
Tufts University, Delta of Massachusetts, Medford	1892
Tulane University, Alpha of Louisiana, New Orleans	1909
Tulsa, University of, Beta of Oklahoma, Tulsa	1989
Union College, Alpha of New York, Schenectady	1817
Utah, University of, Alpha of Utah, Salt Lake City	1935
Vanderbilt University, Alpha of Tennessee, Nashville	1901
Vassar College, Mu of New York, Poughkeepsie	1899
Vermont, University of, Alpha of Vermont, Burlington	1848
Villanova University, Sigma of Pennsylvania, Villanova	1986
Virginia, University of, Beta of Virginia, Charlottesville	1908
Virginia Polytechnic Institute and State University, Mu of Virginia, Blacksburg	1977
Wabash College, Beta of Indiana, Crawfordsville	1898
Wake Forest University, Delta of North Carolina, Winston-Salem	1941
Washington, University of, Alpha of Washington, Seattle	1914
Washington and Jefferson College, Kappa of Pennsylvania, Washington	1937
Washington and Lee University, Gamma of Virginia, Lexington	1911
Washington State University, Gamma of Washington, Pullman	1929
Washington University, Beta of Missouri, St. Louis	1914
Wayne State University, Gamma of Michigan, Detroit	1953
Wellesley College, Eta of Massachusetts, Wellesley	1904
Wells College, Xi of New York, Aurora	1932
Wesleyan University, Gamma of Connecticut, Middletown	1845
West Virginia University, Alpha of West Virginia, Morgantown	1910
Western Maryland College, Delta of Maryland, Westminster	1980
Wheaton College, Kappa of Massachusetts, Norton	1932
Whitman College, Beta of Washington, Walla Walla	1920
William and Mary, College of, Alpha of Virginia, Williamsburg	1776
Williams College, Gamma of Massachusetts, Williamstown	1864
Wilson College, Nu of Pennsylvania, Chambersburg	1950
Wisconsin, University of	
Madison, Alpha of Wisconsin	1899
Milwaukee, Eta of Wisconsin	1974
Wofford College, Beta of South Carolina, Spartanburg	1941
Wooster, College of, Kappa of Ohio, Wooster	1926
Wyoming, University of, Alpha of Wyoming, Laramie	1940
Yale University, Alpha of Connecticut, New Haven	1780

Notes

1. Beginnings in Revolutionary Virginia

1. Herbert B. Adams, *The College of William and Mary* (Washington, 1887), 20, 36–39; J. E. Morpugo, *Their Majesties' Royall Colledge: William and Mary in the Seventeenth and Eighteenth Centuries* (Washington, 1976), 180–82; Wilford Kale, *Hark upon the Gale: An Illustrated History of the College of William and Mary* (Norfolk, 1985), 51, 55–60; Gilbert Chinard, *Thomas Jefferson: The Apostle of Americanism* (Boston, 1929), 8–9, 98–100; Louis F. Snow, *The College Curriculum in the United States* (New York, 1907), 74, quoting James Madison to Ezra Stiles, Aug. 27, 1780.

2. Jane Carson, *James Innes and His Brothers of the F. H. C.* (Williamsburg, 1965), 1 n, 5–7, 60–61, 67–69; George P. Coleman, ed., *The Flat Hat Club and the Phi Beta Kappa Society: Some New Light on Their History* (Richmond, 1916), an unpaged collection of a few documents, among them an F. H. C. "catalogue of books" and two letters of Thomas Jefferson; William T. Hastings, *Phi Beta Kappa as a Secret Society, with Its Relation to Freemasonry and Antimasonry; Some Supplementary Documents* (Washington, 1965), 84, quoting William Short to Edward Everett, July 8, 1831. Hastings says, pp. 2–3, that "in three respects" Phi Beta Kappa at William and Mary "developed uniqueness": it was highly selective, was secret, and put out branches. But the F. H. C. and the P. D. A. were also secret and selective.

3. Quotations of and references to the society's proceedings, except where otherwise noted, are derived from the manuscript record in the William and Mary library, which provided a xerographic copy of the manuscript. Heath's role in the founding is recalled by Short in his letter of July 8, 1831, in Hastings, *Phi Beta Kappa*, 84.

4. By 1935 the William and Mary librarian, E. G. Swem, had deciphered most of the crossed-out description of the salutation. He then sent the manuscript to Bert C. Farrar, examiner of questioned documents for the federal government. With the aid of infrared light, Farrar reproduced the undeciphered words as follows: "back of the same hand, and a return with the last (_____) by the saluted." (Farrar to Swem, Aug. 14, 1935, William and Mary file, United Chapters archives.) Hastings, *Phi Beta Cappa*, 59 n, gives for "last

(_____)" a conjectural reading of "hand used," which is undoubtedly correct.

The pronunciation "Fie Beeta Kappa" is indicated by the clerk's writing (and then crossing out) "Fi be Ca" for "Φ Β Κ" in the minutes for June 5, 1779. This must be a "phonetic representation of the name," as Hastings, *Phi Beta Kappa*, 68 n, suggests.

5. *Catalogue of the Harvard Chapter of Phi Beta Kappa: Alpha of Massachusetts; with the Constitution, the Charter, Extracts from the Records, Historical Documents, and Notes* (Cambridge, Mass., 1912), 91–93.

6. The quotation regarding "irregular" students is from Short's letter to Everett, in Hastings, *Phi Beta Kappa*, 84.

7. Allan B. Magruder, *John Marshall* (Boston, 1895), 18–24.

8. The minutes state that three anniversary celebrations and a farewell party took place in Raleigh Tavern, and another anniversary was celebrated in Davenport's tavern. The minutes do not specify where the ordinary (regular and "call") meetings were held but make it obvious that most if not all of them were held at the college. As the record shows, the society appointed a vice president residing at the college so that he would be on hand to take charge during the frequent absences of a president who lived off campus. The society purchased candles, which the members would have used to light a college room, not a tavern room. Again and again the society elected and initiated a new member at one and the same meeting; this would hardly have been feasible if the meeting had not been held close to where the initiate stayed, that is, at the college. Very likely the minutes omitted mentioning the place of the ordinary meetings simply because it was taken for granted.

Nevertheless, the tradition persists that the very first of these meetings, the one of December 5, 1776, was held elsewhere. The earliest known source for this story is a William and Mary catalog of 1859, which—83 years after the event—stated on page 15: "The first meeting was held in the Apollo Hall of the old Raleigh Tavern of Williamsburg, the room in which the first revolutionary spirit was breathed in the burning words of [Patrick] Henry." That sentence was repeated verbatim in *The History of the College of William and Mary from Its Foundation, 1660, to 1874* (Richmond, 1874), 50–51. The author of the statement could not have written it from personal knowledge, nor could he have derived it from the society's minutes, the one authentic source available to him.

Not only does the story lack documentary support, but it also raises questions of credibility. Would a group of only five, planning a sober and secret society, be likely to meet in a tavern ballroom and banquet hall that was one of the most commodious, conspicuous, and boisterous places in town? If the clerk in his minutes noted the Raleigh Tavern (though not the Apollo Room) as the scene of anniversary celebrations and a farewell party, would he be likely to omit mentioning it as the scene of the historic first meeting?

The Phi Beta Kappa historian Oscar M. Voorhees came to disbelieve the story, and in 1934 he wrote out an argument against it. (The 6-page typescript of the paper "Was the Phi Beta Kappa Organized in Raleigh Tavern?" is in the William and Mary file, United Chapters archives, and a carbon copy is in the William and Mary library.) His argument received a courteous but cool reception in Williamsburg. (W. A. R. Goodwin to Voorhees, June 5, 1934, William and Mary file, United Chapters archives.) In *The History of Phi Beta Kappa* (New York, 1945) Voorhees refrained from discussing the question and merely observed, on page 5, that the traditional account "was first set forth" in 1859.

9. George E. Kidd, *Early Freemasonry in Williamsburg, Virginia* (Richmond, 1957), 2, 59–90; Hastings, *Phi Beta Kappa*, 4–7.

10. Short to Everett, July 8, 1831, in Hastings, *Phi Beta Kappa*, 84; Edward Everett Hale, "A Fossil from the Tertiary," *Atlantic Monthly*, XLIV (July 1879), 100–102.

11. Jonathan Leavitt, 1787 memorandum, Connecticut Alpha archives, in Hastings, *Phi Beta Kappa*, 86–88.

12. Short to Abraham Bishop, Jan. 23, 1781, Connecticut Alpha archives, in Voorhees, *History*, 28.

13. Beckley to Elizur Goodrich, July 1, 1782, Connecticut Alpha archives, typed copy in possession of the United Chapters.

14. Lyon G. Tyler, "Brief Personal Sketches," *William and Mary College Quarterly*, IV (Apr. 1896), 245–54; Arthur T. Vanderbilt, "An Example to Emulate," *South Atlantic Quarterly*, LII (Jan. 1953), 5–6.

15. Voorhees, *History*, 26–31, 220–21.

2. New Beginnings in New England

1. Voorhees, *History*, *26–27, 36–37;* proceedings of the Massachusetts Alpha, July, Sept. 1781, *Catalogue of the Harvard Chapter* (1912), 99–100.

2. Voorhees, *History*, 40–41, 45–46; Samuel Kendal to Yale Brothers, Mar. 23, 1782, and to Henry T. Channing, Oct. 16, 1782, *Catalogue of the Harvard Chapter* (1912), 103–4, 108–9. In this and most other correspondence between branches a simple cipher was used.

3. Proceedings of the Massachusetts Alpha, Sept. 5, 1786, May 24, June 21, 1787; Henry Ware to Barna Bidwell, Mar. 8, 1787; Bidwell to Ware, May ?, 1787, *Catalogue of the Harvard Chapter* (1912), 110–15; Frederick Chase, "The Phi Beta Kappa Society," in John K. Lord, *A History of Dartmouth College, 1815–1909* (Concord, N. H., 1913), 540–41.

4. Jesse Appleton to Asa McFarland, July 25, 1796; Joseph McKean to the Connecticut Alpha, Sept. 4, 1806; Asa Peabody to the New Hampshire Alpha, Sept. 11, 1806, in Hastings, *Phi Beta Kappa*, 20, 33–35.

5. Voorhees, *History*, 34, 41–46, 68; Chase, "Phi Beta Kappa Society," 541; *Catalogue of the Harvard Chapter* (1912), 107.

6. Voorhees, *History*, 83–84; Chase, "Phi Beta Kappa Society," 544; Sydney H. Gay, *James Madison* (Boston, 1898), 297–301.

7. Voorhees, *History*, 39, 53–56; Chase, "Phi Beta Kappa Society," 541.

8. Louis F. Snow, *The College Curriculum in the United States* (New York, 1907), 37, 79–82; Russel B. Nye, *The Cultural Life of the New Nation, 1776–1830* (New York, 1960), 184–87, 396.

9. Voorhees, *History*, 42, 56–59, 68; *Catalogue of the Harvard Chapter* (1912), 101–3; Lewis R. Packard, "The Phi Beta Kappa Society," in William L. Kingsley, ed., *Yale College: A Sketch of Its History*, vol. I (New York, 1879), 326; Reginald H. Phelps, "Phi Beta Kappa at Harvard: A Bicentennial History," in *Phi Beta Kappa: Alpha of Massachusetts, 1781–1981, Bicentennial Exercises, December 11, 1981* (n. p., n. d.), 3–5; *Eighty Years' Progress of the United States*, vol. II (Hartford, Conn., 1869), 254.

10. Voorhees, *History*, 72–73, 93–98, 109–14; *Catalogue of the Harvard Chapter* (1912), 108–9, 144; Phelps, "Phi Beta Kappa," 8.

11. William R. Thayer, *An Historical Sketch of Harvard University, from Its Foun-*

dation to May, 1890 (Cambridge, Mass., 1890), 59–60; David Potter, *Debating in the Colonial Chartered Colleges: An Historical Survey, 1642 to 1900* (New York, 1944), 66–67; Oscar M. Voorhees, "College Societies that Antedate Phi Beta Kappa," *Phi Beta Kappa Key*, V (Mar. 1925), 679–80.

12. Voorhees, *History*, 43, 45; Hastings, *Phi Beta Kappa*, 14–21; *Catalogue of the Harvard Chapter* (1912), 108–9, 120–21; Josiah Quincy, *The History of Harvard University*, 2d ed., vol. I (Boston, 1860), 397–99.

13. Voorhees, *History*, 31–32; Voorhees, "College Societies," 676–79; *Catalogue of the Harvard Chapter* (1912), 112–13; Packard, "Phi Beta Kappa Society," 326.

14. Voorhees, *History*, 51–52, 59–62; Voorhees, "College Societies," 682; Chase, "Phi Beta Kappa Society," 539, 543–44.

15. Hastings, *Phi Beta Kappa*, 23–37; *Catalogue of the Harvard Chapter* (1912), 142; Packard, "Phi Beta Kappa Society," 326; Vernon Stauffer, *New England and the Bavarian Illuminati* (New York, 1918), 246–52, 283, 355–60.

16. "Alexander H. Everett," *United States Magazine and Democratic Review*, X (1842), 462–64, quoted in Marta Wagner, "The American Scholar in the Early National Period: The Changing Context of College Education, 1782–1837" (Ph.D. dissertation, Yale University, 1983), 258.

17. Chase, "Phi Beta Kappa Society," 545; *Catalogue of the Members of the New-Hampshire Alpha of the φBK Society, Dartmouth University* (Andover, Mass., 1815), listing Woodward at the head of the "Officers for the Years 1814–15"; unidentified typescript, quoting minutes of the New Hampshire Alpha, in the Dartmouth file, United Chapters archives.

18. Voorhees, *History*, 105–8.

3. Limiting the Fraternity's Growth

1. Donald G. Tewksbury, *The Founding of American Colleges and Universities Before the Civil War* (New York, 1932), 75, quoting an 1856 pamphlet by E. N. Kirk; Nye, *Cultural Life of the New Nation*, 177–78.

2. Colin B. Burke, *American Collegiate Populations: A Test of the Traditional View* (New York, 1982), 18–19, 26.

3. Tewksbury, *Founding of American Colleges*, 28, 31, and passim, grossly exaggerates the number of colleges that were founded and the number that failed. He has misled many writers on the subject. Burke, *American Collegiate Populations*, 14 and passim, provides a corrective. Burke's estimates seem much more reliable than Tewksbury's, but the figures given in all sources are more or less suspect.

4. Russel B. Nye, *Society and Culture in America, 1830–1860* (New York, 1974), 180–81; Merle Curti, *The Growth of American Thought* (New York, 1943), 226; John S. Brubacher and Willis Rudy, *Higher Education in Transition: A History of American Colleges and Universities, 1636–1976*, 3d ed. (New York, 1976), 355–56. The enrollment figures are from Clarence F. Birdseye, *Individual Training in Our Colleges* (New York, 1907), 135.

5. R. Freeman Butts, *The College Charts Its Course* (Philadelphia, 1939), 100–06, 118–25, 129; George P. Schmidt, *The Liberal Arts College* (New Brunswick, N. J., 1957), 39; Robert S. Fletcher, *A History of Oberlin College from Its Foundation Through the Civil War*, vol. I (Oberlin, Ohio, 1943), 434–36; Snow, *College Curriculum*, 141, 146–48; Curti, *Growth of American Thought*, 362.

6. Voorhees, *History*, 76–77; minutes of the Massachusetts Alpha, Sept. 8,

1789, Sept. 1, 1790; minutes of the Connecticut Alpha, Feb. 22, Mar. 8, Sept. 9, 1790; Thomas Thompson to the president of the Connecticut Alpha, Sept. 2, 1790, in William T. Hastings, ed., *A Century of Scholars: Rhode Island Alpha of Phi Beta Kappa, 1830–1930* (Providence, 1932), 173–78.

7. Voorhees, *History*, 77–79; minutes of the Connecticut Alpha, Sept. 14, Dec. 5, 1797, Sept. 13, Dec. 3, 5, 24, 1798, Feb. 22, 1799, in Hastings, *Century of Scholars*, 178–84.

8. Voorhees, *History*, 128–32, 158–62. The enrollment figures are from Birdseye, *Individual Training*, 135.

9. Edward Everett to John Pickering, July 14, 1828; Samuel Deane to Francis Wayland, Apr. 26, 1829; Wayland to the Massachusetts Alpha, June 25, 1829; Joseph Story to Wayland, July 20, 1829; James Gould to Wayland, July 30, 1829, in Hastings, *Century of Scholars*, 184–87; *Catalogue of the Rhode Island Alpha of Phi Beta Kappa, with the History of the Founding of the Chapter and Its Present Constitution and By-Laws* (Providence, 1914), 4–5; Voorhees, *History*, 170–78; Birdseye, *Individual Training*, 135.

10. Letters of Roebet Elfe et al., Abraham Nott, Jonathan Maxcy, John C. Calhoun, John Quincy Adams, and Samuel Gilman, and minutes of the New Hampshire Alpha, quoted in Voorhees, *History*, 79–81; additional quotation from the May 12, 1818, letter of application in *The Phi Beta Kappa Key*, V (May 1923), 235.

11. Application from Hampden-Sydney, July 9, 1837, quoted in Voorhees, *History*, 197–98, minutes of the Massachusetts Alpha, Aug. 26, 1841, quoted in *Catalogue of the Harvard Chapter* (1912), 161.

12. Voorhees, History, 198–204, quoting correspondence and minutes; Birdseye, *Individual Training*, 135.

4. Transition to an Honor Society

1. Thomas McAuley to Jefferson, May 19, 1819, and Jefferson to John D. Taylor, n. d., in Coleman, ed., *Flat Hat Club*, unpaginated.

2. James Gould, *An Oration Pronounced at New-Haven, Before the Connecticut Alpha of the Phi Beta Kappa* (New Haven, 1825), 18, quoted in Wagner, "American Scholar," 15; [John Mitchell,] *Reminiscences of Scenes and Characters in College* (New Haven, 1847), 106–7.

3. Thaddeus Stevens to Samuel Merrill, Jan. 5, 1814, quoted in Richard N. Current, *Old Thad Stevens: A Story of Ambition* (Madison, Wis., 1942), 6–7.

4. Samuel L. Knapp, *The Genius of Masonry, or a Defence of the Order . . . with Some Notice of Other Secret Societies in the United States* (Providence, 1828), 91; *Rochester Republican*, n. d., reprinted in the *American Masonic Record and Albany Saturday Magazine*, Apr. 11, 1829, and quoted in Hastings, *Phi Beta Kappa*, 43.

5. Avery Allyn, *A Ritual of Freemasonry, Illustrated by Numerous Engravings; with Notes and Remarks, to Which Is Added a Key to the Phi Beta Kappa* (Boston, 1831), quoted in Voorhees, *History*, 184–87. See also Hastings, *Phi Beta Kappa*, 43–45 and frontispiece reproducing Allyn's illustrations of the society's medal, grip, and sign.

6. Everett to Short, July 5, 1831, and Short to Everett, July 8, 1831, in Hastings, *Phi Beta Kappa*, 83–84.

7. Charles Francis Adams, ed., *Memoirs of John Quincy Adams*, vol. VIII (Philadelphia, 1874–77), 383–87, 389–92, 394–400, 405–7, 408–10.

8. Voorhees, *History,* 188–92; Hastings, *Phi Beta Kappa,* 50–51, quoting the recollection of Charles Tracy, Yale '32; *Catalogue of the Rhode Island Alpha* (1914), 9; Chase, "Phi Beta Kappa Society," 541–42.

9. Voorhees, *History,* 143–44, 188–89; Adams, *Memoirs,* VIII, 386, 389–91, 398, 406–7, 409.

10. Voorhees, *History,* 162–63; David F. Allmendinger, Jr., *Paupers and Scholars: The Transformation of Student Life in Nineteenth Century New England* (New York, 1975), 121–24.

11. Voorhees, *History,* 149–50, 153–54; Quincy, *History of Harvard University,* II, 398.

12. Voorhees, *History,* 196–97; Hastings, *Phi Beta Kappa,* 49; Chase, "Phi Beta Kappa Society," 546–47; Harold G. Rugg to Voorhees, Dec. 8, 1933, quoting minutes of the New Hampshire Alpha for Apr. 19 and June 30, 1845, in Dartmouth file, United Chapters archives.

13. William R. Baird, *American College Fraternities: A Descriptive Analysis of the Society System of the United State with a Detailed Account of Each Fraternity* (Philadelphia, 1879), 14–15; John Robson, ed., *Baird's Manual of American College Fraternities,* 14th ed. (Menasha, Wis., 1977), 5–6; Oscar M. Voorhees, "Kappa Alpha at Its Centennial," *Phi Beta Kappa Key,* VI (Jan. 1926), 85–87; Hastings, *Phi Beta Kapppa,* 51.

14. Voorhees, *History,* 147–48; [Lyman H. Bagg,] *Four Years at Yale,* (New Yaven, 1871), 51–52, 111, 142–44, 146, 160.

15. Voorhees, *History,* 181–83; *Catalogue of the Harvard Chapter* (1912), 160, 162; Charles P. Curtis, "Liquor and Learning in Harvard College, 1792–1846," *New England Quarterly,* XXV (Sept. 1952), 344–53.

16. Voorhees, *History,* 69–70, 149–50; Phelps, "Phi Beta Kappa at Harvard," 5–6; Hastings, *Phi Beta Kappa,* 49–50.

17. Voorhees, *History,* 153–54, 163–64, 179; Chase, "Phi Beta Kappa Society," 542, 546.

18. Voorhees, *History,* 194–96, 206–7; Mitchell, *Reminiscences,* 105–9.

19. Voorhees, *History,* 69–70, 149–50; Phelps, "Phi Beta Kappa at Harvard," 10–11; Clarence C. Mondale, "Gentlemen of Letters in a Democracy: Phi Beta Kappa Orations, 1788–1865" (Ph.D. dissertation, University of Minnesota, 1960), 36–37.

20. Clark S. Northup, William C. Lane, and John C. Schwab, eds., *Representative Phi Beta Kappa Orations* (Boston, 1915), vii, 24–42; Clark S. Northup, ed., *Representative Phi Beta Kappa Orations: Second Series* (New York, 1927), v, 4–7, 9; Richard Beale Davis, ed., "Edward Tyrrel Channing's 'American Scholar' of 1818," *Key Reporter,* XXVI (Spring 1961), 1–4, 8; Voorhees, *History,* 122.

21. Quincy, *History of Harvard University,* II, 397; Knapp, *Genius of Masonry,* 92; Mitchell, *Reminiscences,* 106–7.

5. Expansion, Disunification, and Decline

1. Voorhees, *History,* 207–8, 212, 223–26, 230; Clement R. Wood, "Alabama's Chapter of Phi Beta Kappa," *Phi Beta Kappa Key,* I (May 1912), 21–23.

2. Voorhees, *History,* 222–23; M. I. Smead to "Dear Sir," Aug. 8, 1852, and minutes of the Maine Alpha, Sept. 2, 1852, typed copies in the William and Mary file, United Chapters archives; Adams, *College of William and Mary,* 61–62, quoting a 42d Cong., 2d sess. House Report (1872); *The War of the Rebellion: A Compila-*

tion of the Official Records of the Union and Confederate Armies, series I, vol. XVII (Washington, 1887), 203–7.

3. Voorhees, *History,* 209–11, 224–25, 233, 244; Chase, "Phi Beta Kappa Society," 542–43, 547.

4. Voorhees, *History,* 217–18, 260, 262–64; Raymond Du Bois Cahall, "The Beta Chapter of Ohio at Kenyon College," *Phi Beta Kappa Key,* V (Jan. 1923), 120 n; Allan Nevins, *The Emergence of Modern America, 1865–1878* (New York, 1927), 273–74, 280–81.

5. Voorhees, *History,* 214, 218–20, 229, 231, 239, 254–55, 257, 260; Cahall, "Beta Chapter of Ohio," 124–25.

6. Voorhees, *History,* 211, 232–33, 242–43; Mondale, "Gentlemen of Letters," 36; Cahall, "Beta Chapter of Ohio," 124–26; *Phi Beta Kappa Gamma Chapter of New York: Centennial Directory, 1867–1967* (New York, 1967), 20.

7. Potter, *Debating in the Colonial Chartered Colleges,* 122–23; [Bagg], *Four Years at Yale,* 51, 148, 200, 228, 231; Packard, "Phi Beta Kappa Society," 327.

8. Voorhees, *History,* 207–10, 212–13, 223–26, 240–42.

9. Voorhees, *History,* 215–16, 227–28, 230, 234–35, 238, 244–46; Cahall, "Beta Chapter of Ohio," 120.

10. Voorhees, *History,* 247, 251, 252–55, 258–60; *Phi Beta Kappa Gamma Chapter of New York,* 17–18; *Harvard University Bulletin,* II (Oct. 1, 1881), 264–65.

11. [Bagg], *Four Years at Yale,* 233–34.

12. Baird, *American College Fraternities* (1879), 77–79.

13. Hale, "A Fossil from the Tertiary," 98, 100, 102–3, 105–6; Voorhees, *History,* 254–55, 266.

6. Organizing the United Chapters

1. Ida M. Tarbell, *The Nationalizing of Business, 1878–1898* (New York, 1936), 76–82, 147–67; Arthur M. Schlesinger, *The Rise of the City, 1878–1898* (New York, 1933), 221–22, 311–12, 410.

2. *Harvard University Bulletin,* II, 264–65; Voorhees, *History,* 219–20, 253, 265–67.

3. *Harvard University Bulletin,* II, 264–65. The minutes of the convention, kept by Justin Winsor and published in the *Bulletin,* list 56 delegates by name and affiliation. Then the minutes note: "Twenty-nine delegates present, representing six alphas and six other branches." This is confusing, but it probably means that 56 delegates were appointed and only 29 of them attended.

4. George William Curtis, "The Editor's Easy Chair," *Harper's Magazine,* LXIII (Sept. 1881), 625–26; Northup et al., *Representative Phi Beta Kappa Orations,* 191–215.

5. One of the delegates wrote much later in regard to the Harvard convention: ". . . my recollections have a tragic tinge. I reached New York a day or two after the meeting and that morning the news came that President Garfield had been assassinated." Joseph Ullman to Oscar M. Voorhees, Nov. 26, 1919, United Chapters archives.

6. Justin Winsor, "University Notes," *Harvard University Bulletin,* III (Jan. 1, 1882), 300–301.

7. Voorhees, *History,* 233, 274; *Catalogue of the Rhode Island Chapter* (1914), 10.

8. Voorhees, *History,* 275–80, 285–86.

7. Keeping Up with the Colleges

1. Voorhees, *History,* 284–85, 301, 310; *Phi Beta Kappa Key,* I (Jan. 1911), 23.

2. Eben B. Parsons, *Phi Beta Kappa Handbook and General Address Catalogue of the United Chapters* (North Adams, Mass., 1900). According to Voorhees, *History,* 292–93, Parsons also published the information other than the membership list in a separate pamphlet, of which "only one copy is known to be in existence."

3. Voorhees, *History,* 308, 313; *Phi Beta Kappa Key,* I (Nov. 1910), 3–4; Northup et al., *Representative Phi Beta Kappa Orations,* iv–v.

4. Voorhees, *History,* 316; *Phi Beta Kappa Key,* II (Oct. 1913), 24–25; III (Oct. 1916), 21.

5. James Bryce, *The American Commonwealth,* vol. II (New York, 1889), 550; Curti, *Growth of American Thought,* 468, 512–14.

6. Brubacher and Rudy, *Higher Education in Transition,* 357–58; Bureau of the Census, *Historical Statistics of the United States* (Washington, D.C., 1960), 210–11; Birdseye, *Individual Training,* 134–39.

7. Bryce, *American Commonwealth,* II, 529–30, 550–51; Schlesinger, *Rise of the City,* 202–4; Daniel J. Boorstin, *The American: The National Experience* (New York, 1965), 155.

8. Bryce, *American Commonwealth,* II, 534–35, 541–42, 549–50; Brubacher and Rudy, *Higher Education in Transition,* 356–59; Birdseye, *Individual Training,* 137; Laurence R. Veysey, *The Emergence of the American University* (Chicago, 1965), 358–59.

9. Voorhees, *History,* 290–91, 304. As an example of the application form used in the early 1900s, see the 1907 Oberlin application in the Oberlin file, United Chapters archives.

10. Robson, *Baird's Manual of American College Fraternities,* 14th ed., 183; C. G. Rockwood, "historical statement," Sept. 22, 1909, Princeton file, United Chapters archives.

11. Quotations from Parsons's letters and other information in "Beta of California," an unidentified page conjecturally dated 1909, and "The Golden Key" by Raymond Macdonald Allen (1905), in the Stanford file, United Chapters archives.

12. *Report of the Dean* (1909?), Oberlin file, United Chapters archives. See also John M. Poor to Voorhees, July 23, 1905, discussing the Oberlin application, in the Dartmouth file, United Chapters archives.

13. W. A. Chamberlin, ed., *Phi Beta Kappa, Theta of Ohio: History and Membership* (Granville, Ohio, 1914), 9–13, 16–18, 26–30; Shepardson to Voorhees, Dec. 23, 1912, University of Chicago file, United Chapters archives.

14. Voorhees, *History,* 256–57, 277–78, 290, 314; *Phi Beta Kappa Key,* VI (Oct. 1927), 559–60.

8. Classicism or Eclecticism

1. Bryce, *American Commonwealth,* II, 536–38.

2. *History of the College of William and Mary . . . to 1874,* 170–71, 174; William R. Thayer, *An Historical Sketch of Harvard University, From Its Foundation to May, 1890* (Cambridge, Mass., 1890), 25–26, 37.

3. Butts, *College Charts Its Course,* 239, 244–47; Chamberlin, *History of Theta Chapter of Ohio,* 31–36; Birdseye, *Individual Training,* 122–23.

4. Charles Francis Adams, Jr., "A College Fetich," *The Independent*, XXV (Aug. 9, 1883), 997–1000; Daniel Henry Chamberlain, *Not "a College Fetich"* (Boston, 1884), 1–29, and "The Harvard Elective System," *New Englander and Yale Review*, LXV (Apr. 1886), 359–72.

5. R. H. Soper, ed., *The Complete Prose Works of Matthew Arnold*, vol. X (Ann Arbor, Mich., 1974), 56, 59.

6. Andrew P. Peabody, "A Liberal Education," *New Englander and Yale Review*, XLV (Mar. 1886), 193–207.

7. Thayer, *Historical Sketch of Harvard*, 37; Veysey, *Emergence of the American University*, 248–51.

8. Cahall, "Beta Chapter of Ohio," 126–27; Parsons, *Phi Beta Kappa Handbook and General Address Catalogue*, 251–52; Voorhees, *History*, 305, 315, 317; Calvin O. Davis, "Courses Pursued by Members of Phi Beta Kappa," *School and Society*, V (June 9, 1917), 686–90.

9. Bernard C. Steiner, "Historical Sketch of the Alpha of Maryland," typescript dated April 1909; Wilfred P. Mustard to Voorhees, Nov. 30, 1914; Voorhees to Mustard, Dec. 2, 1914, John Hopkins file, United Chapters archives.

10. *Phi Beta Kappa Gamma Chapter of New York*, 22–23; Chase, "Phi Beta Kappa Society," 542; Chamberlin, *History of Theta Chapter of Ohio*, 19; Voorhees, *History*, 290.

11. Parsons, *Phi Beta Kappa Handbook and General Address Catalogue*, 252–53; Hastings, *Century of Scholars*, 21–22; *Phi Beta Kappa Gamma Chapter of New York*, 24–25.

12. *Harvard Crimson*, n. d., quoted in *Phi Beta Kappa Key*, I (Mar. 1911), 28–30; Edwin E. Slosson, *Great American Universities* (New York, 1910), 192; Parsons, *Phi Beta Kappa Handbook and General Address Catalogue*, 252.

13. Voorhees, *History*, 291–92, 316; Parsons, *Phi Beta Kappa Handbook and General Address Catalogue*, 253–55; *Phi Beta Kappa Gamma Chapter of New York*, 22–23; *Phi Beta Kappa Key*, I, 29–30; Cahall, "Beta Chapter of Ohio," 125–26; *Catalogue of the Rhode Island Alpha* (1914), 9–10.

14. Bryce, *American Commonwealth*, II, 544–46; Schlesinger, *Rise of the City*, 207–11; Voorhees, *History*, 301.

15. Robson, *Baird's Manual of American College Fraternities*, 19th ed., 707–8; Bryce, *American Commonwealth*, II, 545; Schlesinger, *Rise of the City*, 316–18; Allison Danzig, *The History of American Football* (New York, 1956), 29.

16. George Birkbeck Hill, *Harvard by an Oxonian* (New York, 1894), 116–19; Slosson, *Great American Universities*, 237; Birdseye, *Individual Training*, 211–12; Chamberlin, *History of Theta Chapter of Ohio*, 37; Voorhees, *History*, 307; *Phi Beta Kappa Key*, I, 30.

9. Money and the American Scholar

1. Voorhees, *History*, 317–18; secretary's report to the 13th Council, *Phi Beta Kappa Key*, IV (Oct. 1919), 32; Albert Shaw to Edwin A. Grosvenor, Dec. 21, 1916, and Voorhees to Shaw, Jan. 12, 1917, Stanford file, United Chapters archives.

2. Voorhees, *History*, 318–20, 322–25; secretary's report to the 14th Council and his article "Permanent Phi Beta Kappa Headquarters," *Phi Beta Kappa Key*, V (Oct. 1922), 45, and VI (Mar. 1928), 697–98.

3. "The Proposed Sesquicentennial Fund," *Phi Beta Kappa Key,* V (Oct. 1922), 44–45; Voorhees, *History,* 327–28, 331.

4. Elizabeth A. Smith to Voorhees, Apr. 30, 1923, Wisconsin file, United Chapters archives; Kirk L. Cowdery to Voorhees, Mar. 6, 1924, Oberlin file, United Chapters archives.

5. William C. Lane, report, June 24, 1926, and Voorhees to Lane, July 2, 1926, Harvard file, United Chapters archives; Charles W. Hendel to Voorhees, Apr. 8, 1927, Princeton file, United Chapters archives.

6. Voorhees, *History,* 323–24, 333–35; William H. Winters to Voorhees, Dec. 20, 1911; J. Leslie Hall to Edwin A. Grosvenor, Aug. 21, 1919; Robert M. Hughes to Charles F. Thwing, Sept. 23, 1925, William and Mary file, United Chapters archives.

7. Roy B. Varnado, "Mr. Rockefeller's Other City: Background and Response to the Restoration of Williamsburg, Virginia, 1927–1929" (master's thesis, College of William and Mary, 1974), 2–3, 8–11, 18–20, 22–25. The poet was William M. Davidson, superintendent of Pittsburgh public schools, and a copy of the poem, which consists of many stanzas, is in the William and Mary file, United Chapters archives. The editor of the fraternity magazine *Greek Exchange* said he had "some marvelous photographs showing the reconstructed Raleigh Tavern in which ΦBK is supposed to have been founded." He invited Voorhees's successor William A. Shimer to write an "article concerning this reconstruction and working in the ΦBK and the fraternity angle prominently." Shimer did so. Leland F. Leland to Shimer, Feb. 7, 20, Mar. 13, 17, May 15, June 16, 21, 1933, and Shimer to Leland, Feb. 10, Mar. 15, May 4, June 14, 19, 1933, *American Scholar* records, Library of Congress. But the secretary of the Virginia Alpha later wrote Shimer: "In connection with the use of the cut of the Raleigh, may I suggest that the founders of 1776 to 1780 held in that building three of their four anniversary celebrations." Donald W. Davis to Shimer, Nov. 19, 1937, William and Mary file, United Chapters archives. Significantly, the chapter secretary did *not* say the meeting of December 5, 1776, was held in the Raleigh. For further discussion of this question, see note 8 of Chapter 1 above.

8. *Phi Beta Kappa Key,* VI (Mar. 1928), 697–98; Voorhees, *History,* 331–32, 340–41.

9. Voorhees, *History,* 343–46; Clark S. Northup, ed., *Representative Phi Beta Kappa Orations, Second Series* (New York, 1927), iv; Shepardson to Shimer, Aug. 20, 1931, Shimer to Shepardson, Aug. 21, 1931, and Treasurer to Charles Heebner, Feb. 17, 1933, *American Scholar* records.

10. Erskine to Ruth E. Campbell, Jan. 29, 1933, Campbell to Ada L. Comstock, Jan. 29, 1932, Comstock to Campbell, Feb. 1, 1932, and F. J. E. Woodbridge to Campbell, Mar. 14, 1932, *American Scholar* records.

11. Maurice Holland to Ruth E. Campbell, Feb. 2, 1933, Shimer to Holland, Feb. 3, 1933, Angela Melville to Robert M. Henry, Apr. 6, 1933, Henry to Melville, May 28, 1933, and Melville to Henry, June 5, 1933, *American Scholar* records.

12. Voorhees, *History,* 346; *ΦBK Annals 1934* (New York, 1934), 50–52, 61–62.

13. Shimer to Ellen Glasgow, Aug. 20, 1936, to John Erskine, Nov. 10, 1937, June 20, 29, 1938, and to Christian Gauss, May 27, 1938; Dorothy Blair (assistant to Shimer) to Philip F. Myers, Sept. 8, 1938, *American Scholar* records.

14. *ΦBK Annals 1934,* 51; Shimer to Franklin Dunham, Dec. 9, 1938, and

Information for Authors of Articles, Jan. 15, 1943, *American Scholar* records; *New York Times,* Apr. 9, 1939, sec. X, p. 12; *Key Reporter,* V (Spring 1940), 3; minutes of the Senate, Dec. 17, 1941, United Chapters archives.

15. *ΦBK Annals 1934,* 44, 48–49, 57, 67; *Key Reporter,* V (Summer 1940), 1–2; Voorhees, *History,* 351–53.

16. *ΦBK Annals 1934,* 45; Angela Melville to Bruce Barton, Mar. 24, Apr. 6, 13, 1933, and Barton to Melville, Apr. 14, 1933, *American Scholar* records.

17. George Milton Janes, "The Future of Phi Beta Kappa," and William A. Shimer, "Some Phi Beta Kappa Purposes," *School and Society,* XL (Sept. 1, 29, 1934), 281–82, 421–23.

18. *Time,* XXXIII (Feb. 27, 1939), 58; *Key Reporter,* IV (Spring 1939), 1–2; IX (Spring 1944), 4; XXVI (Winter 1960–61), 2–3; Voorhees, *History,* 354–55; minutes of the Senate, Dec. 13, 1939, United Chapters archives.

10. Applications and Qualifications

1. *Historical Statistics of the United States: Colonial Times to 1957,* 210–11; Rita S. Halle, *Which College?* (New York, 1928), 3, quoted in David O. Levine, *The American College and the Culture of Aspiration, 1915–1940* (Ithaca, N. Y., 1986), 113–14.

2. Brubacher and Rudy, *Higher Education in Transition,* 358–59; Preston W. Slosson, *The Great Crusade and After, 1914–1928* (New York, 1930), 329–30; Frank Aydelotte, *Breaking the Academic Lock Step: The Development of Honors Work in American Colleges and Universities* (New York, 1944), 102–3 (a book consisting of lectures that Aydelotte, ex-president of Swarthmore College, delivered at Teachers College, Columbia University, in 1942).

3. Secretary's reports to the 12th, 13th, and 14th Councils and proceedings of the 14th Council, *Phi Beta Kappa Key,* III (Oct. 1916), 21; IV (Oct. 1919), 25–26; V (Oct. 1922), 43–46, 61–62.

4. Clyde S. Atchison, "The Mission of Phi Beta Kappa," and Voorhees on the society's expansion, *Phi Beta Kappa Key,* V (Mar. 1924), 445–48; VI (Oct. 1927), 555–63.

5. Charles N. Cole to Voorhees, Nov. 27, 1923, and Voorhees to Cole, Dec. 20, 1923, Oberlin file, United Chapters archives.

6. Secretary's report on the 18th Council, *ΦBK Annals 1934,* 58–61; Voorhees, *History,* 347–49. The 18th Council voted down a proposal to refuse a charter to any institution less than fifty years old—fortunately for Connecticut College, which opened in 1915 and was approved for a charter by the Council in 1934. *Key Reporter,* I (Spring, 1936), 35.

7. Edward J. Larson, *Trial and Error: The American Controversy over Creation and Evolution* (New York, 1985), 48–63.

8. J. Gresham Machen to the editor, Sept. 12, 1925, and James Bradbury to the editor, Sept. 18, 1925, in the *New York Times,* Sept. 18, 1925, p. 22, and Sept. 20, 1925, sec. II, p. 6.

9. Cyril A. Nelson to Voorhees, May 12, 1926; Voorhees to Nelson, May 21, 1926; and undated clipping from the *Baltimore Sun,* Johns Hopkins file, United Chapters archives; Larson, *Trial and Error,* 65.

10. *ΦBK Annals 1934,* 61; *Phi Beta Kappa: Gamma Chapter of New York,* 25–26; Voorhees, *History,* 350.

11. Secretary's report to the 16th Council, *Phi Beta Kappa Key,* VII (Oct. 1928), 32.

12. Voorhees to Shepardson, May 11, 1925; Shepardson to Voorhees, May 16, 1925, with a copy of Shepardson to Scott, May 15, 1925; Scott to Shepardson, May 18, 1925; A. R. Ellingwood to Shimer, Nov. 11, 1932; Melville to Ellingwood, Nov. 15, 1932, Northwestern file, United Chapters archives; Shailer Mathews, "Chicago Association Organizes," *The Key Reporter,* I (Summer 1936), 77–78.

13. *ΦBK Annals 1934,* 15–16, 37–38, 60; Eustace Percy to Shimer, Dec. 16, 1936, and Frank Darvall to Shimer, Oct. 24, 1938, English Association file, United Chapters archives; Voorhees, *History,* 350–51.

11. Merit, Marks, and Membership

1. Levine, *American College,* 94; Butts, *College Charts Its Course,* 266–67, 358, 408–16; Aydelotte, *Breaking the Academic Lock Step,* ix.

2. News release for Spring 1937 number of the *American Scholar,* "Hutchins' Educational Views Called an Escape from Reality," *American Scholar* records, Library of Congress; Butts, *College Charts Its Course,* 297.

3. Wilfred P. Mustard to Voorhees, Nov. 30, 1914, and Voorhees to Mustard, Dec. 2, 1914, Johns Hopkins file, United Chapters archives.

4. William A. Shimer, "Higher ΦBK Standards," *American Scholar,* III (May 1934), 367–68; *ΦBK Annals 1934,* 43, 62; Voorhees, *History,* 350.

5. Walter A. Montgomery to Voorhees, Nov. 30, 1911, with clipping from the *Richmond Times-Dispatch,* Nov. 28, 1911; Robert M. Hughes to Voorhees, Nov. 11, 1912, June 24, 1927; James S. Wilson to Voorhees, Feb. 14, 1915; Voorhees to Hughes, Dec. 4, 1922, Nov. 29, 1927; Voorhees to J. A. C. Chandler, Apr. 16, 1928; Chandler to Voorhees, Apr. 21, 1928, William and Mary file, United Chapters archives.

6. Robert Scoon to Voorhees, Sept. 27, 1920; Voorhees to Edward S. Worcester, Apr. 10, May 10, 1924; Worcester to David Layton, May 2, 1924; Voorhees to Robert G. Albion, Mar. 30, 1934, Princeton file, United Chapters archives.

7. Clipping from the *New York Times,* Mar. 24, 1927; Voorhees to Charles W. Hendel, Jr., Mar. 29, May 13, 1927; Hendel to Voorhees, Apr. 19, May 9, 1927, Princeton file, United Chapters archives.

8. John C. French to Voorhees, July 3, 1928; Voorhees to French, July 6, 1928, Johns Hopkins file, United Chapters archives.

9. Voorhees to Henry Churchill King, Dec. 8, 1925; Karl F. Geiser to Shimer, Dec. 10, 17, 1932, Feb. 12, 20, 1933; Donald M. Love to Shimer, Jan. 19, 1933, Oberlin file, United Chapters archives.

10. Voorhees, *Phi Beta Kappa General Catalog, 1776–1922* (Somerville, N. J., 1923), xxix; Voorhees, *History,* 326–27, 341, 350; *ΦBK Annals 1934,* 66.

11. Robert M. Hughes to Caroline R. Fletcher, Apr. 2, 1926, and to Voorhees, June 24, 1927; Voorhees to Hughes, July 7, Nov. 29, 1927, William and Mary file, United Chapters archives; Voorhees, *History,* 350; *ΦBK Annals 1934,* 66.

12. Maynard M. Metcalf to Voorhees, Nov. 1, 1927, Feb. 21, 1928, Voorhees to Metcalf, Nov. 11, 1927, Feb. 29, 1928, Johns Hopkins file, United Chapters archives.

13. *New York Times,* Mar. 6, 1922, p. 1, col. 2; Feb. 22, 1928, p. 9, col. 2; Feb. 11, 1934, sec. II, p. 1, col. 2; Feb. 25, 1934, sec. VIII, p. 4, col. 5; Voorhees, "Is the

Key a 'Badge of Grinds'?" *Phi Beta Kappa Key,* VI (Mar. 1928), 706–8; Shimer, "Higher ΦBK Standards," *American Scholar,* III (May 1934), 367–68; *ΦBK Annals 1934,* 66.

14. C. S. Boucher, "Phi Beta Kappa Prospects at Chicago under the New Plan," *American Scholar,* III (Oct. 1934), 487–93; A. C. Hanford, "New Basis of Harvard Elections," *Key Reporter,* I (Summer 1936), 73–74, 97–99. See also the *Key Reporter,* II (Winter 1937), 2, 6, and V (Spring 1940), 3.

15. Aydelotte, *Breaking the Academic Lock Step,* 56, 58–59, 61–62; *Key Reporter,* I (Summer 1936), 79; Shimer to O. E. Albrecht, July 30, 1936, Pennsylvania file, United Chapters archives.

16. *New York Times,* Feb. 25, 1934, sec. VIII, p. 4, col. 5; *ΦBK Annals 1934,* 15.

17. *Phi Beta Kappa Key,* I (Nov. 1910), 28–29; *New York Times,* Jan. 13, 1928, p. 5, col. 3; Chamberlin, *History of Theta Chapter of Ohio,* 54–55; *Key Reporter,* I (Summer 1936), 99; VI (Winter 1940–41), 1–2.

12. Great Men—or Grinds?

1. William T. Foster, *Should Students Study?* (New York, 1917), 3–11; *Liberty,* July 17, 1926, quoted in Slosson, *Great Crusade and After,* 275–76; John P. Gavit, *College* (New York, 1925), 116–19, quoted in Levine, *American College,* 121–22.

2. *New York Times,* Aug. 25, 1925, p. 12, col. 5; May 23, 1928, p. 24, col. 4; May 15, 1930, p. 14, col. 2; May 19, 1930, p. 20, col. 7; May 21, 1930, p. 30, col. 2; May 23, 1930, p. 22, col. 6.

3. *New York Times,* Apr. 11, 1926, p. 14, col. 3; Mabel Newcomer, "The Phi Beta Kappa Student," *School and Society,* XXV (Jan. 1, 1927), 24.

4. Frank C. Ewart, "Does Scholarship Pay?" *Phi Beta Kappa Key,* VII (Mar. 1930), 454–55; Voorhees, "Phi Beta Kappa and the Beginnings of Football," *Phi Beta Kappa Key,* VI (Jan. 1926), 89–91; *Key Reporter,* I (Summer 1936), 79; II (Autumn 1937), 2; IV (Winter 1938–39), 2.

5. *New York Times,* July 20, 1935, sec. II, p. 3, col. 3; Dec. 15, 1935, sec. II, p. 8, col. 7; "Thorndike Intelligence Scores of Columbia College Seniors Elected to Phi Beta Kappa in 1928," *School and Society,* XXVIII (July 7, 1928), 10–11; *Key Reporter,* I (Summer 1936), 76, quoting Crile, *The Phenomena of Life,* 89, summarizing a CCNY study.

6. Edwin G. Dexter, "High-Grade Men: In College and Out," *Popular Science Monthly,* LXII (Mar. 1903), 429–35; Foster, *Should Students Study?,* 37–40; John C. Minot, "What Door Does the Phi Beta Kappa Key Open?" *North American Review,* CCXXIV (Nov. 1927), 532–36; *New York Times,* Mar. 28, 1926, sec. IX, p. 11, *ΦBK Annals 1934,* 77.

7. Slosson, *Great American Universities,* 67–69; Foster, *Should Students Study?,* 34–36; Stephen S. Visher, "Phi Beta Kappa and Scientific Research" and "Phi Beta Kappa Members of the National Academy of Science," *School and Society,* XX (Aug. 16, 1924), 216, and XXI (Mar. 28, 1925), 390.

8. *New York Times,* Dec. 15, 1923, p. 4, col. 3; Dec. 17, 1923, p. 16, col. 5; Walter S. Gifford, "Does Business Want Scholars?" *Harper's Magazine,* CLVI (Apr. 1928), 669–74; Ewart, "Does Scholarship Pay?" 456; Harold A. Larrabee, "Was ΦBK Worth While?" *Key Reporter,* II (Winter 1937), 1, 6.

9. *New York Times,* Sept. 11, 1925, p. 22, col. 5.

10. Edward L. Thorndike, "The Careers of Scholarly Men in America," *Century Magazine,* LXVI (May 1903), 153–55; Charles F. Thwing, "The Phi Beta

Kappa Society as a World-Force," *American Review of Reviews*, LXIX (June 1924), 629–33. The *New York Times*, Apr. 15, 1934, sec. II, p. 8, col. 7, reported a study of 88 Knox College ΦBK graduates of the years 1917–1927 and observed: "As would be expected, teaching is the profession which has drawn the greatest number."

11. *Phi Beta Kappa Key*, IV (Oct. 1921), 536; IV (Jan. 1922), 589; V (Mar. 1923), 184–86; V (Oct. 1923), 322–24; V (Jan. 1924), 367, 372; VI (Mar. 1928), 706–8; *Key Reporter*, I (Summer 1936), 75–76.

12. Mark Sullivan, *Our Times: The United States, 1900–1925*, vol. VI (New York, 1935), 398–400; Voorhees to J. A. C. Chandler, Mar. 15, 1923, William and Mary file, United Chapters archives.

13. Sullivan, *Our Times*, VI, 623–28; W. S. Mays to Voorhees, June 4, 1924; Voorhees to Mays, June 5, 1924; Lois F. Madsen to Carl Billman, June 9, 1961; Billman to Madsen, June 13, 1961, Chicago file, United Chapters archives.
The society continued to be judged by its notorious as well as its illustrious members. In 1987, after the revelation of Gary Hart's relationship with Donna Rice and his withdrawal from the presidential race, one reader of *People Weekly* thanked the magazine for "exposing lying politicians and stupid Phi Beta Kappas," and another reader said: "I think the Phi Beta Kappa society ought to demand the return of Donna Rice's key." *People Weekly*, XXVIII (July 6, 1987), 2.

14. *New York Times*, Apr. 30, 1930, p. 4, col. 5; May 25, 1930, sec. III, p. 10, col. 2; Sept. 18, 1931, p. 22, col. 4.

15. George A. Coe, "The Future of Phi Beta Kappa," and William A. Shimer, "The Philosophy of Phi Beta Kappa," *School and Society*, XXXIX (May 5, 1934), 568–70; (June 2, 1934), 703.

16. *Key Reporter*, I (Autumn 1936), 5; III (Winter 1938), 5.

13. Sisters in the Brotherhood

1. Nye, *Society and Culture in America*, 393; Wagner, "American Scholar in the Early National Period," 255–56, quoting *Columbian Centinel*, Aug. 30, 1817, and Mary Sophia Quincy, "The Harvard Commencement of 1829," *Harvard Graduates' Magazine*, XXVI (1918), 580.

2. Fletcher, *History of Oberlin College*, I 375–81; Allen M. Bailey to Dorothy Blair, Dec. 18, 1937, Oberlin file, United Chapters archives; Bryce, *American Commonwealth*, II, 548–49; Slosson, *Great Crusade and After*, 204–5; *Historical Statistics of the United States* (1957), 211–12.

3. Cahall, "Beta Chapter of Ohio," 126–27; Herbert Tuttle to Robert Roberts, n. d., *Key Reporter*, III (Spring 1938), 2, 5; Voorhees, *History*, 262–63.

4. *Catalogue of the Rhode Island Alpha* (1914), 9, 10; Hastings, *Century of Scholars*, 22–23; Voorhees, "Romantic Elements in Phi Beta Kappa History," *Phi Beta Kappa Key*, IV (May 1922), 715–16; Voorhees, *History*, 291, 294, 310–12, 315.

5. Voorhees, "Secretary's Report to the 12th National Council," *Phi Beta Kappa Key*, III (Oct. 1916), 20–21; Voorhees, *History*, 264–296; Hastings, *Century of Scholars*, 22–23.

6. Slosson, *Great American Universities*, 191–92, 274–75. It does not appear that at the University of Michigan the women Phi Betes were taking easier courses than the men or greatly outnumbering them. A study of those elected from the classes of 1905 to 1915 led to the following conclusions: "there is no pronounced difference in the work chosen by the men and by the women";

"there has been an approximately equal number of persons elected from each sex." Davis, "Courses Pursued by Members of Phi Beta Kappa," 690.

7. *Phi Beta Kappa Key,* III (Oct. 1916), 20–21; quotation from the *New York Evening Post,* n. d., and letter from Leta S. Hollingworth, n. d., *School and Society,* IV (Dec. 9, 1916), 901–2.

8. Northup, *Representative Phi Beta Kappa Orations, Second Series,* 312–19; Hastings, *Century of Scholars,* 24; *Phi Beta Kappa Key,* IV (Oct. 1919), 33, 44–45; Chamberlin, *Phi Beta Kappa, Theta of Ohio,* 44.

9. *New York Herald Tribune,* Apr. 15, 1936, quoted in *Key Reporter,* I (Summer 1936), 73.

10. J. H. Doyle, "The Phi Beta Kappa Tempest," *Education,* XLI (Oct. 1920), 86–94.

11. John W. Harrington, "'No' to Phi Beta Kappa: First Refusal of Much-Coveted Honor and Why Vassar Girl Did So," *New York Times,* Apr. 2, 1922, sec. VII, p. 10, col. 4.

12. *New York Times,* Apr. 15, 1934, sec. II, p. 8, col. 7; *Key Reporter,* VII (Summer 1942), 3.

13. *Phi Beta Kappa Key,* IV (May 1922), 714–15; VI (Jan. 1927), 375–79; *New York Times,* Apr. 2, 1922, sec. VII, p. 10, col. 4.

14. J. A. C. Chandler to Voorhees, June 24, 1924, William and Mary file, United Chapters archives.

15. Mrs. J. Scott Anderson to Voorhees, Jan. 5, 1921, Voorhees to Mrs. Anderson, Jan. 6, 1921; Shimer to Helen O. Shollenberger, Apr. 13, 1932, Pennsylvania file, United Chapters archives.

16. Voorhees, *History,* 311, 358–59; Shimer to Otto E. Albrecht, May 13, 1938; Albrecht to Shimer, telegram, May 18, 1938, Pennsylvania file, United Chapters archives.

17. *Newsweek,* XV (Mar. 4, 1940), 35; Eunice Fuller Barnard, "Let the Fair Creatures In," *Key Reporter,* III (Spring 1938), 1–2, 5.

14. To Quicken the Activities

1. *Key Reporter,* VIII (Autumn 1942), 2; XI (Spring 1946), 7; XXXIV (Autumn 1968), 1; Voorhees, *History,* 355–56.

2. Hiram Haydn, *Words & Faces* (New York, 1974), 3–5; *Key Reporter,* IX (Spring 1944), 1; XI (Winter 1945), 5.

3. Billman obituary, *Key Reporter,* XXXIX (Winter 1973–74), 3; Haydn, *Words & Faces,* 200; *New York Times,* Dec. 18, 1948, p. 27, col. 5; "Over Mrs. Altman's Strudels," *New Yorker,* XXVI (June 17, 1950), 19–20.

4. *New York Times,* Sept. 22, 1951, p. 18, col. 7; May 12, 1957, sec. X, p. 23, col. 3; *Key Reporter,* XIX (Feb. 1954), 1, 6; (Sept. 1954), 1, 7; XX (Nov. 1954), 1; (Feb. 1955), 1; XXXI (Autumn 1975), 1; Billman to Mary Alice Newman, Feb. 1, 1963, Chicago file, United Chapters archives; records of income and expenses for 1940–41 and 1975–76, United Chapters archives.

5. *New York Times,* Dec. 25, 1974, p. 20, col. 3; *Key Reporter,* XXXIX (Winter 1973–74), 3; XL (Winter 1974–75), 1.

6. Haydn, *Words & Faces,* 165–66; typescript of report [c. April 1946], *American Scholar* records, Library of Congress.

7. Haydn to Editorial Board, Jan. 22, 1945, and to Henry Simon, Jan. 31, 1947, *American Scholar* records.

8. Haydn, *Words & Faces*, 165–67; Haydn to Robeson, Oct. 19, 25, 1944; Robeson to Haydn, telegram, Oct. 21, 1944, Mar. 8, 1945; D. Sommers to Haydn, Apr. 26, 1945; Celia Lewis to Robeson, June 5, 1945, *American Scholar* records.

9. Matthew Page Andrews to G. A. Works, Nov. 14, 1946, and to Haydn, Dec. 6, 27, 1946; Haydn to Andrews, Dec. 2, 24, 1946, *American Scholar* records.

10. Haydn, *Words & Faces*, 181–83; Edwin J. Akutowicz to "Gentlemen," Apr. 30, 1944, and to Haydn, received May 18, 1944; Haydn to Akutowicz, May 8, 19, 1944, *American Scholar* records.

11. Typescript of report [c. Apr. 1946]. *American Scholar* records; Haydn, *Words & Faces*, 197–99; *Key Reporter*, XVIII (Feb. 1953), 1; XL (Autumn 1974), 1; records of income and expenses, 1941 to 1975; minutes of the executive committee, Apr. 6, 1966, United Chapters archives; Kenneth M. Greene to the author, July 1, 1988.

12. *Key Reporter*, VII (Spring 1942), 5; (Summer 1942), 1; (Autumn 1942), 3; VIII (Winter 1942–43), 3.

13. *Phi Beta Kappa Bulletin*, X (Dec. 1946), 13–14; *Key Reporter*, VII (Autumn 1942), 3; XI (Summer 1946), 1; XVI (Summer 1951), 3; XVII (Aug. 1952), 5; XXVII (Winter 1961–62), 1, 8; XXVIII (Spring 1963), 1; (Summer 1963), 1; XXXII (Summer 1967), 1, 8; XLII (Summer 1976), 1, 8; *Minutes of the Twenty-Ninth Triennial Council Meeting* (1970), 10–11; William T. Hastings, ed., *Man Thinking: Representative Phi Beta Kappa Orations, 1915–1959* (Ithaca, N. Y., 1962); Alpha Association of Phi Beta Kappa in Southern California *News Letter*, IV (Summer 1979), 1–4.

14. Mrs. Ewart K. Lewis, 1956 ΦΒΚ talk at Oberlin, *Key Reporter*, XXXV (Summer 1970), 1–3; G. C. Nearing to Haydn, Feb. 19, 1944; Haydn to Nearing, Mar. 6, 1944, *American Scholar* records.

15. *Key Reporter*, XX (Feb. 1955), 1; XXIII (Jan. 1958), 1; XXIV (Jan. 1959), 1; XXV (Winter 1959–60), 1; *Minutes of the Twenty-Ninth Triennial Council Meeting* (1970), 5–6.

16. *Key Reporter*, XXI (Nov. 1955), 2; John W. Dodds, "8 Years of the Visiting Scholar Program," *Key Reporter*, XXIX (Summer 1964), 1–3; XXXV (Spring 1970), 1; XLII (Spring 1977), 1; Hallett D. Smith, "The Visiting Scholar Program: A Quarter-Century Overview," *Key Reporter*, XLIV (Spring 1979), 1, 4.

17. William T. Hastings, "Phi Beta Kappa Today," *Key Reporter*, XXII (July 1957), 1–4; XXVIII (Winter 1963–64), 1, 4; XXX (Autumn 1964), 2–3; XXXI (Summer 1966), 1–3; *New York Times*, June 2, 1963, sec. IV, p. 9, col. 1; June 24, 1964, p. 27, col. 6; June 28, 1964, sec. IV, p. 7, col. 1.

18. *Key Reporter*, XXXIII (Autumn 1967), 1; (Spring 1968), 1; XXXIV (Spring 1969), 3; XXXVI (Summer 1971), 1, 8; XXXIX (Winter 1973–74), 4; *New York Times*, Sept. 10, 1972, p. 64, col. 7.

19. Voorhees, *History*, 352; *Key Reporter*, V (Autumn 1940), 6; XXXVI (Winter 1970–71), 1; XXXVII (Spring 1972), 1.

15. Academics, Athletics, Ecclesiastics

1. *Key Reporter*, VII (Autumn 1942), 5; XI (Winter 1945), 1; Brubacher and Rudy, *Higher Education in Transition*, 360, 406.

2. *Key Reporter*, XIII (Spring 1948), 1; XV (Winter 1949–50), 1; XXIII (July

1958), 4; XXVII (Autumn 1961), 1–2; Joyce S. Steward to Billman, Oct. 20, 1971, and Billman to Steward, Oct. 27, 1971, Wisconsin file, United Chapters archives.

3. *New York Times,* Nov. 5, 1961, sec. IV, p. 7, col. 6; Nov. 6, 1970, p. 43, col. 6; *Minutes of the Twenty-Ninth Triennial Council,* 8; *Phi Beta Kappa Bulletin,* XXVIII (Sept. 1973), 12. For a list of chapters chartered by 1979, see *Phi Beta Kappa: A Handbook for New Members,* edition for the triennium 1982–85, pp. 24–29.

4. William T. Hastings, "Phi Beta Kappa Today," *Key Reporter,* XXII (July 1957), 1–4.

5. *Key Reporter,* XIV (Jan. 1959), 4; XXIV (Autumn 1960), 1; Arthur E. Gordon to Secretary, United Chapters, May 10, 1963; Billman to Gordon, May 14, 1963, Berkeley file, United Chapters archives.

6. *Phi Beta Kappa Bulletin,* XIX (Feb. 1956), 10–11; *Key Reporter,* XX (May 1955), 1–2; XXIII (July 1958), 4; XXIV (Oct. 1958) 2–3.

7. *Key Reporter,* VII (Spring 1942), 5; XII (Winter 1946), 1–2; W. A. Shimer to Emily Moore Hall, July 10, 1952, William and Mary file, United Chapters archives; minutes of the Senate, Sept. 9–10, 1946, United Chapters archives; Haydn, *Words & Faces,* 159–62.

8. *Key Reporter,* XVI (Spring 1951), 1; Helen R. MacGregor, Governor Warren's private secretary, to Billman, Feb. 7, 1951; *Phi Beta Kappa Bulletin,* XXVIII (Sept. 1973 [misdated 1974]), 12–13, 21.

9. Neil G. McCluskey, "Phi Beta Kappa and Catholic Colleges," *America,* XCVIII (Feb. 22, 1958), 597–99. In a later issue of the same magazine, XCIX (Apr. 5, 1958), 18–20, some Catholic educators disagreed with McCluskey. "Phi Beta Kappa cannot welcome all comers . . . other colleges are passed over too," Brother Dionysius, C. F. X., pointed out. Two laymen at Holy Cross said Catholic colleges should hire more lay professors and should "treat them as full-fledged members of the faculty and not as mere adjuncts to the religious faculty."

10. Nina Grun Shafer to Phi Beta Kappa, June 8, 1961, Georgetown file, United Chapters archives. "I can assure you," Billman replied to Shafer, June 13, 1961, in regard to Georgetown, "that its prospects for inspection will not be affected, either favorably or unfavorably, by the fact that it happens to be a Roman Catholic institution." Georgetown file, United Chapters archives.

11. Bryce, *American Commonwealth,* II, 546–48; Brubacher and Rudy, *Higher Education in Transition,* 361–62; Slosson, *Great American Universities,* 430.

12. William J. Martin to Voorhees, July 29, 1914, and Davidson application for a ΦBK charter, 1914, Davidson file, United Chapters archives; Davidson College, *Self-Study Report for the Southern Association of Colleges and Schools* (Davidson, N. C., 1965), 84–85; Suhail Hanna, "Calling for a Yardstick for Excellence," *Christianity Today,* XXIII (Nov. 2, 1979), 17; Report to the Committee on Qualifications, Furman University, Dec. 6–7, 1971, United Chapters archives; *Phi Beta Kappa Bulletin,* XXVIII, 13, 15, 21.

13. Stephen Steinberg, *The Academic Melting Pot: Catholics and Jews in American Higher Education* (New York, 1974), 33, 52–53.

14. James Brown Scott to Voorhees, Feb. 18, 1926; Voorhees to Scott, Feb. 27, 1926; Walter J. O'Connor to Voorhees, Jan. 25, 1929; O'Connor to Clark S. Northup, Feb. 14, 1933; Joseph Q. Adams to Northup, Feb. 15, 1933, Shimer to O'Connor, Feb. 23, 1933, Georgetown file, United Chapters archives.

15. Roy J. Deferrari to Voorhees, Feb. 16, 1929, and to Shimer, Jan. 6, 1931;

David A. Robertson to Deferrari, Jan. 23, 1937, Catholic University file, United Chapters archives; E. G. Swem to Voorhees, Sept. 6, 1929, William and Mary file, United Chapters archives; Joseph M. Corrigan, "The Future of the Catholic University of America," *School and Society,* XLV (Jan. 9, 1937), 44.

16. Lowell J. Ragatz, *The Founding of the Alpha Chapter of Phi Beta Kappa in the District of Columbia and a History of Its First Two Years, 1937–1939* (n. p., 1939), 3–10; Deferrari to Shimer, Jan. 21, 1937; Shimer to Harold G. Rugg, Mar. 26, 1937, Catholic University file, United Chapters archives.

17. Andrew J. Torrielli to Shimer, Mar. 16, 1938; Robert I. Gannon to Billman, Jan. 29, 1948, and Billman to Gannon, Feb. 3, 1948; Billman to Jean Misrahi, Feb. 10, 1956, and Misrahi to Billman, Feb. 25, 1956; E. C. Kirkland, report on a Fordham visit of Jan. 7–8, 1960; *New York Times* clipping, Dec. ?, 1960, Fordham file, United Chapters archives; Danzig, *History of American Football,* 337–39.

18. Shimer to O'Connor, Mar. 22, 1935, and O'Connor to Shimer, Apr. 3, 1935; Franklin B. Williams to Billman, Oct. 23, 1958; E. C. Kirkland, report on 1964 Georgetown visit, Georgetown file, United Chapters archives; Danzig, *History of American Football,* 347–48.

19. For the list of chapters, see the *Phi Beta Kappa Handbook* for 1982–1985, pp. 24–29.

16. Race, Ethnicity, and Scholarship

1. President's Commission on Higher Education, *Higher Education for American Democracy,* vol. I (Washington, D. C., 1947), 27; vol. II, 1; Levine, *American College and the Culture of Aspiration,* 136–37; Steinberg, *Academic Melting Pot,* 9, 19–20.

2. Dan W. Dodson, "College Quotas and American Democracy," *American Scholar,* XV (Summer 1946), 267–76; *Key Reporter,* XIII (Winter 1947–48), 3.

3. Clarence E. Josephson to Shimer, Feb. 2, 1940, and Shimer to Josephson, Feb. 17, 1940, Pennsylvania file, United Chapters archives.

4. Mrs. B. Milton Garfinkle to Billman, June 20, 1949; Billman to John E. Pomfret, May 7, 14, 1951; Pomfret to Billman, May 10, 1951, Pennsylvania file, United Chapters archives.

5. Broun and Britt, p. 54, cited in Levine, *American College and the Culture of Aspiration,* 156; Arthur F. Payne, "Who's Like Us?" *Key Reporter,* I (Summer 1936), 89.

6. Nathaniel Weyl, "National Origins of the Phi Beta Kappa Membership," *Names: Journal of the American Name Society,* XII (June 1964), 119–22.

7. Everett Carll Ladd, Jr., and Seymour Martin Lipset, *The Divided Academy: Professors and Politics* (New York, 1975), 150; Steinberg, *Academic Melting Pot,* 109.

8. Harry W. Roberts to United Chapters, Aug. 5, 1968; Evelyn Greenberg to Nancy Zintbaum, Feb. 19, 1969, miscellaneous file, United Chapters archives; *Key Reporter,* XXXIV (Autumn 1968), 1.

9. Chamberlin, *History of Theta Chapter of Ohio,* 45; *Phi Beta Kappa Key,* VII (Mar. 1929), 173.

10. Harold G. Rugg to Shimer, Nov. 1, 18, 1938, Dartmouth file, United Chapters archives; *Key Reporter,* V (Spring 1940), 1–2; XII (Summer 1947), 3.

11. Donald W. Davis to Shimer, Aug. 9, 1936, and Shimer to Davis, Aug. 12,

1936, William and Mary file, United Chapters archives; Dorothy E. Blair to Harold G. Rugg, Nov. 23, 1938, Dartmouth file, United Chapters archives; Ramona S. Barth, "Miami ΦBKs Survey Negro Education," *Key Reporter,* VII (Summer 1942), 2–3.

12. Sarah Martin Eason to Marjorie Hope Nicholson, Apr. 24, 1944, *American Scholar* records, Library of Congress. At Hiram Haydn's invitation, Mereb Mossman of the Woman's College of the University of North Carolina wrote a dissenting letter: "Miss Cromwell's assumption seems to be that if we could but eliminate segregation we would eliminate much of the inequality. This may or may not be true." Mossman to the Editor, n. d., and Haydn to Mossman Aug. 23, 1944, *American Scholar* records.

13. "Phi Beta Kappa and Segregated Education," *Key Reporter,* X (Autumn 1945), 3; Evelyn Tedford to the Editor, n. d., and Clarence Lohman to the Editor, n. d., *Key Reporter,* XI (Spring 1946), 3; (Summer 1946), 3.

14. *Key Reporter,* XXI (Nov. 1955), 6; XXIV (Oct. 1958), 3–4; *New York Times,* Apr. 25, 1963, p. 14, col. 3.

15. Henry F. and Katharine Pringle, "America's Leading Negro University," *Saturday Evening Post,* CCXXI (Feb. 19, 1945), 95.

16. George Hedley and Elizabeth Pope to the Committee on Qualifications, Mar. 10, 1949; Edna A. Merson to United Chapters, Mar. 14, 1949; Mrs. G. R. B. Symonds to Phi Beta Kappa, Mar. 16, 1949; Bernard A. Shepard to Phi Beta Kappa, Mar. 30, 1949, Howard file, United Chapters archives.

17. Billman to Pope, Mar. 15, 1949; to Merson, Mar. 21, 1949; to Symonds, Mar. 21, 1949; and to Shepard, Apr. 5, 1949, Howard file, United Chapters archives.

18. Horace Mann Bond, "A Century of Negro Higher Education," in William W. Brickman and Stanley Lehrer, eds., *A Century of Higher Education: Classical Citadel to Collegiate Colossus,* (New York, 1962), 190–92.

19. Letters to Voorhees from Ernest E. Just, Aug. 30, Sept. 10, 1911; from F. R. Letcher, Nov. 13, 1923; from Lawrence E. Wilson, Nov. 9, 1926; from James E. Walker, Mar. 11, 1927; Voorhees to Letcher, Nov. 21, 1923; to Wilson, Nov. 13, 1926, to Walker, Mar. 16, 1927, Howard file, United Chapters archives.

20. Edward L. Parks to Voorhees, Jan. 9, 1924, Feb. 27, 1924; Voorhees to Parks, Jan. 14, 1924; "Secretary to Dr. Voorhees" to Parks, Mar. 3, 1924, Howard file, United Chapters archives.

21. E. P. Davis to Secretary, United Chapters, June 10, 1930; Voorhees to Davis, June 11, 1930; Shimer to Davis, May 21, June 7, 1934; Davis to Shimer, June 5, 14, 1934, Howard file, United Chapters archives.

22. Pringle to Shimer, Oct. 18, 1948; Shimer to Pringle, Oct. 22, 1948, Howard file, United Chapters archives; Pringle, "America's Leading Negro University," 92–93, 97.

23. William T. Hastings, Report on Howard University visit of Apr. 26–27, 1951; Mordecai W. Johnson to Billman, Feb. 28, 1952, Howard file, United Chapters archives; McCluskey, "Phi Beta Kappa and Catholic Colleges," 597–99.

24. Billman to George St. John, Jr., and Charles S. Johnson, Feb. 2, 1952; Theodore S. Currier to Billman, Feb. 27, 1962, Fisk file, United Chapters archives.

25. Haydn, *Words & Faces,* 180, 184–85, 192; John Hope Franklin to Billman, Mar. 2, 1955, Howard file, United Chapters archives; *Phi Beta Kappa Bulletin,*

XXVIII (Sept. 1973), 20–21; Northup, *Representative Phi Beta Kappa Orations,* 195.

17. The Flattery of Imitation

1. William T. Hastings, *The Insignia of Phi Beta Kappa* (Washington, D.C., 1964), 5–13, 24–26; Parsons, *Phi Beta Kappa Handbook,* 256.

2. Voorhees, *History,* 210; Hastings, *Insignia of Phi Beta Kappa,* 13–14, 27–31; Voorhees, "The Phi Beta Kappa Badge," *Phi Beta Kappa Key,* I (Jan. 1911), 19–22.

3. Voorhees, *History,* 222; Haydn, *Words & Faces,* 5; *ΦBK Annals 1934,* 62; *Key Reporter,* I (Spring 1936), 43–44; (Winter 1936), 10; III (Winter 1938), 8. Dictionaries continued, as before, to offer a variety of pronunciations.

4. Robson, *Baird's Manual of American College Fraternities,* 19th ed., 20–21, 42–43; *Phi Beta Kappa Key,* VI (Oct. 1925), 9–10; VII (Oct. 1928), 32–33; *New York Times,* Nov. 7, 1946, p. 33, col. 5.

5. Chamberlin, *History of Theta Chapter of Ohio,* 52–54; Billman to Herbert R. Sensenig, Oct. 10, 1958, Dartmouth file, United Chapters archives.

6. Marsa W. White to Lloyd W. Taylor, May 6, 1928; Taylor to White, May 18, 1928, Oberlin file, United Chapters archives. Taylor sent a copy of the correspondence to Voorhees's office, and Voorhees thoroughly approved of Taylor's stand. Taylor to ΦBK Headquarters, May 18, 1928, and Voorhees to Taylor, June 6, 1928, Oberlin file, United Chapters archives.

7. Aileen Wells to Voorhees, Mar. 3, 1931, and Shimer to Wells, Mar. 5, 1931, Infringements file, United Chapters archives.

8. Noel L. Keith to Billman, Oct. 24, 1952, and Billman to Keith, Oct. 30, 1952, Texas Christian file, United Chapters archives.

9. Announcement of joint meeting, May 1, 1918, Pennsylvania file, United Chapters archives; ΦBK Senate, minutes of the Executive Committee, Nov. 30, 1962; Robson, *Baird's Manual of American College Fraternities,* 19th ed., vii–viii.

10. This list is based on correspondence in the Infringements file, United Chapters archives.

11. *Key Reporter,* I (Summer 1936), 85; *New York Times,* Feb. 25, 1954, p. 18, col. 5.

12. *Phi Beta Kappa Key,* I (Jan. 1911), 21; VI (Oct. 1925), 9–10; VII (Oct. 1928), 33; *Key Reporter* I (Summer 1936), 85.

13. Record of registration of "Phi Beta Kappa," July 6, 1954, and of the registration of the key with inscriptions, Feb. 15, 1955; Billman to John N. Williams, Dec. 9, 1961, Infringements file, United Chapters archives.

14. Diane Threlkeld to Mrs. Royal A. Stone, Feb. 28, 1967, and Fybate Lecture Notes catalog, Infringements file, United Chapters archives.

15. Paul Gerhardt to Diane Threlkeld, Oct. 15, 1963, and clipping of "Phi Beta Jantzen" newspaper ad, Infringements file, United Chapters archives.

16. Ralph E. Lazarus to Billman, July 23, 1964, Infringements file, United Chapters archives.

17. Minutes of the Executive Committee, Dec. ?, 1978, Oct. ?, 1979.

18. Paul V. Claudon to Billman, Oct. 4, 1962; Billman to Claudon, Oct. 10, 1962; Frank Piombo to Billman, Oct. 30, 1962, Infringements file, United Chapters archives.

19. John N. Williams to Billman, July 13, 1962, Infringements file, United Chapters archives.

18. "Relevance"—or Rationality

1. *Key Reporter,* XXII (July 1957), 2; Aydelotte, *Breaking the Academic Lock Step,* 1–2.

2. Christian Gauss, "Wartime Policy of Phi Beta Kappa," *The American Scholar,* XII (Apr. 1943), 143–52.

3. Press release for Oct. 3, 1944, *American Scholar* records, Library of Congress.

4. Harlow Shapley to Hiram Haydn, Jan. 17, 1945, *American Scholar* records.

5. William T. Hastings, "The Task at Hand," *Key Reporter,* VII (Winter 1951–52), 5, 7, 11; minutes of the Senate, Dec. 5–6, 1969, p. 10.

6. *New York Times,* Dec. 3, 1951, p. 30, col. 3; Vanderbilt, "An Example to Emulate," 19.

7. Denys Smith, "American News: Phi Beta Kappas in Government; or, The American Old School Tie," *National Review,* CXIX (July 1942), 69–73; Robert H. Ferrell, ed., *The Eisenhower Diaries* (New York, 1981), 239.

8. *New York Times,* Feb. 23, 1958, sec. IV, p. 8, col. 5; Nov. 9, 1958, p. 65, col. 3; "The Golden Key," *The Reporter,* XVIII (Mar. 20, 1958), 2; "In Honorable Company," *Key Reporter,* XXIII (July 1958), 4.

9. *New York Times,* Dec. 25, 1960, sec. IV, p. 5, col. 1; Sept. 4, 1961, p. 16, col. 2; Nov. 26, 1964, p. 48, col. 1; "The Phi Beta Kappas on Kennedy's Team," *U.S. News & World Report,* Jan. 30, 1961, p. 30.

10. Ernest Havemann and Patricia Salter West, *They Went to College: The College Graduate in America Today* (New York, 1952), 157–65; G. Armour Craig, review of *They Went to College, Key Reporter,* XVII (Summer 1952), 4; David Boroff, *Campus U.S.A.: Portraits of American Colleges in Action* (New York, 1961), 133.

11. Albert Beecher Crawford, *Phi Beta Kappa Men of Yale, 1780–1959,* reviewed in the *Key Reporter,* XXXIV (Autumn 1968), 1; *New York Times,* June 7, 1970, p. 1, col. 5.

12. Boroff, *Campus U.S.A.,* 47, 106–7, 110, 118.

13. *New York Times,* June 7, 1970, p. 1, col. 5; Melba B. Bowers to the Editor, *Key Reporter,* XXIV (July 1959), 8.

14. Brubacher and Rudy, *Higher Education in Transition,* 79, 234–51, 278–79, 285–86; *Phi Beta Kappa Bicentennial Bulletin,* 44; Dennis Cuddy, "Higher Education Needs Higher Ground," *Chicago Tribune,* Dec. 15, 1986, Sec. I, p. 22, col. 4.

15. Jerrold Tannenbaum to the Editor, *Cornell Daily Sun,* Jan. 16, 1967; *Time,* CXVIII (Dec. 28, 1981), 67.

16. Brian McGuire, "Berkeley's Best Student: 'It Wasn't Worth It,'" *Speaking Out,* p. 27, undated clipping in the Berkeley file, United Chapters archives; *Key Reporter,* XXXIV (Autumn 1968), 3.

17. Joe. E. Cohen, "Phi Beta Kappa: Who Needs It?" *Harvard Crimson,* May 7, 1964, reprinted in *Key Reporter,* XXIX (Summer 1964), 3, 8; Robert Reinhold, "School Ferment Roils Phi Beta Kappa," *New York Times,* June 7, 1970, p. 1, col. 5.

18. "Pass/Fail Study Committee Reports Findings," *Key Reporter,* XXXV (Winter 1969–70), 2–4.

19. Minutes of the Senate, Dec. 6–7, 1968, p. 13; Minutes of the Executive Committee, May 25, 1969, pp. 3–4; *New York Times,* June 7, 1970, p. 1, col. 5.

20. *Minutes of the Twenty-Ninth Triennial Council Meeting* (Sept. 9–11, 1970), 6, 10.

21. Minutes of the Executive Committee, Dec. 4, 1970, p. 2; "ΦΒΚ Special

Committee," *Key Reporter*, XXXVII (Autumn 1971), 1, 4; *New York Times*, Nov. 6, 1970, p. 43, col. 6.

22. Nancy L. Ross, "Phi Beta Kappa: A New Look?" *Washington Post*, Dec. 6, 1971.

23. "Minutes of the Thirtieth Triennial Council Meeting," *Phi Beta Kappa Bulletin*, XXVIII (Sept. 1973), 8–10, 19.

24. *New York Times*, Dec. 6, 1976, p. 35, col. 5; *Phi Beta Kappa Bicentennial Bulletin*, 44–49.

25. Eleanor Blau, "Why Women Seldom Wear Their Phi Beta Kappa Keys," *New York Times*, Feb. 24, 1971, p. 46, col. 1; *Key Reporter*, XXXV (Summer 1970), 3–4; XLII (Winter 1976–77), 1.

26. *Time*, Dec. 28, 1981, p. 67.

27. Frederick J. Crosson to the 34th Council, Nov. 2, 1985, *Phi Beta Kappa Bulletin* (1985), 19.

Bibliography

Manuscript sources for this history consist of the following. 1) The minutes of the original society, 1776–1781, in the Earl Gregg Swem Library, College of William and Mary. 2) The chapter correspondence and miscellaneous records of the United Chapters office, which are located in the Phi Beta Kappa Society's building in Washington, D.C. Very few of these materials are dated before 1910, but they become increasingly rich from that time to the 1970s, beyond which they were not systematically consulted. 3) The correspondence and other records of the *American Scholar* editors, 1931–1976, in the manuscripts division of the Library of Congress.

The *New York Times* was searched, with the aid of its index, from the 1880s to the 1970s. Other printed contemporary sources include articles in popular magazines and in educational periodicals, but most important are the publications of the Phi Beta Kappa Society itself, among them the *Phi Beta Kappa Key* (1910–1931), *ΦBK Annals 1934,* the *American Scholar* (1932–1935, when it served partly as a house organ), the *Key Reporter* (1935–1976), and the *Phi Beta Kappa Bulletin* (1940s–1970s).

The most helpful secondary works have been those by Oscar M. Voorhees and by William T. Hastings as cited below. Indeed, the present history would scarcely have been possible without the pioneering work of those two scholars, both of whom uncovered and made available a good deal of original source material. Following is a list of the most useful books, articles, theses, and dissertations.

Adams, Charles F., ed. *Memoirs of John Quincy Adams,* vol. VIII. Philadelphia, 1877.

Adams, Charles F., Jr., "A College Fetich." *The Independent,* XXV (Aug. 9, 1883), 997–1000.

Adams, Herbert B. *The College of William and Mary: A Contribution to the History of Higher Education, with Suggestions for Its National Promotion.* Washington, D.C., 1887.

Allmendinger, David F., Jr. *Paupers and Scholars: The Transformation of Student Life in Nineteenth-Century New England.* New York, 1975.

Aydelotte, Frank. *Breaking the Academic Lock Step: The Development of Honors Work in American Colleges and Universities.* New York, 1944.

[Bagg, Lyman H.] *Four Years at Yale.* New Haven, 1871.

Baird, Raymond W. *American College Fraternities: A Descriptive Analysis of the Society System of the United States, with a Detailed Account of Each Fraternity.* Philadelphia, 1879.

Birdseye, Clarence F. *Individual Training in Our Colleges.* New York, 1907.

Boorstin, Daniel J. *The Americans: The National Experience.* New York, 1965.

Boroff, David. *Campus U. S. A.: Portraits of American Colleges in Action.* New York, 1961.

Boucher, Chauncey S. "Phi Beta Kappa Prospects at Chicago under the New Plan." *The American Scholar,* III (Oct. 1934), 487–93.

Brickman, William W., and Stanley Lehrer, eds. *A Century of Higher Education: Classical Citadel to Collegiate Colossus.* New York, 1962.

Brubacher, John S., and Willis Rudy. *Higher Education in Transition: A History of American Colleges and Universities, 1636–1976.* 3d ed. New York, 1976.

Bryce, James. *The American Commonwealth.* London and New York, 1889.

Burke, Colin B. *American Collegiate Populations: A Test of the Traditional View.* New York, 1982.

Butts, R. Freeman. *The College Charts Its Course.* Philadelphia, 1939.

Cahall, Raymond D. B. "The Beta Chapter of Ohio at Kenyon College." *The Phi Beta Kappa Key,* V (Jan. 1923), 116–27.

Carson, Jane. *James Innes and His Brothers of the F. H. C.* Williamsburg, Va., 1965.

Chamberlin, Willis A. *History of Theta Chapter of Ohio Phi Beta Kappa, 1911–1936.* Granville, Ohio, 1937.

———, ed. *Phi Beta Kappa, Theta of Ohio: History and Membership.* Granville, Ohio, 1914.

Chase, Frederick. "The Phi Beta Kappa Society." In John K. Lord. *A History of Dartmouth College, 1815–1909.* Concord, N.H., 1912.

Curti, Merle. *The Growth of American Thought.* New York, 1943.

Curtis, Charles P. "Liquor and Learning in Harvard College, 1792–1846." *The New England Quarterly,* XXV (Sept. 1952), 344–53.

Curtis, George W. "The Editor's Easy Chair." *Harper's New Monthly Magazine,* LXIII (Sept. 1881), 625–26.

Davidson College. *Self-Study Report for the Southern Association of Colleges and Schools.* Davidson, N.C., 1965.

Davis, Calvin O. "Courses Pursued by Members of Phi Beta Kappa [at Michigan]." *School and Society,* V (June 9, 1917), 686–90.

Dexter, E. G. "High-Grade Men: In College and Out." *Popular Science Monthly,* LXII (Mar. 1903), 429–35.

Doyle, J. A. "Phi Beta Kappa Tempest." *Education,* XLI (Oct. 1920), 86–94.

Ferrell, Robert H., ed. *The Eisenhower Diaries.* New York, 1981.

Fish, Carl R. *The Rise of the Common Man, 1830–1850.* New York, 1927.

Fivehouse, Janice. "The History of the Alpha of Virginia Chapter of Phi Beta Kappa." Master's thesis, College of William and Mary, 1968.

Fletcher, Robert S. *A History of Oberlin College from Its Foundation through the Civil War.* Oberlin, Ohio. 1943.

Foster, William T. *Should Students Study?* New York, 1917.

Gauss, Christian. "Wartime Policy of Phi Beta Kappa." *The American Scholar*, XII (Apr. 1943), 143–52.

Gifford, Walter S. "Does Business Want Scholars?" *Harper's Magazine*, CLVI (Apr. 1928), 669–74.

"Golden Key." *The Reporter*, XVIII (Mar. 1920), 2.

Hale, Edward E. "A Fossil from the Tertiary." *Atlantic Monthly*, XLIV (July 1879), 98–106.

Hanna, Suhail. "Calling for a Yardstick for Excellence." *Christianity Today*, XXIII (Nov. 2, 1979), 17.

Harvard Chapter. *Catalogue of the Harvard Chapter of Phi Beta Kappa: Alpha of Massachusetts. With the Constitution, the Charter, Extracts from the Records, Historical Documents, and Notes.* Cambridge, 1912.

Hastings, William T., ed. *A Century of Scholars: Rhode Island Alpha of Phi Beta Kappa, 1830–1930.* Providence, 1932.

———. *The Insignia of Phi Beta Kappa.* Washington, D.C., 1964.

———. *Phi Beta Kappa as a Secret Society, with Its Relation to Freemasonry and Antimasonry.* Washington, D.C., 1965.

———. "Phi Beta Kappa Today." *The Key Reporter*, XXII (July 1957), 1–4.

Havemann, Ernest, and Patricia S. West. *They Went to College: The College Graduate in America Today.* New York, 1952.

Haydn, Hiram. *Words & Faces.* New York, 1974.

Hill, G. Birkbeck. *Harvard by an Oxonian.* New York, 1894.

History of the College of William and Mary from Its Foundation, 1660, to 1874. Richmond, 1874.

Hollingworth, Leta. "Phi Beta Kappa and Women Students." *School and Society*, IV (Dec. 16, 1916), 932–33.

Kale, Wilford, *Hark upon the Gale: An Illustrated History of the College of William and Mary.* Norfolk, Va., 1985.

Kidd, George E. *Early Freemasonry in Williamsburg, Virginia.* Richmond, 1957.

Ladd, Everett C., Jr., and Seymour M. Lipset. *The Divided Academy: Professors and Politics.* New York, 1975.

Levine, David O. *The American College and the Culture of Aspiration, 1915–1940.* Ithaca, N.Y., 1986.

McCluskey, Neil G. "Phi Beta Kappa and Catholic Colleges." *America: National Catholic Weekly Review*, XCVIII (Feb. 22, 1958), 597–99. Discussion of this article in ibid., XCIX (Apr. 5, 1958), 18–20.

McCord, David T. "An Essay on the Orations and Poems, with a Handful of Each." In *Catalogue of the Harvard College Chapter of ΦBK.* Lunenburg, Vt., 1970.

McLachlan, James. "The *Choice of Hercules:* American Student Societies in the Early Nineteenth Century." In Lawrence Stone, ed. *The University in Society.* Vol. II. Princeton, N.J., 1974.

Minot, John C. "What Door Does the Phi Beta Kappa Key Open?" *North American Review*, CCXXIV (Nov. 1927), 531–36.

[Mitchell, John.] *Reminiscences of Scenes and Characters in College.* New Haven, 1847.

Mondale, Clarence C. "Gentlemen of Letters in a Democracy: Phi Beta Kappa Orations, 1788–1865." Ph.D. dissertation, University of Minnesota, 1960.

Morpugo, J. E. *Their Majesties' Royall Colledge: William and Mary in the Seventeenth and Eighteenth Centuries.* Washington, D.C., 1976.

[Munro, Wilfred H.] *Catalogue of the Rhode Island Alpha of Phi Beta Kappa, with the History of the Founding of the Chapter and Its Present Constitution and By-Laws.* Providence, 1914.

Newcomer, Mabel. "The Phi Beta Kappa Student." *School and Society,* XXV (Jan. 1, 1927), 24.

Northup, Clark S. *A Bibliography of the Phi Beta Kappa Society.* New York, 1928.

———, William C. Lane, and John C. Schwab, eds. *Representative Phi Beta Kappa Orations.* Boston, 1915.

Nye, Russel B. *The Cultural Life of the New Nation, 1776–1830.* New York, 1960.

———. *Society and Culture in America, 1830–1860.* New York, 1974.

"Over Mrs. Altman's Strudels." *The New Yorker,* XXVI (June 17, 1950), 19–20.

Packard, Lewis R. "The Phi Beta Kappa Society." In William L. Kingsley, ed. *Yale College: A Sketch of Its History.* Vol. I. New York, 1879.

Parsons, Eben B. *Phi Beta Kappa Handbook and General Address Catalogue of the United Chapters.* North Adams, Mass., 1900.

Peabody, Andrew P. "A Liberal Education." *New Englander and Yale Review,* XLV (Mar. 1886), 193–207.

Phelps, Reginald H. "Phi Beta Kappa at Harvard: A Bicentennial History." In *Phi Beta Kappa: Alpha of Massachusetts, 1781–1981, Bicentennial Exercises, December 11, 1981.* N. p., n. d.

"Phi Beta Kappa and Women." *School and Society,* IV (Dec. 9, 1916), 901–2.

"Phi Beta Kappas on Kennedy's Team." *U.S. News & World Report,* L (Jan. 30, 1961), 30.

Potter, David. *Debating in the Colonial Chartered Colleges: An Historical Survey, 1642 to 1900.* New York, 1944.

Pringle, Henry F. and Katharine. "America's Leading Negro University." *Saturday Evening Post,* CCXXI (Feb. 19, 1949), 36–37, 92–93, 95–98.

Quincy, Josiah. *The History of Harvard University.* 2d ed. Boston, 1860.

Ragatz, Lowell J. *The Founding of the Alpha Chapter of Phi Beta Kappa in the District of Columbia and a History of Its First Two Years, 1937–1939.* N. p., 1939.

Robson, John, ed. *Baird's Manual of American College Fraternities.* 19th ed. Menasha, Wis., 1977.

Sanford, Nevitt, ed. *The American College: A Psychological and Social Interpretation of the Higher Learning.* New York, 1962.

[Sas, Louis F., and others.] *Phi Beta Kappa, Gamma Chapter of New York, The City College: Centennial Directory, 1867–1967.* New York, 1967.

Schlesinger, Arthur M. *The Rise of the City, 1878–1898.* New York, 1933.

Schmidt, George P. *The Liberal Arts College.* New Brunswick, N.J., 1957.

Shimer, William A. "Higher ΦBK Standards." *The American Scholar,* III (May 1934), 367–68.

Slosson, Edwin E. *Great American Universities.* New York, 1910.

Smith, Denys. "American News: Phi Beta Kappas in Government; or, The American Old School Tie." *National Review,* CXIX (July 1942), 69–73.

Snow, Louis F. *The College Curriculum in the United States.* New York, 1907.

Stauffer, Vernon. *New England and the Bavarian Illuminati.* New York, 1918.

Steinberg, Stephen. *The Academic Melting Pot: Catholics and Jews in American Higher Education.* New York, 1974.

Stone, Lawrence, ed. *The University in Society.* Princeton, 1974.

Tewksbury, Donald G. *The Founding of American Colleges and Universities Before the Civil War.* New York, 1932.

Thayer, William R. *An Historical Sketch of Harvard University from Its Foundation to May 1890.* Cambridge, 1890.

Thorndike, E. L. "Careers of Scholarly Men in America." *Century Magazine,* LXVI (May 1903), 153–55.

"Thorndike Intelligence Scores of Columbia College Seniors Elected to Phi Beta Kappa in 1928." *School and Society,* XXVIII (July 7, 1928), 10–11.

Thwing, Charles F. "Phi Beta Kappa Society As a World Force." *American Review of Reviews,* LXIX (June 1924), 629–33.

Touraine, Alain. *The Academic System in American Society.* New York, 1974.

Tyler, Lyon G. "Original Records of the Phi Beta Kappa Society," "Brief Personal Sketches," "William Short." *William and Mary College Quarterly,* IV (Apr. 1896), 213–63.

U.S. Bureau of the Census. *Historical Statistics of the United States, Colonial Times to 1957.* Washington, D.C., 1960.

Vanderbilt, Arthur T. "An Example to Emulate." *The South Atlantic Quarterly,* LII (Jan. 1953), 5–19.

Varnado, Roy B. "Mr. Rockefeller's Other City: Background and Response to the Restoration of Williamsburg, Virginia, 1927–1939." Master's thesis, College of William and Mary, 1974.

Veysey, Laurence R. *The Emergence of the American University.* Chicago, 1965.

Visher, Stephen S. "Phi Beta Kappa and Scientific Research." *School and Society,* XX (Aug. 16, 1924), 216.

———. "Phi Beta Kappa Members of the National Academy of Science." *School and Society,* XXI (Mar. 28, 1925), 390.

Voorhees, Oscar M. *The History of Phi Beta Kappa.* New York, 1945.

———, ed. *Phi Beta Kappa General Catalog, 1776–1922.* Somerville, N.J., 1923.

———. "College Societies That Antedate Phi Beta Kappa." *The Phi Beta Kappa Key,* V (Mar. 1925), 672–82.

———. "The Phi Beta Kappa Badge." *The Phi Beta Kappa Key,* I (Jan. 1911), 16–25.

Wagner, Marta. "The American Scholar in the Early National Period: The Changing Context of College Education, 1782–1837." Ph.D. dissertation, Yale University, 1983.

Weyl, Nathaniel. "National Origins of Phi Beta Kappa Membership." *Names: Journal of the American Name Society,* XII (June 1964), 119–22.

Winsor, Justin. "Φ. B. K. (Records of the Centennial Convention, June 30, 1880 [i.e., 1881])." *Harvard University Bulletin,* II (Oct. 1, 1881), 264–65.

———. "Records of the Convention of October 18, 1881." *Harvard University Bulletin,* III (Jan. 1, 1882), 300–301.

Index